VOLUME TWO HUNDRED AND FOURTEEN

ADVANCES IN IMAGING AND ELECTRON PHYSICS

VOLUME TWO HUNDRED AND FOURTEEN

ADVANCES IN IMAGING AND ELECTRON PHYSICS

Computer Techniques for Image Processing in Electron Microscopy

Edited by

MARTIN HŸTCH and PETER W. HAWKES
CEMES-CNRS
Toulouse, France

ELSEVIER

ACADEMIC PRESS
An imprint of Elsevier

Cover photo credit:
The iterative transform algorithm.

Academic Press is an imprint of Elsevier
125 London Wall, London EC2Y 5AS, United Kingdom
525 B Street, Suite 1650, San Diego, CA 92101, United States
50 Hampshire Street, 5th Floor, Cambridge, MA 02139, United States
The Boulevard, Langford Lane, Kidlington, Oxford OX5 1GB, United Kingdom

ISBN: 978-0-12-820999-8
ISSN: 1076-5670

For information on all Academic Press publications
visit our website at https://www.elsevier.com/books-and-journals

Publisher: Zoe Kruze
Acquisitions Editor: Jason Mitchell
Editorial Project Manager: Hilal Johnson
Production Project Manager: James Selvam
Designer: Christian J. Bilbow

Typeset by VTeX

Contents

Contributor

W. Owen Saxton
Murray Edwards College, University of Cambridge, Cambridge, United Kingdom

Foreword☆

Ever since the beginnings of electron microscopy, image interpretation has been one of the most challenging tasks. A necessary condition for any interpretation is the knowledge of the steps taken in processing the image: be it the chemical composition or the temperature of the solution used in developing a photographic emulsion, as in the simplest case, or the very sophisticated computer generated processes discussed in the present volume. The essential requirement is to extract from the image all the information that it may contain: Dr. Saxton's monograph shows us the possibilities at our disposal at the present time, leading to more reliable interpretations than those available in the not so distant past.

One important aspect of this monograph is that it serves as a clear demonstration of the interdisciplinary aspects of modern science. No specialist of any branch of science can extract all information from the data obtained without recourse to adjacent and, in many cases, far from adjacent branches of science or technology.

To the thanks expressed by the author we would like to add our appreciation to the staff of Academic Press for all the assistance in both the editorial and production aspects of this book.

L. Marton
C. Marton

Editorial Preface

This volume of the Advances marks a departure from regular volumes in the series. *Advances in Imaging and Electron Physics* is the successor to *Advances in Electronics and Electron Physics*, which comprised regular volumes and Supplements. All the regular volumes are available on ScienceDirect but the supplements were never digitized. Some of these are extremely important and routinely cited as monographs. Elsevier have generously agreed to reprint a selection of these supplements as volumes of these Advances and this volume is the first of these reprinted texts.

The first practical solution to the phase problem—the fact that there are many situations in which information is coded in both the amplitude and the phase of a wave but only the amplitude can be recorded—was obtained by Ralph Gerchberg and Owen Saxton in 1972. A very full account of this work was later prepared by Saxton and published as Supplement 10 in 1978. We have chosen this seminal text as the first of these reprinted volumes. It will be followed by *Morphological Image Operators* by H.J.A.M. Heijmans (Supplement 25, 1994), *Coulomb Interactions in Charged Particle Beams* by G.H. Jansen (Supplement 21), *Quadrupoles in Electron Lens Design* by one of us (Supplement 7) and several other supplements that deserve to be accessible on-line.

We are very grateful to Elsevier for agreeing to rekey the text and undertake many other production tasks.

Martin Hÿtch
Peter W. Hawkes

Preface

Conventional electron microscopes have for some time been capable of imaging structures only slightly larger than atoms. The potential importance of reliable imaging at the atomic level is obvious, but attempts to achieve this have so far been frustrated, partly by the susceptibility of some specimens to electron exposure, but also by the nature of the imaging process. The inability of electron detectors to record the phase of the wave as well as the intensity often necessitates the deliberate exploitation of lens aberrations to force the image information from the former into the latter; the process can only be partially successful, and the images formed cannot be easily interpreted except when the specimen is known to interact only weakly with the electron beam. The development of fast Fourier transform algorithms, which has made possible so much image processing in other fields, has led naturally to a corresponding interest in the possibility of achieving a more reliable interpretation of electron micrographs by digital means, and it is this hope that is the subject of the present volume.

At the outset of the decade, these hopes were largely invested in the class of weak (i.e., thin) specimens previously mentioned. More recently, a number of attempts have been made, with some success, to increase the specimen thickness and/or atomic number for which image interpretation in this way is possible, and this work is reported and reviewed here at length (Chapters 3–7). Special attention is given to the earliest of these proposals, namely, the combined use of an image and diffraction pattern of a given object area; but the most important alternatives (defocus pairs and one-sided image pairs) are also treated, and the various sets of data are compared. The constraints imposed on the image wave function by the objective lens aperture are established (Chapter 3), and the special case of periodic wave functions is treated fully.

The hardware available and the basic mathematical techniques of image processing in electron microscopy are reviewed (Chapters 8–9), particular attention being given to transformation, correlation, alignment, and filtering in the linear approximation. One particular processing system is described in detail (Chapter 10). The properties of discrete Fourier transforms are collected together in Chapter 2. Chapter 1 gives an introductory account of image formation in which the properties of the linear (weak object) approximation are included and the circumstances for which these fail are made clear.

It is of course impossible to acknowledge all the generous advice that has been of help to me in the preparation of this volume. I owe a substantial debt, however, to Dr. R. W. Gerchberg whose enthusiasm stimulated my own interest in the image and diffraction pattern problem, and to Dr. P. W. Hawkes whose interest initiated the pursuit of image processing within the Electron Microscopy Section of the Cavendish Laboratory, and to whom I am indebted in innumerable ways. I am grateful also to M. Horner for sharing his knowledge of computers with me and to Dr. J. Frank whose practical experience of processing helped me to retain some sense of realism. Finally, I acknowledge the financial assistance of the Science Research Council of Great Britain for both personal support and the generous provision of equipment.

W. Owen Saxton

CHAPTER ONE

Image formation theory☆

W. Owen Saxton
Murray Edwards College, University of Cambridge, Cambridge, United Kingdom
e-mail address: wos1@cam.ac.uk

Contents

1.1. Electron optics

W. Hoppe has referred to "the interference patterns known as micrographs"; this introductory chapter gives a brief account of the considerations reflected in this remark. We begin not at the beginning but in the middle, outlining first the propagation of electron waves through the optical system of the microscope and turning later to how the object information is imposed on the beam and how it may be extracted from it. Electron optics is an interesting blend of classical (relativistic) dynamics, wave mechanics, and light optics. A comprehensive and authoritative account of the fundamental theory is given by Glaser (1952) and also, in abbreviated form, by Glaser (1956); the transition to the Fourier theory is concisely expressed by Hawkes (1973b).

The principal elements in the optical systems of electron microscopes are magnetic lenses with rotational symmetry; the symmetry property means that the field is determined everywhere by its value on the axis, $B(z)$. If the electrons have been accelerated through a potential Φ, the equations of motion can be conveniently separated in the "paraxial" approximation (all trajectories near and at small angles to the axis) by the introduction of a rotating Cartesian coordinate system (x, y, z), the angle of rotation varying

☆ Reprinted from Advances in Electronics and Electron Physics, Supplement 10, 1978.

Advances in Imaging and Electron Physics, Volume 214
ISSN 1076-5670
https://doi.org/10.1016/bs.aiep.2020.04.001

with z according to

$$d\theta/dz = B(e/8mV)^{1/2}, \tag{1.1.1}$$

V being the "relativistically corrected" accelerating voltage $\Phi(1+e\Phi/2mc^2)$. In this system, the x and y equations both take the form

$$x'' + (e/8mV)B^2x = 0 \tag{1.1.2}$$

(prime denoting differentiation with respect to z). The solutions of this are obviously oscillatory, so that all lenses of this type are converging. More important is the fact that the equation is linear. Choosing an arbitrary "object" plane z_o, we choose the two fundamental solutions to be $g(z)$ and $h(z)$, satisfying

$$\begin{aligned} g(z_o) &= h'(z_o) = 1, \\ g'(z_o) &= h(z_o) = 0. \end{aligned} \tag{1.1.3}$$

An electron leaving the object plane at x_o in a direction x'_o then follows the trajectory

$$x = x_o g + x'_o h. \tag{1.1.4}$$

If in some plane z_i, $h(z)$ again vanishes, then $x(z_i)$ will be $x_o g(z_i)$ regardless of the initial direction of the electron: i.e., an "image" will be formed with a magnification $g(z_i)$. In these circumstances, $g(z)$ vanishes at an intermediate plane z_d, and $x(z_d)$ will be $x'_o h(z_d)$ regardless of the initial position of the electron—a basic requirement for a diffraction pattern.

The transition to the wave mechanical form is now made. The Lagrangian describing the paraxial motion (1.1.2) is

$$L = \frac{p}{2}(x'^2 + y'^2) - \frac{e^2B^2}{8p}(x^2 + y^2), \tag{1.1.5}$$

where $p = (2emV)^{1/2}$ is the electron momentum; the independent variables are x, x', y, y' and z. The generalized momenta are px' and py'; the Hamiltonian is thus

$$H = \frac{1}{2p}\{(px')^2 + (py'^2)\} + \frac{e^2B^2}{8p}(x^2 + y^2), \tag{1.1.6}$$

and the z-dependent Schrödinger equation follows at once as

$$-\frac{\hbar^2}{2p}\nabla^2\psi + \frac{e^2B^2}{8p}(x^2 + y^2)\psi = i\hbar\frac{\partial\psi}{\partial z}, \tag{1.1.7}$$

the ∇^2 operator being two-dimensional. A solution in the form $b(z)\exp(iW/\hbar)$ can be obtained with b and W real; the equation for W separates, being in fact the classical Hamilton–Jacobi equation. Adopting vector notation with $\mathbf{x} = (x, y)$, it is easily confirmed by trial that

$$W = \frac{p}{2h}(h'|\mathbf{x}|^2 - 2\mathbf{x} \cdot \mathbf{x}_o + g|\mathbf{x}_o|^2) \tag{1.1.8}$$

is a solution, g and h being the classical trajectories satisfying (1.1.2) and (1.1.3) and $\mathbf{x}_o$ being an arbitrary constant of integration. This established, the equation for b is easily solved to give $b = 1/h$. By superposition we can build the solution

$$\psi(\mathbf{x}, z) = \frac{1}{i\lambda h}\exp\left(\frac{ikh'}{2h}|\mathbf{x}|^2\right)\int \psi_o(\mathbf{x}_o)\exp\left(\frac{ikg}{2h}|\mathbf{x}_o|^2\right)\exp\left(-\frac{ik}{h}\mathbf{x} \cdot \mathbf{x}_o\right)d^2\mathbf{x}_o, \tag{1.1.9}$$

where $k = p/\hbar = 2\pi/\lambda$. This integral is the propagation law for the electron wave function between different planes of the microscope, for it is easy to show that $\psi(\mathbf{x}, z_o) = \psi_o(\mathbf{x})$ since as $z \to z_o$, $h \to 0$, and the rapidly oscillating exponential factors tend to delta functions. In the same way, images of $\psi_o(\mathbf{x})$ are formed in any subsequent planes z_i for which h vanishes. To treat the effects of lens apertures and aberrations conveniently, however, we shall obtain the images in two stages, considering initially the intermediate plane z_d (subsequently referred to variously as the diffraction plane, the aperture plane, the back focal plane, and the Fourier plane) for which g vanishes:

$$\psi(\mathbf{x}, z_d) = \frac{1}{i\lambda h}\exp\left(\frac{ikh'}{2h}|\mathbf{x}|^2\right)\int \psi_o(\mathbf{x}_o)\exp\left(-\frac{ik}{h}\mathbf{x} \cdot \mathbf{x}_o\right)d^2\mathbf{x}_o. \tag{1.1.10}$$

If we define the Fourier transform $\tilde{\psi}_o(\mathbf{t})$ of $\psi_o(\mathbf{x})$ by the expression

$$\tilde{\psi}_o(\mathbf{t}) = \int \psi_o(\mathbf{x})\exp(-2\pi i\mathbf{t} \cdot \mathbf{x})\,d^2\mathbf{x}$$

(cf. Chapter 2: tildes are used throughout this work to denote transforms), it will be noted that, apart from a constant and a quadratic phase factor, the diffraction plane wave function (1.1.10) is simply $\tilde{\psi}_o(\mathbf{x}/\lambda h_d)$ or $\tilde{\psi}_o(\mathbf{x}/\lambda f)$ if we identify h_d with the focal length f of the lens, being thus a Fraunhöfer diffraction pattern of the object wave as in light optics.

The presence of an aperture in the diffraction plane is now described by multiplying (1.1.10) by a masking function $A(\mathbf{x})$, after which the image wave function can be obtained by a second application of (1.1.9) if g and h are replaced by those linear combinations which satisfy (1.1.3) in the *diffraction* plane. The quadratic phase factor in (1.1.10) cancels, leaving

$$\Psi(\mathbf{x}, z_i) = \frac{-1}{\lambda^2 h_d^2 g_i} \exp\left(\frac{ikg_i'}{2g_i}|\mathbf{x}|^2\right) \int A(\mathbf{x}_d)\tilde{\psi}_o\left(\frac{\mathbf{x}_d}{\lambda h_d}\right) \exp\left(\frac{ik}{h_d g_i}\mathbf{x} \cdot \mathbf{x}_d\right) d^2\mathbf{x}_d \tag{1.1.11}$$

(with g_i denoting $g(z_i)$, etc.). If we define $\psi_i(\mathbf{x})$ as the inverse Fourier transform of $\tilde{\psi}_o(\mathbf{x})A(\lambda f\mathbf{x})$, this image wave function, apart from a constant and a quadratic phase factor, is simply $\psi_i(\mathbf{x}/g_i)$ or $\psi_i(\mathbf{x}/M)$ if we identify g_i with the magnification M of the image. Because of this simple relationship between $\tilde{\psi}_o$ and ψ_i and the *actual* wave functions in the diffraction and image plane, we shall loosely refer to them throughout this work as if they were identical; it is obviously preferable to treat $\tilde{\psi}_o$ and ψ_i because of their relative mathematical simplicity. In terms of these, the process of image formation is thus concisely described in three steps: the transformation of $\psi_o(\mathbf{x})$ to yield $\tilde{\psi}_o(\mathbf{t})$; the multiplication of $\tilde{\psi}_o(\mathbf{t})$ by an aperture function to yield $\tilde{\psi}_i(\mathbf{t})$; and the inverse transformation of $\tilde{\psi}_i(\mathbf{t})$ to yield $\psi_i(\mathbf{x})$.

The next step is the inclusion of lens aberrations in the scheme, i.e., the removal of the restriction to the paraxial approximation. This is achieved by including an appropriate perturbation term in the classical equations of motion, which leads to the modification of the eikonal function W (1.1.8) by the addition of several small (real) terms of the fourth order in $\mathbf{x}_o$ and $\mathbf{x}$ whose coefficients, the aberration coefficients, are given by various integrals involving the classical ray paths and the axial magnetic field distribution $B(z)$. The most serious of these, the spherical aberration term, is independent of $\mathbf{x}_o$, and in consequence may be conveniently incorporated into the double Fourier transform scheme above as a phase shift to be applied in the diffraction plane, i.e., multiplying $\tilde{\psi}_o(\mathbf{t})$. In terms of the spherical aberration coefficient C_s, the phase shifting factor is

$$\exp(-\pi i\lambda^3 C_s|\mathbf{t}|^4/2). \tag{1.1.12}$$

Two other effects can also be usefully included in this way, though they are not strictly aberrations. Defocusing the system slightly reintroduces the quadratic phase factor in (1.1.10) which previously cancelled; an object space defocus D is thus described by a phase shifting factor

$$\exp(\pi i\lambda D|\mathbf{t}|^2), \tag{1.1.13}$$

(positive D being described as *under* focus), while an obviously related phase shifting factor

$$\exp(\pi i\lambda A(2(\mathbf{t}\cdot\mathbf{a})^2 - |\mathbf{t}|^2)/2) \tag{1.1.14}$$

describes first order (axial) astigmatism, as defined by the astigmatism coefficient A (the defocus difference between maximum and minimum effective defocus directions) and a unit vector $\mathbf{a}$ in the direction of maximum effective defocus.

The image formation scheme is thus extended to include the multiplication of $\tilde{\psi}_o$ not merely by an aperture function but also by the three factors (1.1.12)–(1.1.14) before inverse transformation. The neglect of aberrations other than spherical is known as the isoplanatic assumption since it implies that the image of a point object has the same form wherever the point object may be; the area within which the approximation to reality is close is limited by the next most serious aberration, namely coma, but is normally quite large enough to be useful for high resolution purposes.

The phase shifting factors broaden the image of a point source considerably by comparison with the diffraction limited image. Defocus is of course easily varied through the lens currents; axial astigmatism can be removed with a stigmator (a small quadrupole lens), though a small amount is normally present in practical micrographs; the spherical aberration cannot be eliminated however without the use of quadrupole/octopole lens combinations, microwave cavity lenses with pulsed beams, or electron mirror systems, all of which involve great technical difficulties and none of which has yet been successfully demonstrated in practice. It is of some consolation to reflect that, although its removal would undoubtedly improve our chances of imaging single heavy atoms, for a large class of objects we depend on positive exploitation of the spherical aberration to provide interpretable micrographs at all (see Section 1.3).

The image recorded on a fluorescent screen, photographic plate, or camera tube system, is proportional, of course, not to the wave function ψ_i but to the electron flux or to $|\psi_i|^2$, and the ambiguity arising is the subject of much discussion in Section 1.4 and succeeding chapters. The imperfect spatial and temporal coherence of a practical illuminating beam is treated by superposing the image $|\psi_i|^2$ predicted by the coherent theory previously discussed for each source element and each energy (wavelength) present in turn; this is in general a prohibitively lengthy calculation, but circumstances in which it reduces to reasonable proportions are discussed in Section 1.3.

1.2. Beam-specimen interaction

Having seen how information is carried by an electron beam through the microscope, we examine now what information about an object is imposed on the beam passing through it. This is a wide subject; useful accounts are given by Reimer and Gilde (1973) for isolated atoms and thin films, by Misell (1973c), Burge (1973), and Misell and Burge (1973) with particular reference to inelastic scattering events, and by Hirsch et al. (1965) for crystals. Here we are concerned only to indicate some basic features of the interactions that take place, including their relative importance under various conditions, and how far they allow the specimen structure to be deduced from the form of the scattered waves.

Isolated atoms are effectively transparent to the electron wave; they form a potential well $\phi(\mathbf{x}, z)$ which shortens the wavelength thus advancing the phase (retarding the wave front). Recalling that $p = \hbar k = \sqrt{2emV}$ (cf. (1.1.1) and (1.1.5)), the change in wave vector is

$$\delta k = em(1 + e\Phi/mc^2)\phi/\hbar p,$$

and the phase shift is thus

$$\eta(\mathbf{x}) = (em/\hbar p)(1 + e\Phi/mc^2) \int \phi(\mathbf{x}, z)\, dz, \tag{1.2.1}$$

being proportional to the z-projection of the atomic potential. The form of this result is in sharp contrast to that of most scattering calculations in which it is the far-field diffracted wave (the Fourier transform of $\exp(i\eta)$) which is obtained in the first instance from a Born series or partial wave analysis. We have made in effect a WKB approximation, treating the potential as slowly varying compared with the electron wavelength: this is a good approximation except near the nucleus, and we may ignore that small region since the scattering from it is predominantly at large angles while only electrons scattered at small angles contribute to the image. It is frequently convenient to break the resulting wave function $\exp(i\eta)$ into two terms, namely, the incident or primary beam (sometimes called "unscattered") represented by unity and a scattered wave represented by the rest.

If the phase shift η is small, (1.2.1) is equivalent to the first Born approximation, for the far-field diffraction pattern is then the two-dimensional Fourier transform of

$$\exp(i\eta(\mathbf{x})) \simeq 1 + i\eta(\mathbf{x}), \tag{1.2.2}$$

and is thus a section through the three-dimensional transform of the scattering potential ϕ. In this very important limit (the "weak phase approximation" in electron microscopy) the scattered wave is symmetric and purely imaginary in the diffraction plane. The limit is approached closely by light atoms at fairly high voltages, the phase of the diffracted wave for the carbon atom deviating by no more than 3° from $\pi/2$ over the range of angles normally contributing to the image. With heavier atoms however the next term in (1.2.2), namely $-\eta^2(\mathbf{x})/2$, becomes significant; its transform is symmetric but real, so that the diffraction plane wave ceases to be imaginary; since $\eta^2(\mathbf{x})$ is narrower than $\eta(\mathbf{x})$, its transform is wider, and the phase of the diffraction plane wave accordingly exceeds $\pi/2$ by an amount which grows with off-axis distance. In general, all the terms of (1.2.2) are retained, but the general behavior of the scattered wave is by now established. The familiar "atomic scattering factor" $f(\mathbf{t})$ of the atom for electrons is obtained from the diffraction plane wave

$$f(\mathbf{t}) = (1/i\lambda) \int \exp(i\eta(\mathbf{x})) \exp(-2\pi i\mathbf{t} \cdot \mathbf{x})\, d^2\mathbf{x}. \tag{1.2.3}$$

Various models have been used by different authors for the scattering potentials ϕ, the screened Coulomb potential $Z\exp(-r/r_0)/r$ being a popular simple model, with Z being the atomic number and r_0 proportional to $Z^{-1/3}$ (Lenz, 1954). This predicts that the total scattering cross section varies as $Z^{4/3}$, so that the scattering factor amplitude is roughly proportional to $Z^{2/3}$; the scattering factor phase is roughly proportional to $Z^{5/6}$ (Frank, 1973c), so that its imaginary (or "anomalous") part is much more sensitive to atomic number than the real part. This separation of the scattering into two components will reappear subsequently.

For thin specimens, we can simply extend the potential in (1.2.1) to describe the whole object film instead of a single atom. In the "tight-binding" approximation, in which the atomic potentials are assumed to be unmodified by their neighbors, this amounts simply to adding the scattering factors for individual atoms, each multiplied by a linear phase factor describing the position of the atom concerned, the sum being called the "structure factor" of the object. At the other extreme, crystals are often represented by a truncated Fourier series potential. In this thin-specimen limit, the object structure could be recovered, in projection, from the electron wave function. The objects are said to be "multiplicative," i.e., capable of full description by a transparency function which depends only on object position and not on the direction or energy of the illumination.

Numerous problems arise as the specimen thickness increases. Atoms at different heights contribute at different levels of focus, thus having their scattering factors modified by a quadratic phase factor (1.1.13) and appearing to have a much larger anomalous scattering factor than an in-focus atom. The "depth of focus" can be assessed from (1.1.13); allowing the scattering factor phases to vary by ± 0.2 rad at the relatively high angle of 0.01 rad gives a limiting thickness of 5 nm, for example. For light atoms, a higher thickness can be tolerated since it is the normal part of the scattering factor which is chiefly responsible for the image formation (cf. Section 1.1), and this is relatively little affected by such phase shifts.

Multiple scattering probabilities become significant above 50 nm for carbon or 3 nm for gold as the scattered wave ceases to be negligible compared with the primary wave, and in the diffraction plane convolutions of the scattered wave with itself appear. In thick perfect crystals, the problem finds a very elegant treatment in the matching of the incident wave to a number of Bloch waves (the electron states within the crystal); the differences between the z-components of these give rise to interference effects so that periodicities may appear and disappear as the crystal thickness varies. With the appearance of crystal defects, it is necessary to effect approximate integration of the Schrödinger equation over the faulty regions at least. In all these cases, the object structure could not be deduced from a knowledge of the electron wave function beneath the specimen, though in the case of thick crystals, the variations of the image intensity with the crystal orientation or thickness, the accelerating voltage, or the particular diffracted beams used in image formation can be of great assistance in characterizing it. We shall not subsequently be concerned with this range of specimen thickness.

For biological objects stained with heavy atom salts, a different view of the scattering process is useful in view of the low resolution normally permitted by the tendency of the salts to form small crystals 1–2 nm in size under the electron beam. Erickson (1971) [see also Erickson & Klug (1970)] uses the same approach as that we have discussed to derive the normal part of the scattering factors and takes the projected potential of the specimen to be proportional to the projected mass (of stain). The anomalous part, however, is derived by treating the interception of the higher angle scattered electrons at the (relatively small) objective aperture as equivalent to an absorption of electrons at the specimen, and so equivalent to a term in (1.2.2) playing the role of the term $-\eta^2(\mathbf{x})/2$ and also proportional to the projected mass of the specimen by way of the atomic cross section for scattering outside the aperture. The contrast produced in the image has traditionally been called "amplitude contrast," but the term "scattering contrast" is preferable and is in increasing use. Erickson develops an attractive object model in attempting—perhaps not wholly successfully—to treat scattering contrast more rigorously, in which a high frequency phase object due to individual heavy atoms is modulated by a slowly varying function describing the stain density.

The interactions considered so far have been purely elastic collisions (in the center of mass frame) imparting negligible energy to the target atoms. Inevitably however, a certain number of inelastic scattering events also occur, and useful images can be formed with a narrow energy band of inelastically scattered electrons in a microscope with a suitable energy filter. More commonly the inelastic scattering is unwanted, reducing the contrast in the elastic image and in fact determining the image resolution for thick biological specimens through the chromatic aberration of the objective lens. The most significant loss processes for our present purposes are core-electron excitation and plasmon excitation.

In contrast to the well-defined plasmon energy in metals and the correspondingly delocalized nature of the excitation process, "plasmon" excitation in disordered materials such as biological specimens, in which the electron states extend over much smaller areas, is localized perhaps to 0.5 nm, and the energy spectrum is accordingly broad, with a peak at 20–25 eV. Lenz (1954) gives the cross section for core excitation as proportional to $Z^{1/3}$ only; Misell and Burge find very little variation with Z in a combined scattering model. The balance between the elastic and inelastic electrons thus swings heavily towards the elastic component as the atomic number rises, but for atoms as light as carbon there may easily be twice or four times as many inelastic electrons contributing to the image as elastic ones. The angular distribution of the inelastic electrons is more strongly peaked in the forward direction than that for the elastic electrons, implying a lower resolution image. When multiple scattering effects are not large, the combined energy-angle distribution can be approximately separated into a product of angular and energy distributions, easily observed experimentally; for a given energy loss range and a given number of scattering events, the scattered wave appears to have a reasonable coherence area and is usually treated as a coherent wave, incoherent with the primary beam and with waves scattered inelastically a different number of times. In the absence of reliable information on the subject, the phase of the scattered wave is usually taken to be independent of scattering angle so that it cancels in the image formation theory.

In view of the complex nature of the inelastic interactions, it is not clear precisely how the object structure and the inelastic waves are related: Misell and Burge present theoretical "point images" formed by the inelastic electrons, but do not state explicitly with what object function these should be convolved. Moreover, an opposite view has been used recently with some success [e.g., Krakow, Ast, et al. (1976)] to account for certain features of micrograph diffraction patterns at medium specimen thickness. In this, the inelastic scattering process is taken to be completely delocalized, even in disordered materials, resulting merely in the provision of new "incident" beams, spatially similar to the primary beam, but at a range of energies, and these beams are then supposed elastically scattered in their turn like the primary beam.

An indirect effect of inelastic scattering that can be of some importance is the removal of electrons from the elastic beam. This really does correspond to an attenuation of the elastic wave function at the object and makes a further contribution to the anomalous part of the object structure factor. The implication is that even if the image formed by the inelastically scattered electrons is blurred beyond recognition, or is completely removed with an energy filter, the inelastic scattering events are still capable of producing high resolution image contrast.

1.3. Linear imaging

We have seen in Section 1.1 that the image wave function $\psi_i(\mathbf{x})$ is related to the object wave function $\psi_o(\mathbf{x})$ by a simple linear filter, denoted by the equation

$$\tilde{\psi}_i(\mathbf{t}) = \tilde{T}(\mathbf{t})\tilde{\psi}_o(\mathbf{t}) \tag{1.3.1}$$

applied to their Fourier transforms, $\tilde{T}(\mathbf{t})$, the wave function transfer function, is the product of an aperture function and a phase shifting factor $\exp(-i\gamma(\mathbf{t}))$, with

$$\gamma(\mathbf{t}) = \pi\lambda\{\tfrac{1}{2}C_s\lambda^2|\mathbf{t}|^4 - D|\mathbf{t}|^2 - \tfrac{1}{2}A(2(\mathbf{t}\cdot\mathbf{a})^2 - |\mathbf{t}|^2)\}, \tag{1.3.2}$$

which combines the effects of spherical aberration, defocus, and axial astigmatism. This satisfactory linear relationship is lost as soon as it established by the recording of $|\psi_i|^2$ instead of ψ_i; however, a linear image formation theory can be preserved under bright-field imaging conditions for a wide class of objects, namely "weak" objects, which scatter only a small fraction of the incident beam, and this theory is the subject of the present section. Detailed accounts of image formation are given by Lenz (1971) and Hanszen (1971), while Zeitler (1968) gives a looser, more widely ranging discussion; Misell (1973c) pays particular attention to the inelastic component.

The linearity of the theory depends on the most significant term in the image intensity being the interference term between the primary beam and the scattered wave. With a suitable normalization, we may write the wave function leaving a weak object, for axial plane coherent illumination, in the form

$$\psi_o(\mathbf{x}) = 1 + a(\mathbf{x}) + i\eta(\mathbf{x}),$$

with $|a| \ll 1$ and $|\eta| \ll 1$, and both a and η real. From (1.3.1),

$$\tilde{\psi}_i(\mathbf{t}) = \tilde{T}(\mathbf{t})(\delta(\mathbf{t}) + \tilde{a}(\mathbf{t}) + i\tilde{\eta}(\mathbf{t})). \tag{1.3.3}$$

The transform $\tilde{I}(\mathbf{t})$ of the image intensity $I(\mathbf{x}) = |\psi_i(\mathbf{x})|^2$ is the autocorrelation of (1.3.3), which is

$$\tilde{I}(\mathbf{t}) = \delta(\mathbf{t}) + \tilde{T}(\mathbf{t})(\tilde{a}(\mathbf{t}) + i\tilde{\eta}(\mathbf{t})) + \tilde{T}^*(-\mathbf{t})(\tilde{a}^*(-\mathbf{t}) - i\tilde{\eta}^*(-\mathbf{t})), \tag{1.3.4}$$

since $T(0) = 1$ and second order terms are being neglected. It is convenient to remove the constant background from the image intensity, defining the contrast $c(\mathbf{x}) = I(\mathbf{x}) - 1$. From (1.3.4) its transform is

$$\tilde{c}(\mathbf{t}) = (\tilde{T}(\mathbf{t}) + \tilde{T}^*(-\mathbf{t}))\tilde{a}(\mathbf{t}) + i(\tilde{T}(\mathbf{t}) - \tilde{T}^*(-\mathbf{t}))\tilde{\eta}(\mathbf{t}) \tag{1.3.5}$$

since $\tilde{a}$ and $\tilde{\eta}$ are conjugate symmetric. Within a symmetric aperture, this may conveniently be reexpressed in the form

$$\tilde{c} = 2\tilde{a}\cos\gamma + 2\tilde{\eta}\sin\gamma \tag{1.3.6}$$

describing the independent transfer of the real and imaginary ("amplitude" and "phase" or "normal" and "anomalous") parts of the object into the image intensity by simple spatial frequency filters known as contrast transfer functions, involving attenuation and/or sign reversal. The oscillating form of the frequency response implies an irretrievable loss of information about periodicities near the filter zeros, considered further in Section 1.5.

If the aperture obscures one half of the diffraction plane (passing the primary beam) instead of being centrosymmetric, only one of $\tilde{T}(\mathbf{t})$ and

$\widetilde{T}(-\mathbf{t})$ is nonzero at any given point, and (1.3.5) becomes

$$\tilde{c} = \tilde{a}\exp(+i\gamma) \pm i\tilde{\eta}\exp(\mp i\gamma), \tag{1.3.7}$$

the upper signs applying over the open, and the lower over the obscured half plane. Different object periodicities, instead of being attenuated, are now shifted laterally by varying amounts; however, there is no irretrievable loss of information as there is with a symmetric aperture.

The form of $\gamma(\mathbf{t})$ and its sine deserve a little study in view of the wide use of the weak phase object approximation in studies of thin films. The use of the dimensionless variables

$$D' = D/\sqrt{C_s\lambda}, \qquad A' = A/\sqrt{C_s\lambda}, \qquad \mathbf{t}' = \mathbf{t}C_s^{1/4}\lambda^{3/4} \tag{1.3.8}$$

(the *normalized* or *generalized* defocus, astigmatism, and spatial frequency) and the introduction of ϕ as the angle between $\mathbf{t}$ and $\mathbf{a}$ allows us to write

$$\gamma(\mathbf{t}') = \pi\{\tfrac{1}{2}|\mathbf{t}'|^4 - (D' + \tfrac{1}{2}A'\cos 2\phi)|\mathbf{t}'|^2\}. \tag{1.3.9}$$

Allowing the astigmatism to be absorbed into an azimuthal dependence of D', we can consider a radial section of this function,

$$\gamma(t') = \pi(t'^4/2 - D't'^2), \tag{1.3.10}$$

from which vector notation has been dropped. This expression is dominated by the negative second order term at low t' values (low resolution) and by the positive fourth order term at high t' values. The extremum falls at $t' = \sqrt{D'}$, where $\gamma = -\pi D'^2/2$. Around the extremum, $\sin\gamma$ like γ itself varies relatively slowly, giving an extended range of t' with the same imaging conditions; a suitable choice of D' allows $|\sin\gamma|$ to be unity at this extremum, thereby giving "optimal" phase contrast transfer conditions. We require that

$$\pi D'^2/2 = n\pi/2,$$

with n odd, so that $D' = \sqrt{n}$. With $n = 1$, the defocus is that widely associated with the name of Scherzer. The extended transfer intervals form bands about $t' = n^{1/4}$ so that they move out rather slowly as the defocus increases, and they become narrower at the same time since the zeros of $\sin\gamma$ adjacent to the broad band fall at $t' = n^{1/4}(1 \pm 1/\sqrt{n})^{1/2}$. The value $t' = \sqrt{2}$ is thus the highest directly interpretable spatial frequency in the broad spectrum Scherzer defocus image and thus defines a nominal instrumental resolution

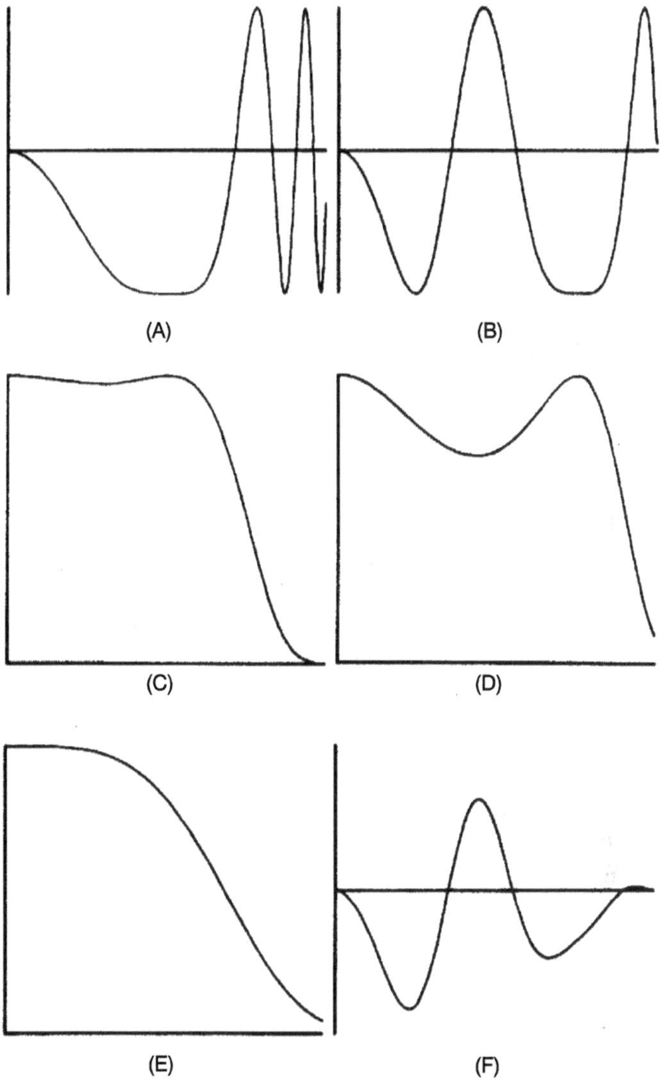

Figure 1.3.1 Weak object transfer functions: (A) $\sin\gamma$ for $D' = 1$; (B) $\sin\gamma$ for $D' = \sqrt{5}$; (C) $\widetilde{B}$ for $D' = 1$; (D) $\widetilde{B}$ for $D' = \sqrt{5}$; (E) $\widetilde{E}$; (F) $\widetilde{B}\widetilde{E}\sin\gamma$ for $D' = \sqrt{5}$.

of about $0.7C_s^{1/4}\lambda^{3/4}$. Better values can refer only to the imaging of narrow bands of spatial frequencies, such as isolated components of crystal lattices. The above expressions with *even n*, incidentally described optimized conditions for weak amplitude objects (which do not exist at high resolution) by way of $\cos\gamma$. Fig. 1.3.1 shows the transfer function $\sin\gamma(t)$ for $D' = 1$ and

$\sqrt{5}$ at the top left and top right, respectively, in order to illustrate the general nature of these functions. (The other curves in this figure are explained subsequently.)

If subsequent processing of the micrographs is envisaged, either optically or digitally, reversed sign transfer and any significant transfer further out than the broad band provide valuable information capable of being rendered interpretable. If this is not to be performed however there is some advantage, in terms of the shape of the point image associated with the $\sin\gamma$ transfer functions, in excluding the rapid oscillations of the functions beyond the broad band which only lead to uninterpretable fine detail, by an aperture cutting off at $t' = n^{1/4}(1 + 1/\sqrt{n})^{1/2}$. It is also helpful to exclude all those regions for which the transfer has the sign opposite to that in the broad band; since the point image is the Fourier transform of the transfer function, its central contrast is the area under the transfer function, and this is clearly increased greatly by removing negative contributions. Suitable apertures, called "zone plates," are difficult to manufacture since the diffraction plane is typically less than 100 μm across at 100 kV, but they have nevertheless been used successfully (Hoppe, 1971b). Their suitability for more general objects not consisting largely of isolated heavy atoms is debatable however, and they are clearly inappropriate to periodic objects.

Since it is capable of removing transfer function zeros, a special device of value, even if subsequent processing *is* anticipated, is the "phase plate" (Willasch, 1975), consisting of a carbon film placed in the back focal plane of the objective, the thickness of which varies (up to 100 nm) so as to provide phase shifts according to (1.2.1) which cancel those contained in γ. Though this remains difficult to manufacture and causes a contrast loss through further inelastic scattering in the phase plate, it does appear viable, and further improvements in microfabrication techniques may make it an attractive proposition. A compromise phase plate consisting of a single filament spanning the diffraction plane aperture, and therefore much easier to manufacture, has also been used (Unwin, 1972, 1974; Krakow & Siegel, 1975); this depends on the charging of the filament in the primary beam which is relatively difficult to control, but it provides an additional degree of freedom in the form of a phase shift proportional to

$$q \ln t/t_0$$

to be subtracted from γ, q being the charge developed on the filament and t_0 a parameter dependent on the angular spread of the illumination; the phase shift is, in practice, surprisingly near to being circularly symmetric.

Some attention is now given to the consequences of the imperfect coherence of the illumination, which means that the object is not illuminated by a single plane wave but by an incoherent superposition of waves over a small range of angles and a small range of energies (wavelengths). Since this commits us to relatively lengthy expressions in any case, we shall lengthen them slightly further so as to treat tilted illumination modes, in which the high resolution obtainable in one direction (double that of the axial mode) is bought at the price of the one-sided nature of the transfer with the consequent lateral shifts in the high spatial frequencies. In the axial mode, the effects of the partial coherence have been studied in detail by Hanszen and Trepte (1971a, 1971b) and Frank (1973b), while the characteristics of the tilted beam modes have been the subject of numerous recent papers, such as Hoppe et al. (1975), McFarlane and Cochran (1975), and Krakow, Ast, et al. (1976).

For brevity only, we shall restrict the treatment to weak *phase* objects. A plane wave directed at $\mathbf{t}_0$ in the diffraction plane then produces an object wave function

$$(1 + i\eta(\mathbf{x}))\exp(2\pi i\mathbf{t}_0 \cdot \mathbf{x}) \tag{1.3.11}$$

and the corresponding diffraction plane wave function is

$$\delta(\mathbf{t} - \mathbf{t}_0) + i\tilde{\eta}(\mathbf{t} - \mathbf{t}_0). \tag{1.3.12}$$

From (1.3.1) the aberrated diffraction plane wave function is

$$\delta(\mathbf{t} - \mathbf{t}_0)\widetilde{T}(\mathbf{t}_0) + i\tilde{\eta}(\mathbf{t} - \mathbf{t}_0)\widetilde{T}(\mathbf{t}), \tag{1.3.13}$$

and the transform of the image contrast is the autocorrelation of this (less the central delta function), namely

$$\tilde{c}(\mathbf{t}) = i\tilde{\eta}(\mathbf{t})\{\widetilde{T}(\mathbf{t}_0 + \mathbf{t})\widetilde{T}^*(\mathbf{t}_0) - \widetilde{T}^*(\mathbf{t}_0 - \mathbf{t})\widetilde{T}(\mathbf{t}_0)\}. \tag{1.3.14}$$

We describe the effective electron source by an intensity distribution $b(\mathbf{t})$ assumed to be symmetric about some central point $\mathbf{s}_0$ and normalized to unit total intensity; contributions for different $\mathbf{t}_0$ being mutually incoherent, the image intensity, and hence the contrast transform, is obtained as a superposition integral

$$\tilde{c}(\mathbf{t}) = i\tilde{\eta}(\mathbf{t})\int b(\mathbf{t}_0)\{\widetilde{T}(\mathbf{t}_0 + \mathbf{t})\widetilde{T}^*(\mathbf{t}_0) - \widetilde{T}^*(\mathbf{t}_0 - \mathbf{t})\widetilde{T}(\mathbf{t}_0)\}\, d^2\mathbf{t}_0. \tag{1.3.15}$$

Fluctuations in accelerating voltage and lens currents are taken together with the spread of emission energy for electrons at the filament and expressed in terms of an equivalent defocus distribution $E(\delta)$, which we shall again assume symmetric and normalized to unit total intensity. Writing $\widetilde{T}(\mathbf{t},\delta)$ for the transfer function with additional defocus δ, the contrast transform becomes

$$\tilde{c}(\mathbf{t}) = i\tilde{\eta}(\mathbf{t}) \int\int b(\mathbf{t}_0)E(\delta)\{\widetilde{T}(\mathbf{t}_0+\mathbf{t},\delta)\widetilde{T}^*(\mathbf{t}_0,\delta) - \widetilde{T}^*(\mathbf{t}_0-\mathbf{t},\delta)\widetilde{T}(\mathbf{t}_0,\delta)\}\, d^2\mathbf{t}_0\, d\delta, \qquad (1.3.16)$$

which is sufficient to demonstrate the continued separation of object and instrumental factors in a theory which remains linear.

For near coherent conditions (small source sizes and narrow energy spreads), the transfer function of (1.3.16) can be separated into three factors, one relating to the wave aberration, one to the finite spatial extent of the effective source, and one to the finite energy spread in the beam. We replace $\mathbf{t}_0$ by $\mathbf{s}_0+\mathbf{s}$, with $|\mathbf{s}|$ small, and expand γ with respect to $\mathbf{s}$, obtaining from (1.3.2)

$$\widetilde{T}(\mathbf{t}_0+\mathbf{t},\delta) = \widetilde{T}(\mathbf{s}_0+\mathbf{t})\exp(-i\mathbf{s}\cdot\nabla\gamma(\mathbf{s}_0+\mathbf{t}))\exp(i\pi\lambda\delta|\mathbf{s}_0+\mathbf{t}|^2). \qquad (1.3.17)$$

This separation allows us to rewrite (1.3.16) in the form

$$\tilde{c}(\mathbf{t}) = i\tilde{\eta}(\mathbf{t})\widetilde{T}(\mathbf{s}_0+\mathbf{t})\widetilde{T}^*(\mathbf{s}_0)\int B(\mathbf{s})\exp(-i\mathbf{s}\cdot(\nabla\gamma(\mathbf{s}_0+\mathbf{t})-\nabla\gamma(\mathbf{s}_0))\, d^2\mathbf{s} \times \int E(\delta)\exp(i\pi\lambda\delta(|\mathbf{s}_0+\mathbf{t}|^2-|\mathbf{s}_0|^2))\, d\delta, \qquad (1.3.18)$$

minus another term which is the complex conjugate of this with $-\mathbf{t}$ for $\mathbf{t}$; $B(\mathbf{s})$ has been written for $b(\mathbf{s}_0+\mathbf{s})$, and is therefore symmetric. Both of the integrals in (1.3.18) take the form of Fourier transforms of the relevant source profiles, evaluated at $(\nabla\gamma(\mathbf{s}_0+\mathbf{t})-\nabla\gamma(\mathbf{s}_0))/2\pi$ and $\lambda(|\mathbf{s}_0+\mathbf{t}|^2-|\mathbf{s}_0|^2)/2$, respectively, and are real (since $B(\mathbf{s})$ and $E(\delta)$ are real and symmetric). $\nabla\gamma$ is easily obtained from (1.3.2) in the form

$$\nabla\gamma(\mathbf{t}) = 2\pi\lambda\{(C_s\lambda^2|\mathbf{t}|^2-(D-\tfrac{1}{2}A))\mathbf{t}-A(\mathbf{t}\cdot\mathbf{a})\mathbf{a}\}. \qquad (1.3.19)$$

Note that the approximation (1.3.17) is adequate only to describe the finite energy spread of the primary beams; if, for example, plasmon loss electrons are to be included in the theory, the factor $\exp(i\pi\delta|\mathbf{s}+\mathbf{t}|^2)$ must also be expanded with respect to $\mathbf{s}$, which adds a term $-2\pi\lambda\delta\mathbf{t}$ to $\nabla\gamma$ and prevents separation of the coherence factors (though not analytic evaluation of (1.3.16)).

Adopting now the notation

$$\widetilde{T}_0(\mathbf{t}) = \widetilde{T}(\mathbf{s}_0 + \mathbf{t}),$$
$$A_0(\mathbf{t}) = \widetilde{B}((\nabla\gamma(\mathbf{s}_0 + \mathbf{t}) - \nabla\gamma(\mathbf{s}_0))/2\pi)\widetilde{E}(\lambda(|\mathbf{s}_0 + \mathbf{t}|^2 - |\mathbf{s}_0|^2)/2),$$

the full expression for the contrast transform may be written as

$$\tilde{c}(\mathbf{t}) = i\tilde{\eta}(\mathbf{t})\{\widetilde{T}_0(\mathbf{t})\widetilde{T}_0^*(0)A_0(\mathbf{t}) - \widetilde{T}_0^*(-\mathbf{t})\widetilde{T}_0(0)A_0(-\mathbf{t})\} \tag{1.3.20}$$

since $\tilde{\eta}(\mathbf{t})$ is conjugate symmetric. Note that neither $\widetilde{T}_0$ nor A_0 is symmetric; the choice of A to denote the filters due to imperfect coherence is by no means fortuitous for these filters are commonly far more effective in eliminating high spatial frequencies from the image than any physical aperture in the objective back focal plane, and it is these which ultimately limit the resolution of a microscope. Two particular features of (1.3.20) deserve comment. First $\widetilde{E}$ is unity on the circle defined by $|\mathbf{s}_0 + \mathbf{t}|^2 = |\mathbf{s}_0|^2$, i.e., centered at $-\mathbf{s}_0$ and passing through the origin; there is therefore no chromatic attentuation of the transfer of spatial frequencies falling on this "achromatic circle"; moreover, for defocus $C_s\lambda^2|\mathbf{s}_0|^2$ and zero astigmatism, $\nabla\gamma(\mathbf{s}_0 + \mathbf{t})$ and $\nabla\gamma(\mathbf{s}_0)$ are zero on the circle, so that there is no attenuation by $\widetilde{B}$ either. The circle is accordingly a prominent feature of diffraction patterns of micrographs made with broad energy spreads; and this property is sometimes referred to as a "correction" of chromatic aberration in the tilted beam mode, though it is of course only a small proportion of spatial frequencies that benefit from the correction. Second, the presence of a central (physical) aperture makes $\widetilde{T}(\mathbf{t})$ zero beyond the aperture radius from the origin, and hence $\widetilde{T}_0(\mathbf{t})$ and $\widetilde{T}_0(-\mathbf{t})$ zero except within the same radius of $-\mathbf{s}_0$ and $\mathbf{s}_0$, respectively; where these circular regions do not overlap, one of the terms in (1.3.20) vanishes, preventing the presence of transfer function zeros such as plague axial imaging modes.

The diffraction pattern of the micrograph (diffractogram) is easily predicted from (1.3.20) and is of great importance for practical assessment of the transfer conditions in force during the recording of the micrograph. In the regions where the two circles just mentioned overlap so that both terms in (1.3.20) are nonzero, this is

$$|\tilde{c}(\mathbf{t})|^2 = |\tilde{\eta}(\mathbf{t})|^2\{A^2(\mathbf{t}) + A^2(-\mathbf{t}) - 2A(\mathbf{t})A(-\mathbf{t})\cos(\gamma(\mathbf{s}_0 + \mathbf{t}) + \gamma(\mathbf{s}_0 - \mathbf{t}) - 2\gamma(\mathbf{s}_0))\}. \tag{1.3.21}$$

In the nonoverlapping regions, the diffractogram is simply $|\tilde{\eta}(\mathbf{t})|^2 A^2(\mathbf{t})$. With the amorphous objects frequently used as supporting films when not being studied in their own right, the object spectrum $|\tilde{\eta}(\mathbf{t})|^2$ varies relatively slowly, falling steadily with $|\mathbf{t}|$ and exhibiting a broad maximum around the value corresponding to a 0.3 nm object periodicity (normally just excluded in the axial illumination mode) and the relatively rapid modulations due to the cosine term are easily discerned and could be used to establish $\gamma(\mathbf{t})$. In practice this determination has been carried out only for the axial mode when (1.3.20) reduces to

$$\tilde{c}(\mathbf{t}) = 2\tilde{\eta}(\mathbf{t}) \sin \gamma(\mathbf{t}) A(\mathbf{t}), \tag{1.3.22}$$

with a symmetric filter $A(\mathbf{t})$ given by

$$A(\mathbf{t}) = \widetilde{B}(\nabla \gamma(\mathbf{t})/2\pi)\widetilde{E}(\lambda|\mathbf{t}|^2/2), \tag{1.3.23}$$

and the diffractogram becomes the simple expression

$$|\tilde{c}(\mathbf{t})|^2 = 4|\tilde{\eta}(\mathbf{t})|^2 \sin^2 \gamma(\mathbf{t}) A^2(\mathbf{t}) \tag{1.3.24}$$

which reveals $\gamma(\mathbf{t})$ comparatively easily. The effects of imperfect coherence on one-sided axial imaging (1.3.7) are also simple to obtain from (1.3.20) as only one term is nonzero at a time. The diagnosis of mechanical instability from axial mode diffractograms is treated by Frank (1969) who treats the object (and therefore image) movement as a convolution with a vibration spread function the squared transform of which modulates the diffractogram.

Adequate experimental data is not yet available on the shape of $E(\delta)$, so a Gaussian profile (normal distribution) is commonly assumed; the form $(1 - \delta^2)^{-1/2}$ would be appropriate to sinusoidal energy fluctuation. Frank (1975c) gives a method for assessing at least a single width parameter for it. The use of charged-wire biprism interference experiments (Hibi & Takahashi, 1969; Burge, Dainty, et al., 1976) has allowed a better assessment of $B(\mathbf{s})$, and Frank's method allows assessment of the width of this too. It seems clear from results obtained by a new technique (Saxton, 1977) based solely on conventional diffractogram analysis, that the mathematically convenient Gaussian profile is an adequate model, being indistinguishable from the more natural assumption of a rectangular profile under normal operating conditions. The middle row of Fig. 1.3.1 illustrates the spatial coherence filter $\widetilde{B}$ for axial illumination for $D' = 1$ and $\sqrt{5}$ beneath the corresponding functions $\sin \gamma$, taking a Gaussian source profile with a generalized rms width of 0.1; the temporal coherence filter $\tilde{E}$ is shown bottom right for a Gaussian equivalent defocus distribution with a generalized rms width of 0.2: these values are typical of good 100 kV microscopes. The combined transfer function $\tilde{B}\tilde{E}\sin\gamma$ is shown bottom right for $D' = \sqrt{5}$ only.

A brief reference should be made to the fact that we have assumed a range of *plane* waves incident on the specimen, i.e., the effective source is imaged by the condenser system at infinity, whereas in practice the source is at a finite distance and produces a range of *spherical* waves illuminating the object; in fact this makes no difference to the results achieved above. For each component wave in the illumination, $\psi_o(\mathbf{x})$ as used in the preceding is simply multiplied by a quadratic phase factor describing the illuminating wave; consideration of (1.1.9) shows that the factor can be cancelled by an appropriate choice of g/h so that the diffraction pattern of the object is formed in a different plane from that assumed hitherto (namely, the conjugate to the source); onward propagation restores the Fourier transformation as before and gives in the same image plane the same wave function as before, but with the quadratic phase factor multiplying it. The intensity being unchanged, we can carry the other calculations through as before.

We will also note briefly that (1.3.21) permits us to assess the dependence of the signal-to-noise ratio of the image on the illumination coherence and instrumental defocus (Frank & Al-Ali, 1975), for the noise level is independent of these quantities (cf. Section 1.6) while the total signal power is obtainable via Parseval's theorem as the integral of $|\tilde{c}(\mathbf{t})|^2$, i.e., the integral of (1.3.21), or (1.3.24) in the axial case. Decreasing coherence narrows $A_0(\mathbf{t})$, reducing the signal power rapidly, while varying the defocus causes marked fluctuations in the power.

The expression (1.3.16) for the image contrast transform is valid for any source profile no matter how large, and it is probably worth repeating that a linear transfer function exists no matter how large the source if the object is only weak. For spatially extended sources however, the transfer of $\tilde{\eta}(\mathbf{t})$ into the image becomes very small. The same is not true of the amplitude component of the object, and the use of extended sources may be a good approach to the imaging of heavy "marker" atoms (Nagata et al., 1975; Engel et al., 1974). That this is so is most easily seen by considering the *wave function* point spread function—not the point image mentioned previously, but simply the (complex) inverse transform of $\tilde{T}(\mathbf{t})$, which we naturally denote by $T(\mathbf{x})$. This is the image wave function for a delta function object wave function, as is easily confirmed by Fourier transformation of (1.3.1). The intensity of this image would be $|T(\mathbf{x})|^2$, and an assumption of spatially completely incoherent illumination would permit us to calculate the image intensity for an object with an effective complex transmission factor $\psi_o(\mathbf{x})$ as the convolution of $|\psi_o(\mathbf{x})|^2$ with $|T(\mathbf{x})|^2$, which is not only nonzero but shows a higher theoretical resolution than the coherent illumination mode since $|T(\mathbf{x})|^2$ is narrower than $T(\mathbf{x})$. The class of multiplicative objects with high resolution information in $|\psi_o|^2$ is however limited, and the assumption of *complete* spatial incoherence is in any case unsatisfactory; the usual "incoherent" source is in fact annular, and the nonlinear nature of the resulting imaging process has been pointed out by Hanszen and Ade (1975), although a linear theory is commonly assumed (particularly for the scanning transmission electron microscope or STEM—see Section 1.4).

A brief account is appropriate in this section of the charge density approximation, which has recently been used to estimate crystal lattice images for thicknesses up to 5–7 nm (Lynch et al., 1975) when the object-induced phase shifts in the beam arc no longer weak. The degree of success achieved by the approximation in matching experimental images is perhaps surprising for it is in fact a further weak object approximation and indeed a low resolution approximation besides. It is obtained by retaining only the defocus term in the wave aberration function γ (1.3.2), which is a good approximation for normal levels of astigmatism provided that $|\mathbf{t}|^2 \leqslant D/2C_s\lambda^2$ or so. The transfer function is expanded to two terms,

$$\widetilde{T}(\mathbf{t}) = \exp(-i\gamma(\mathbf{t})) \simeq 1 + \pi i\lambda D|\mathbf{t}|^2,$$

which is adequate if $|\mathbf{t}|^2 \leqslant 1/(4\pi\lambda D)$ or so (Lynch et al. allow twice this limit). The object wave function, taken to be free of modulus variation, is represented by the expansion

$$\psi_o(\mathbf{x}) = \exp(i\eta(\mathbf{x})) \simeq 1 + i\eta(\mathbf{x}).$$

From (1.3.1) the transform of the image wave function thus becomes

$$\psi_i(\mathbf{t}) = (\delta(\mathbf{t}) + i\tilde{\eta}(\mathbf{t}))(1 + \pi i\lambda D|\mathbf{t}|^2),$$

so that the image wave function is

$$\psi_i(\mathbf{x}) = 1 + i\eta(\mathbf{x}) - (\lambda D/4\pi)\nabla^2\eta(\mathbf{x}),$$

(since the transform of $|\mathbf{t}|^2\tilde{\eta}(\mathbf{t})$ is the two-dimensional derivative $-\nabla^2\tilde{\eta}(\mathbf{x})/4\pi^2$) and the image intensity, to first order in η, is

$$I(\mathbf{x}) = 1 - (\lambda D/2\pi)\nabla^2\eta(\mathbf{x}). \tag{1.3.25}$$

Since $\eta(\mathbf{x})$ is proportional to the projected object potential, $\nabla^2\eta(\mathbf{x})$ is proportional to the projected charge density (via the Poisson equation, which survives the projection), and the image intensity, within the limits of the present approximations, may be directly interpreted in terms of the object charge density.

Taken together the two restrictions above on $|\mathbf{t}|^2$ are least stringent for $D = (C_s\lambda/2\pi)^{1/2}$ (a generalized defocus of 0.40) when $|\mathbf{t}|$ is limited to $0.45C_s^{-1/4}\lambda^{-3/4}$, implying an interpretable resolution no higher than 1.2 nm on good 100 kV microscopes. At this resolution, the illumination coherence will normally have little effect on the image quality. The approach can be extended to incorporate the spherical aberration but loses its simplicity and therefore usefulness in the process. It should be noted that the optimum defocus for the use of the approximation is roughly half that normally used in high resolution phase contrast studies of weak objects (the Scherzer defocus).

We have said little so far about inelastically scattered electrons, and it is beyond our present scope to treat the image formed by these in any detail, but the general drift of the current theory can be indicated before we leave the realm of the linear theory. Misell and Burge (1973) treat electrons with a given energy, scattered a given number of times, as forming a coherent beam the image of which can be calculated by the theory given above, with a final average over energy (which needs the extension mentioned following (1.3.19)) and number of collisions. At the optimum elastic defocus, the inelastic image is heavily defocused quite apart from chromatic effects and contributes little more than a constant background for high resolution purposes. However, a much higher theoretical resolution is predicted at the relatively large underfocus needed to refocus on the inelastic component, though it remains to be seen precisely how localized the inelastic scattering events are in disordered materials. Certainly inelastic electrons make a significant contribution to the image contrast well beyond the low resolution range in the tilted beam modes which correct the chromatic aberration.

1.4. Nonlinear imaging

By nonlinear imaging, we mean the general case of (1.3.1) when the object is no longer sufficiently weak for the image intensity to be dominated by the interference between primary and scattered waves, or when the primary wave is removed from the imaging process by a suitable aperture to give "dark-field," as opposed to "bright-field," conditions. If we break the object wave function into two parts

$$\psi_o(\mathbf{x}) = 1 + \psi_{so}(\mathbf{x}),$$

we have for the image intensity under bright-field conditions

$$I_B(\mathbf{x}) = 1 + 2\,\mathrm{Re}\{\psi_{si}(\mathbf{x})\} + |\psi_{si}(\mathbf{x})|^2, \tag{1.4.1}$$

in which $\tilde{\psi}_{si} = \tilde{T}\tilde{\psi}_{so}$ according to (1.3.1); while under dark-field conditions,

$$I_D(\mathbf{x}) = |\psi_{si}(\mathbf{x})|^2. \tag{1.4.2}$$

It was the middle term in (1.4.1) that previously gave rise to the linear theory. We are now concerned either with (1.4.2) or cases of (1.4.1) where the third term is no longer negligible. In contrast to the wealth of satisfactory results obtainable within the linear theory, we are now left with a few bald statements of full but not particularly helpful expressions for the image intensity. These must be given at some point however, and we shall present them in a way that enables us to relate the STEM to the conventional instrument.

Since the above decomposition or the wave is only useful for weak or near weak objects, we revert to the full object and image plane wave functions, $\psi_o(\mathbf{x})$ and $\psi_i(\mathbf{x})$ (for perfectly coherent illumination). Transformation of (1.3.1) yields the convolution

$$\psi_i(\mathbf{x}) = \int \psi_o(\mathbf{x}')\,T(\mathbf{x}-\mathbf{x}')\,d^2\mathbf{x}', \tag{1.4.3}$$

so that the image intensity is

$$I(\mathbf{x}) = \left| \int \psi_o(\mathbf{x}')\,T(\mathbf{x}-\mathbf{x}')\,d^2\mathbf{x}' \right|^2, \tag{1.4.4}$$

the expression being difficult to exploit because of the squaring involved. If we restrict ourselves to objects, which although no longer weak, nevertheless remain multiplicative (see Section 1.2) over the range of incidence angle and energy of the illuminating beam, we can proceed to a further simple expression for the intensity under arbitrary illumination conditions. We allow $\psi_o(\mathbf{x})$ to assume the role of the object transmission function; as in (1.3.16) we generalize $\tilde{T}(\mathbf{t})$ to the form $\tilde{T}(\mathbf{t}_o + \mathbf{t}, \delta)$, this being the transfer function appropriate to a monochromatic point source effectively at $\mathbf{t}_o$ with an energy shift equivalent to an additional instrumental defocus of δ. The quantity

$$\psi_i(\mathbf{x}, \mathbf{t}_o, \delta) = \int \tilde{\psi}_o(\mathbf{t}) \widetilde{T}(\mathbf{t}_o + \mathbf{t}, \delta) \exp(2\pi i \mathbf{t} \cdot \mathbf{x})\, d^2\mathbf{t} \tag{1.4.5}$$

then differs from the image wave function for this source only by a linear phase factor $\exp(2\pi i \mathbf{t}_o \cdot \mathbf{x})$, and we obtain an image intensity $|\psi_i(\mathbf{x}, \mathbf{t}_o, \delta)|^2$. If we take $b(\mathbf{t})$ and $E(\delta)$ to quantify the angular and chromatic illumination profiles as in the previous section (1.3.16), we may add contributions for different source elements incoherently to obtain the required general expression for the image intensity,

$$I(\mathbf{x}) = \int \int b(\mathbf{t}_o) E(\delta) |\psi_i(\mathbf{x}, \mathbf{t}_o, \delta)|^2\, d^2\mathbf{t}_o\, d\delta. \tag{1.4.6}$$

This established, we turn to the STEM, in which a lens characterized by a transfer function $\widetilde{T}(\mathbf{t})$ focuses a "probe" $T(\mathbf{x}' - \mathbf{x})$ on the specimen at a point $\mathbf{x}$ which is scanned across the specimen while a detector, effectively in a subsequent diffraction plane, records the total transmitted flux within its own aperture. For a monochromatic point source at $(\mathbf{x}, \delta)$, the detector plane wave function will be

$$\psi_d(\mathbf{x}, \mathbf{t}_o) = \int \psi_o(\mathbf{x}') T(\mathbf{x}' - \mathbf{x}, \delta) \exp(-2\pi i \mathbf{x}' \cdot \mathbf{t}_o)\, d^2\mathbf{x}', \tag{1.4.7}$$

$\mathbf{t}_o$ being the coordinate in the detector plane. Replacing ψ_o and T by their Fourier transforms, it is easy to recast (1.4.7) in the form

$$\psi_d(\mathbf{x}, \mathbf{t}_o) = \int \tilde{\psi}_o(\mathbf{t}) \widetilde{T}(\mathbf{t} - \mathbf{t}_o, \delta) \exp(2\pi i (\mathbf{t} - \mathbf{t}_o) \cdot \mathbf{x})\, d^2\mathbf{t} \tag{1.4.8}$$

since $\widetilde{T}(\mathbf{t})$ is symmetric. This differs from $\psi_i(\mathbf{x}, -\mathbf{t}_o, \delta)$ only by a linear phase factor; accordingly, a detector with an aperture $b(\mathbf{t}_o)$ records a signal

$$I_d(\mathbf{x}) = \int b(-\mathbf{t}_o) E(\delta) |\psi_i(\mathbf{x}, \mathbf{t}_o, \delta)|^2\, d^2\mathbf{t}_o\, d\delta, \tag{1.4.9}$$

in which the energy spread of the beam has been incorporated and which differs from (1.4.6) only in the trivial reflection of $\mathbf{t}_o$. Accordingly if the source distribution and objective lens of the conventional microscope are identified with the detector aperture and probe-forming lens of the STEM, the two instruments give identical images provided only that the specimen is multiplicative (*not* necessarily weak). This satisfactory result is now known as the reciprocity theorem (Zeitler & Thomson, 1970; Thomson, 1973). The finite size of the STEM source, so far ignored, is simply treated by convolving $I_d(\mathbf{x})$ with the source intensity distribution (Engel, 1974); the STEM source function can therefore be considered analogous to the point response function of the photographic emulsion used to record the conventional image intensity (1.4.6). The two instruments differ of course for nonmultiplicative objects such as thick crystals; other differences include the relative signal-to-noise ratios, the relative convenience of subsequent or real-time processing and of obtaining energy filtered images, and the possibility of exploiting the freedom with which a STEM detector profile may be varied (Rose, 1975; Dekkers & De Lang, 1974) compared with the conventional source distribution.

The relationship between the STEM and the conventional microscope means that for the former (as for the latter) a linear image formation theory exists for weak objects whatever the detector (illumination) conditions; a linear theory exists for strong multiplicative objects also for a large detector (completely incoherent illumination). The popular annular detector has been widely treated as corresponding to the completely incoherent case in the conventional microscope but is really an intermediate case, and Cowley (1975) has recently drawn attention to the corresponding restrictions in the use of the linear theory.

The use of (1.4.4) and (1.4.6) is necessary, even for weak objects, under dark-field illumination conditions such as may be produced by placing a narrow wire across the center of the objective aperture, by tilting the primary beam until it is intercepted by this aperture, or by using an annular condenser aperture to provide a ring source falling just outside the objective aperture (which eliminates the directionality of tilted beam modes). Such modes are attractive because of the high contrast levels in the image, but the loss of a linear image formation theory renders interpretation difficult at high resolution (e.g., Hanszen and Ade, 1975; Ade, 1975; Krakow, 1976) except for isolated atoms. The applicability of a linear theory to the semi-incoherent case of a *large* annular source remains the subject of some debate, as mentioned previously. Linear or not, the interesting comparison images calculated by Engel et al. (1974) for a group of isolated osmium atoms on a carbon supporting film in several different imaging modes confirm that the heavy atoms are most reliably identifiable in modes with an extended source and in the STEM annular detector mode in particular; it seems unlikely the results will be so favorable in this mode for other types of specimen.

1.5. Image analysis and object reconstruction

An x-ray crystallographer might reasonably feel that electron microscopists, able to deflect their scattered beams electromagnetically and thereby allow them to interfere with each other, should have little reason to complain. At moderate resolution, micrographs bear out this view, being capable of imaging crystalline regions far smaller than can be analyzed by x-ray methods, and, even more importantly, of providing information on the local structure of disordered materials, inorganic or biological, for which x-ray methods can provide only a rather unhelpful statistical average. However it will have become clear in the course of the two preceding sections that the circumstances in which *high* resolution micrographs may be safely interpreted directly are limited, while even the highest resolution presently available allows the use of the standard crystallographic constraint (that atoms are isolated point or nearly point objects) only occasionally. Accordingly, this section attempts to survey the expedients that can be adopted to enable a more reliable assessment of the specimen than may be made by simple visual inspection of the micrographs (see also Frank, 1973a; Saxton, 1976).

We shall consider only cases where the object is sufficiently thin to be treated in projection (cf. Section 1.2) and will initially seek to construct the object plane wave function. (Altogether more ad hoc procedures, based on the comparison of many micrographs with the theoretical predictions of different object models, are the only current resorts for thicker specimens.) We shall also assume the availability of digital processing equipment capable of implementing the schemes outlined; the choice and use of such equipment, together with the scope of optical processing methods, is reviewed from Chapter 8 onward.

For specimens sufficiently thin to be weak, the image formation theory (Section 1.3) is linear, and object restoration is correspondingly linear and straightforward in principle. For a weak phase object, the bright-field image contrast is given by (1.3.22), i.e.,

$$\tilde{c} = 2A\tilde{\eta}\sin\gamma \tag{1.5.1}$$

for axial illumination, and the phase η of the object wave function is thus easily extracted in principle by transforming the micrograph, dividing by $2A\sin\gamma$, and performing the inverse transformation (Childs & Misell, 1973; Stroke et al., 1974; also see Section 9.7). The division is however impossible where A is too small (beyond a "cut-off" spatial frequency determining the best resolution possible in the reconstruction) or where $\sin\gamma$ is too small (on circles, or near circles with astigmatism present, at various spatial frequencies which may be well below the cut-off); at these places, the transform $\tilde{\eta}$ has to be set to zero, and the reconstruction accordingly has an incomplete spectrum. While the gaps are unimportant for the recognition of isolated atoms, they are a serious nuisance in most cases and of course can be disastrous for periodic specimens where whole diffracted orders may be lost.

A second micrograph taken at a different defocus value may supply the lost information, since the zeros of $\sin\gamma$ will fall in different places. In principle, such a micrograph permits the determination of the amplitude component of a weak object as well as the phase component (Frank, 1972; Burge & Scott, 1975) from the equations

$$\begin{aligned} \tilde{c}_1 &= 2A(\tilde{a}\cos\gamma_1 + \tilde{\eta}\sin\gamma_1), \\ \tilde{c}_2 &= 2A(\tilde{a}\cos\gamma_2 + \tilde{\eta}\sin\gamma_2) \end{aligned} \tag{1.5.2}$$

(cf. (1.3.6) and (1.3.22)), except where A is too small or where γ_1 and γ_2 differ by a multiple of 2π and the two equations become identical; this latter situation is easily avoided by restricting the defocus difference between the two micrographs, a pair at (generalized) defocus 1 and $\sqrt{3}$, for example, giving no gaps below a (generalized) spatial frequency of 1.65, which is almost the 0.3 nm level on a good 100 kV microscope.

The zeros in $\sin\gamma$ can be removed by using half-plane apertures (1.3.7), one such picture being adequate for the characterization of a weak phase object within the limits set by A, and a complementary pair being required for a mixed amplitude and phase object (Downing & Siegel, 1975; also see Sections 3.5–3.7).

Extending the range of A (short of achieving smaller sources and more stable electrical supplies) necessitates a resort to tilted beam modes (1.3.20). The transfer function is again free of zeros over a substantial part of the spatial frequency spectrum, as discussed in Section 1.3. Unfortunately, the reconstructable region of the object spectrum becomes highly asymmetric, and this is not usually tolerable, so that several reconstructions must be combined to fill in the entire transform plane to a higher resolution than is achieved by axial illumination (Hoppe, 1971a).

If a several member focal series of micrographs is available, the problem is greatly eased since the required transform values are overdetermined (Erickson & Klug, 1970). Schiske (1968, 1973) has presented an elegant optimal solution in these circumstances, equally applicable to micrograph sets of other kinds as well (cf. Section 9.7).

The restorations discussed so far have been purely two-dimensional. Where the beam-object interaction can be described by a projection process (of potential, for thin objects at high resolution, or of mass, with a scattering contrast mechanism for stained biological specimens), a set of two-dimensional object projections in different directions can be combined to reconstruct the three-dimensional object. It is not our intention here to give any real discussion of this controversial field because it is with two-dimensional restoration that we shall be concerned, but a brief survey is appropriate. The basic principle of the reconstruction is straightforward (De Rosier & Klug, 1968). Two-dimensional transformations of the projections give sections through the three-dimensional object transform, normal to the direction of projection; a sufficient number of projections thus enables the entire object transform and hence the object itself to be determined. The practical difficulties arise from the fact that, except for objects with so high a degree of internal symmetry that a sufficient number of projections is provided by a single micrograph or a very small number of micrographs, such as icosahedral virus particles or helical bacteriophage tails, it is not possible to support and tilt an object particle through a full range of angle, and there are consequently large "blind" regions in the object transform where no data are available. The use of conical tilting (Hoppe, 1972) leads to somewhat better coverage than does tilting about a fixed axis, but there are still extensive blind regions in the general case. However, the *a priori* knowledge that the object is of limited spatial extent, and that its density is positive, means that the unknowns are overdetermined, and a full reconstruction is often possible even from the limited data available; several alternative methods have been proposed, often by more than one author, for making effective use of them. With a fixed tilting axis, the reconstruction may obviously be reduced to a series of independent problems involving the two-dimensional reconstruction of slices normal to the tilting axis from one-dimensional projections, which greatly reduces the size of the calculations necessary.

The methods may be grouped in three main categories. The first of these comprises Fourier plane methods such as that used by de Rosier and Klug (for a treatment suitable to more general objects at high resolutions, see Hoppe, 1970). The extrapolation problem for the blind regions has been treated in detail by Crowther et al. (1970) in both Cartesian and cylindrical coordinate systems; the sampling equations being linear (cf. (2.2.9)), the interpolation problem reduces to matrix inversion, and the procedure is limited in principle by the decreasing size of the higher order eigenvalues, and in practice by the size of the matrices involved. Although allowing the best use of any symmetry properties the object is known to possess, and easy addition to a given reconstruction of higher resolution data subsequently obtained, Fourier plane methods do not allow the positivity constraint to be exploited easily, nor such other possible constraints as the precise spatial bounds of the object or a maximum object density. The approach indicates clearly however the number of projections needed for a reconstruction of a general object of diameter D to a resolution d, for the sampling theorem (2.2.9) requires the Fourier plane samples to be no further apart than $1/D$ in these circumstances, while the highest order components in the transform will be $1/2d$ from the origin, giving a maximum angular separation of $2d/D$ between the different views.

The second group of methods involve "back projection," i.e., the simple superposition of the available projections extended along the corresponding direction of projection (e.g., see Vainshtein, 1971). This is almost equivalent to performing the Fourier inversion of the object transform as in the first group of methods, since the inverse transform of a section through the object transform is simply the back projection of the relevant projection. The difference lies in the fact that the increasing separation of the data on transform sections with increasing distance from the origin should be reflected by a radial weighting in the back-projection approach; this corresponds to differentiation and Hilbert transformation of the original projections (Radon's integral). Accordingly, the back projection of these "modified" back projections (Ramachandran & Lakshminarayanan, 1971) is equivalent to the first group of methods without any extrapolation, having the same limitations, though being easier in some cases to implement.

The third group of methods tackles the real space projection equations directly. These being linear, simple matrix inversion is possible in principle for the reconstruction, being limited again by the size and conditioning of the matrix. In an iterative approach to the solution of these equations, widely known as ART, Gordon et al. (1970) simply scale the estimated densities within the projected columns to match the observed projection value, repeating the scaling for each projection angle in turn until the process converges. Numerous variants of this process have been developed that permit easy incorporation of positivity and other similar constraints through which it is hoped that the "blind region" problem of the Fourier plane methods may be at least partly overcome. These three categories by no means exhaust the range of methods that have been suggested, the most serious omission probably being the use of orthogonal function decompositions other than Fourier ones; Budinger and Gullberg (1974) and Gordon and Herman (1974) give useful comparative reviews of the subject.

We have carefully avoided thus far any real mention of the effect of the observational noise level present in practical micrographs. This is commonly far higher than is usual in light optics. Hunt (1975, p. 704) refers in that context to a signal-to-noise power ratio of ten as being a high noise case, for example, whereas a ratio of four would be a low noise case for much higher resolution electron microscopy, and ratios of one or less are not uncommon. The factors controlling the noise level are discussed in Section 1.6 where it is also shown that for bright-field microscopy of weak objects the noise is at least additive and signal-independent. We give some attention to how the adverse effects of noise can be minimized because from this new point of view it is clear that procedures such as that based on a close defocus pair for complex object restoration (previously discussed) are likely to be somewhat unreliable, while matrix inversion schemes for extending three-dimensional object transforms into blind regions must be severely curtailed.

For the special, but very important, case of periodic objects, enormous reductions in noise level can be achieved by superposing the images of many unit cells (or, equivalently, suppressing all contributions to the transform not falling on reciprocal lattice sites) as discussed in Section 9.6, while the presence of periodic structures can be detected underneath very high noise levels by examining the image transform for strong local peaks or through the image autocorrelation function (cf. Section 9.3).

On the other hand, separation of noise from the image signal in the general case is possible only if the noise and the signal have different statistics. For example, they might be separated in a histogram of image intensity values, or more commonly they might be spatially separated in the image power spectrum. With carefully planned microscopy however, the separation is at best partial and usually slight, so there is little advantage in attempting to exploit it. However, procedures involving the superposition of several micrographs will reduce the noise levels somewhat. Section 9.7 reviews optimal restoration procedures of this type in addition to procedures that avoid noise amplification in single micrograph restorations.

Since it is the elastic image component from which we are seeking to reconstruct the object, the inelastically scattered electrons might be regarded as a form of noise. At the highest resolutions, these are sufficiently out of focus to contribute only to very low ($\simeq$2 nm) spatial frequencies in normally focused micrograph. Where the inelastic image component is troublesome however, for example, with biological specimens, two expedients can be adopted for its minimization. One of these (Misell & Burge, 1975; Krakow, Welles, et al., 1976) involves taking the difference between two micrographs in either bright- or dark-field conditions, made at carefully chosen defocus values near the elastic focus; the inelastic image varies little between these two and thus largely disappears on subtraction, and the art is then to secure a useful elastic difference image. In bright-field, this is equivalent to having a transfer function $2(\sin\gamma_1 - \sin\gamma_2)$ in place of $2\sin\gamma_1$, and a reasonable choice is not difficult; in dark-field, the difficulty of interpretation, except for isolated heavy atoms, is at any rate rendered little greater by the subtraction. Misell and Burge give difference point images, though the concept of a point image must be used with care in this nonlinear context. The alternative approach (Burge, 1976) is the subtraction of a bright-field/dark-field (central stop) micrograph pair at the same defocus value; apart from the portion intercepted by the beam stop in dark-field, the inelastic image component is the same in the two micrographs, and it thus disappears to a large extent in the subtraction. It is a further attractive feature of this subtraction that the nonlinear term in the image intensity (1.4.1), being equal to the dark-field intensity (1.4.2), is also removed. Both of these methods, of course, being subtraction methods, may have harmful effects on the level of other types of noise in the images and should accordingly be used with some care.

For many specimens, particularly macromolecules, a supporting film is used as a mounting in the microscope, and this inevitably produces an unwanted image signal too. In the case of periodic objects, the support film structure vanishes with other noise when an average of many cells is taken. In the general case, Langer et al. (1970) evaluated a procedure in which micrographs taken of the same area before and after mounting of the specimen are subtracted but find the support film itself insufficiently stable in structure. Very thin crystalline support films are an attractive possibility (e.g., see Riddle & Siegel, 1971, or Baumeister & Hahn, 1974). Where the diffracted beams are intercepted by the object aperture, which still permits high resolution for graphite support films, for example, the support produces no image contrast at all, while if they are passed by the aperture, the resulting periodic structure can be easily filtered out of the image at little cost.

Those procedures involving the use of more than one micrograph, presuppose of course the possibility of achieving adequate relative alignment.

This problem is elegantly solved by the use of cross-correlation techniques, discussed at length in Section 9.4 and 9.5, which permit extremely accurate alignment of micrographs differing in position, orientation, magnification, and transfer conditions. Section 9.7 also considers how the transfer function in force during the recording of a bright-field micrograph can be deduced from the diffraction pattern of an image area for which the object is amorphous (cf. (1.3.24)).

Thus far, noise apart, the reconstruction process has been promising and relatively straightforward because of the assumed restriction to weak objects and bright-field imaging conditions, i.e., to the linear theory of Section 1.3. If we now turn to the nonlinear cases of Section 1.4, the situation is very far from satisfactory. Of the various schemes proposed at different times, none makes any real attempt to treat other than coherent illumination conditions, so we will limit discussion to the coherent case. The schemes have sometimes been designated methods for "phase determination" because it is our inability to record the image wave function phase that gives rise to the difficulties, and they have been contrasted with the transfer function correction methods reviewed earlier for weak objects. However, it is the object rather than image wave function that is sought, and in view of the diversity of the techniques and data proposed, it is perhaps more useful to view them simply as procedures for object restoration for nonlinear rather than linear images.

Gabor's (1948) now well-known paper "A New Microscopic Principle" represents the first proposals for reconstructing general objects. His aim was not so much the correction of the objective lens observations as the elimination of the lens altogether, the far-field diffraction pattern being recorded when a small specimen was placed in a wide electron beam diverging from a small source, and the "hologram" thus formed being reconstructed optically. While Gabor's idea has been so spectacularly applied in other fields, it has long been frustrated in electron microscopy by the stringent coherence requirements, and an account of the current state of electron holography is given at the end of this section. In a subsequent paper echoing Gabor's title, Cowley (1953) suggested that a specimen might be uniquely characterized from a series of tilted dark-field micrographs with a sufficient number of members to approximate a continuous range of tilt angles. The method does not lead directly to the object wave function, and the analysis is difficult unless the object is treated as a collection of point sources.

Nathan (1971) outlines a scheme for periodic objects in which many micrographs are made in turn while a small objective aperture which passes only a few beams at once scans the entire diffraction plane; if the number of beams passed is sufficiently small and the alignment of the micrographs accurately known, the lateral displacement of the image fringe systems reveal the relative phases of the different diffraction orders. In an analogous proposal, Hoppe (1969) [see also Hoppe & Strube (1969); and Hegerl & Hoppe (1972)] determines the relative phases of the spots by the use of extended coherent sources (nonuniform coherent object illumination) which cause the spots to spread and overlap, revealing their relative phase through the resulting interference patterns. Hoppe has called this approach "ptychography"; extension to nonperiodic objects is possible.

Frank (1974) points out that a bright-field/dark-field pair of micrographs at the same defocus value determines the image wave function immediately to within a twofold choice, since knowledge of $|\psi_{\rm i}|$ and $|\psi_{\rm i} - 1|$ restricts $\psi_{\rm i}$ to the intersection of two circles. He suggests the use of continuity arguments to resolve the ambiguity, at least within contours of zero phase, if the micrographs are sufficiently closely sampled, and he surmises that constraints of the type discussed in Chapter 3 might reduce the ambiguity still further.

The remaining possible schemes for reconstruction in the nonlinear case are treated extensively in Chapters 3–7 of this work and need not be detailed here. That given most attention is the use of combined diffraction plane and image plane micrographs (Gerchberg & Saxton, 1971), but the similar case of a defocus micrograph pair (Misell, 1973a) and (for bright-field only) the use of "one-sided" micrograph pairs made with complementary halves of the diffraction plane obscured (Burge et al., 1974) are also examined, together with one case of a micrograph triplet (bright-field/dark-field and diffraction pattern). Of all the schemes mentioned, only one (the image and diffraction plane pair) has yet led to useful practical results, so the scope for future developments is considerable, and interest in these approaches currently remains high.

It would be surprising if the reconstruction problem stepped abruptly from being straightforward to being nearly intractable as soon as the nonlinear term in the image intensity (1.4.1) ceased to be completely negligible, and this is indeed not the case. For *nearly* weak objects, the weak object restoration procedures can be applied iteratively, the first restoration allowing estimation of the nonlinear terms with a consequent improvement in a second restoration and so on. The rate of convergence of course depends on how large the nonlinear term is relative to the linear interference term; when it is the dominant term, convergence fails completely, and recourse to the more sophisticated general methods becomes necessary.

Some brief remarks on electron holography bring this section to an end. This is again the subject of some current interest, particularly in the single side band mode which removes the unwanted conjugate image from the reconstruction. In the arrangements described by Wahl (1974) and Munch (1974), a specimen occupies one half of the object field, the other half being open, while a charged-wire biprism (Möllenstedt & Düker, 1956) in a lower conjugate plane tilts the waves through the two halves of the object field so that they overlap in a final hologram plane. In this plane the wave function takes the form

$$\psi_i(\mathbf{x}) = \psi_a(\mathbf{x}) \exp(\pi i \mathbf{t}_0 \cdot \mathbf{x}) + \exp(-\pi i \mathbf{t}_0 \cdot \mathbf{x}),$$

in which $\psi_a(\mathbf{x})$ denotes the (aberrated) object transmission function and $\mathbf{t}_0$ quantifies the tilt introduced by the biprism. The corresponding intensity is

$$I(\mathbf{x}) = |\psi_i(\mathbf{x})|^2 = |\psi_a(\mathbf{x})|^2 + 1 + \psi_a(\mathbf{x}) \exp(2\pi i \mathbf{t}_0 \cdot \mathbf{x}) + \psi_a^*(\mathbf{x}) \exp(-2\pi i \mathbf{t}_0 \cdot \mathbf{x}),$$

and its transform is

$$\tilde{I}(\mathbf{t}) = \Psi(\mathbf{t}) + \delta(\mathbf{t}) + \tilde{\psi}_a(\mathbf{t} - \mathbf{t}_0) + \tilde{\psi}_a^*(-\mathbf{t} - \mathbf{t}_0), \qquad (1.5.3)$$

$\Psi(\mathbf{t})$ denoting the autocorrelation of $\tilde{\psi}_a(\mathbf{t})$. If $\mathbf{t}_0$ is sufficiently large, the terms of (1.5.3) are spatially separated so that $\tilde{\psi}_a(\mathbf{t})$ can be obtained easily from the image intensity either optically or digitally; if the object is weak, $\Psi(\mathbf{t})$ may be ignored, reducing the required value of $\mathbf{t}_0$. The field of view is limited by the coherence attainable at the specimen, but the low image contrast levels available appear to be a more serious limitation, particularly in view of the fact that the need to obtain $\tilde{I}(\mathbf{t})$ over a larger region than usual implies an image recorded to rather higher resolution than usual. Munch reports a lateral revolution no better than 5 nm to date and recommends "in-line" holograms instead; these, albeit differently treated, amount to no more than conventional defocused micrographs, the information content of which we have here discussed extensively.

A proposal due to Lohmann (1974), closely related to holography, involves the recording of two micrographs made with small off-axis beam stops in different positions. Although this leads to a simple closed form solution for the wave function, it seems likely to be frustrated in practice by observational noise unless there happen to be particularly strong diffracted beams present which might be intercepted by the stops.

1.6. Resolution, contrast, noise, and radiation damage

The topics covered in this section may at first seem many and unrelated, but it is the object of the section to establish their very close interconnection.

Under high resolution conditions, the most important source of noise in the recorded image is that due to the fact that a finite number of electrons contribute to it, and it is called "quantum" or "shot" noise. For a given exposure time and picture element, for which the number of incident electrons predicted by the image formation theories of Sections 1.3 and 1.4 is p, the number actually incident will be a random variable following a Poisson distribution with a mean p, and consequently a variance p also, so that the rms noise level is $\sqrt{p}$. If the image contrast is small, as in the bright-field imaging of weak objects, the noise level is effectively the same for all picture elements and may be treated as additive and signal-independent. This convenient description is not possible in dark-field imaging modes, however.

If the bright-field image contrast at a given point is c and the number of electrons incident on the specimen per unit area is n, then the signal within a resolution element of linear size d will be cnd^2, while the noise level will be $d\sqrt{n}$. Requiring that the former be about five times the latter for adequate visual recognition (Rose, 1948) thus gives

$$n \geqslant 25/c^2 d^2. \tag{1.6.1}$$

This inequality may be viewed from several angles; since an ultimate limit is placed on n by the rate of destruction of the specimen by the electron irradiation, it may be regarded as fixing the resolution ultimately attainable for a given object. To obtain the corresponding inequality for dark-field conditions, we note that a bright-field contrast c implies that the real part of the scattered wave is a fraction $c/2$ of the primary wave, therefore carrying a fraction $c^2/4$ of the electron flux; allowing an equal flux in the imaginary part (though there will be less for low aberration situations), we take the dark-field signal, i.e., the expected number of electrons in the resolution element, to be $nd^2c^2/2$. The rms noise will be the square root of this, and a five-times criterion for recognizability gives

$$n \geqslant 50/c^2 d^2, \tag{1.6.2}$$

differing from the bright-field case by no more than a factor of two in exposure or $\sqrt{2}$ in resolution. In practice, much larger values of n than this are normally used to achieve a darkening of the photographic plate adequate for visual recognition. The factor of five assumed necessary between signal and noise might be relaxed in particular applications. In particular we might consider the detectability of the presence of an isolated heavy marker atom in the object; if no more than one scattered electron contributes on average to its dark-field image, the probability of a nonzero image is about 63%, and it is about 86% if there are two scattered electrons on average. In practice, the support film normally produces a background signal that prevents recognition of the atom at exposures as low as this, but the use of crystalline support films, which could give a zero *wave function* between the heavy atoms, could yet make detection at these levels possible.

The implication behind the assertion that the electron quantum noise is the most significant of the noise processes is that the great majority of the electrons are successfully detected by the photographic emulsion or whatever other recording device is being used. Electron emulsions have been usefully reviewed by Valentine (1966) and by Farnell and Flint (1975). In contrast to light photography, where the cooperation of many photons is required to render a given silver grain developable, a single electron at normal accelerating voltages typically renders between 2 and 20 grains developable, the number varying with the emulsion thickness, the accelerating voltage, and the grain density. This increase in the mean number of quanta per picture element implies that the signal-to-noise ratio becomes little worse when the electron beam is recorded on a photographic plate, and no reasonable emulsion does more than halve it, while the deterioration can be negligible for some. Moreover, again unlike light photography, the optical density of the developed emulsion grows linearly with the exposure (i.e., the incident electron intensity) for densities lower than one, which is extremely convenient for subsequent interpretation. Typically 1 electron μm^{-2} will render a fast emulsion developable to an optical density of one.

Image intensifier systems cannot offer an ultimate improvement in the image noise statistics since those for the best emulsions are already as good as can be expected, but they may greatly ease the problem of focusing. Typical emulsions have a grain size of less than 1 μm, offering a resolution too small to be visible on a fluorescent screen, which is of course necessary for focusing; increasing the magnification leads to very low screen brightness, requiring considerable dark adaptation, besides involving a higher object exposure than is necessary for a given final optical density in the plate.

Glaeser (1975) recommends the use of larger grain faster emulsions, giving a more suitable resolution level, but an image intensifier might well be necessary to render the image visible to the operator at the high magnification involved.

Biological materials are most sensitive to radiation damage; detailed reviews of the problem are given by Stenn and Bahr (1970) and Glaeser (1975). Heating effects appear to be relatively unimportant, since except for high beam currents, the damage effected appears to depend only on the total electron dose and not on the dose rate; accordingly a "critical dose" for destruction can be defined for a given material. Not surprisingly, in practice, different values for the dose are obtained depending on the damage criterion used; a relatively low value is obtained from the fading of electron diffraction patterns, implying the loss of periodic order in the specimen, but a higher value may be appropriate to mass loss and a higher one still to the loss of the characteristic energy loss spectrum of the material; therefore individual molecules may be substantially more resistant to damage than the lattice forces. There is a wide range in the critical dose for different materials, but the range 1–100 mC cm^{-2} ($\frac{1}{2}$–60 electrons Å^{-2}) embraces most biological macromolecules, the higher figure applying to heavily double-bonded substances and including the DNA bases. Typically the dose increases by a factor of two or three if the specimen is held at liquid helium temperatures and by a similar factor as the accelerating voltage rises from 100 kV to 600 kV–1 MV. Where its influence on the image can be tolerated, a conducting coating (e.g., of gold) around the specimen increases the dose by a rather larger factor.

These considerations make it clear that even assuming the careful design of the experiment and the use of minimum exposure techniques in focusing (e.g., Williams & Fisher, 1970), we are unlikely in practice ever to be able to obtain a single high resolution image of aperiodic biological materials, let alone pairs or series of them, since the contrast levels available from such materials in the absence of heavy atom stains are at most a few percent. The case of periodic specimens is much more promising than the general situation, since high noise levels may be reduced by averaging the images of many unit cells (cf. Section 9.6) and successive micrographs can be obtained from entirely fresh preparations of the specimen. Apart from this case, however, it would seem to be inorganic specimens, disordered or crystalline, for which really high resolution imaging will be possible.

CHAPTER TWO

The discrete Fourier transform☆

W. Owen Saxton
Murray Edwards College, University of Cambridge, Cambridge, United Kingdom
e-mail address: wos1@cam.ac.uk

Contents

2.1. Definition and fundamental properties

The "discrete" Fourier transform appears initially as an approximation to an integral transform

$$\tilde{g}(t) = \int_{-\infty}^{\infty} g(x) \exp(-2\pi ixt)\, dx$$

in which the transform values are calculated at an evenly spaced series of points from a similar set of tabulated values of $g(x)$ by means of the expression

$$\tilde{g}(j\Delta t) = \sum_{k=-N/2}^{N/2} g(k\Delta x) \exp(-2\pi ijk\Delta t\Delta x)\Delta x,$$

the summation being restricted to a finite number of terms for obvious practical reasons. However, a discrete Fourier transform is unlike most such approximations; it represents an exact decomposition of a sequence of values into a series of sampled harmonic terms and is described by a matrix algebra in which many of the results normally associated with the integral transform have exact analogs. This chapter collects together the more important of these as they are used extensively in subsequent chapters; digital methods deal necessarily with "discrete" images, and discrete transforms provide a natural and satisfactory computational tool. In what follows, sequences of N values are parameterized by a subscript running from 0 to

☆ Reprinted from Advances in Electronics and Electron Physics, Supplement 10, 1978.

Advances in Imaging and Electron Physics, Volume 214
ISSN 1076-5670
https://doi.org/10.1016/bs.aiep.2020.04.002

$N-1$. A summation convention is used in which summation is assumed over subscripts appearing two *or more* times in a given expression.

Given a sequence of N values g_k, we define its discrete Fourier transform by

$$\tilde{g}_j = g_k W_{jk}, \tag{2.1.1}$$

where $W_{jk} = W_{kj} = \exp(-2\pi ijk/N)$, the value depending on the *product* only of the subscripts. By inspection we see that W_{jk} is a unitary matrix with the property that

$$W_{jk} W_{lk}^* = N\delta_{jl} \tag{2.1.2}$$

and, in consequence, the unique inverse

$$W_{jk}^{-1} = (1/N) W_{jk}^*.$$

Thus we may retrieve g_k from $\tilde{g}_j$ by inverse transformation as

$$g_k = (1/N)\tilde{g}_j W_{jk}^*. \tag{2.1.3}$$

We note that W_{jk} is periodic in j and k with period N; thus (2.1.1) and (2.1.3) may be taken to define $\tilde{g}_j$ and g_k for any value of subscript as periodic sequences of period N. We may therefore shift any subscript into the range 0 to $N-1$ by addition of a suitable multiple of N without affecting the value of the subscripted quantity. Finally, we note that

$$W_{-j,k} = W_{j,-k} = W_{jk}^*, \tag{2.1.4}$$

that

$$W_{jk} W_{lk} = W_{j+l,k}, \tag{2.1.5}$$

and that

$$W_{jk} W_{lk} W_{mk\ldots} W_{pk}^* W_{qk}^* W_{rk\ldots}^* = N\delta_{j+l+m\ldots,p+q+r\ldots}. \tag{2.1.6}$$

(Note that (2.1.2) is a special case of this result.)

A form of Parseval's theorem can be obtained. Given two sequences f_k and g_k, we may write

$$\tilde{f}_j \tilde{g}_j^* = f_k W_{jk} g_l^* W_{jl}^* = N f_k g_k^*$$

by (2.1.2), giving the result

$$\tilde{f}_j\tilde{g}_j^* = N f_k g_k^*, \tag{2.1.7}$$

of which a special case is the well-known property of unitary matrices that

$$\tilde{g}_j\tilde{g}_j^* = N g_k g_k^*. \tag{2.1.8}$$

The transform of a constant sequence is one with a single nonzero value, a discrete "delta function": if $g_k = c$, its transform $g_k W_{jk}$ may be written as $cW_{0k}^* W_{jk}$ (since $W_{0k} = W_{k0} = 1$) and by (2.1.6) this is simply $cN\delta_{j0}$.

The transforms of real or imaginary sequences have symmetry properties. We have

$$\begin{aligned} \tilde{g}_{-j} &= g_k W_{-jk} = g_k W_{jk}^* \quad \text{by (2.1.4)}, \\ &= (g_k W_{jk})^* \quad \text{if } g_k \text{ is real}, \end{aligned}$$

so that the transform of a real sequence has the property of conjugate symmetry

$$\tilde{g}_{-j} = \tilde{g}_j^* \tag{2.1.9}$$

(implying that $\tilde{g}_0$ and $\tilde{g}_{N/2}$ are real). Similarly, the transform of an imaginary sequence has the property of conjugate antisymmetry

$$\tilde{g}_{-j} = -\tilde{g}_j^*. \tag{2.1.10}$$

Next, we consider reflection, conjugation, and translation of sequences. If $f_k = g_{-k}$, then

$$\tilde{f}_j = g_{-k} W_{jk} = g_k W_{j,-k} = g_k W_{-jk} = \tilde{g}_{-j}, \tag{2.1.11}$$

so that reflecting a sequence also reflects its discrete transform. If $f_k = g_k^*$, then

$$\tilde{f}_j = g_k^* W_{jk} = (g_k W_{jk}^*)^* = (g_k W_{-jk})^* = \tilde{g}_{-j}^*, \tag{2.1.12}$$

so that taking the complex conjugate of a sequence conjugates and reflects the transform. Translating $\tilde{g}_j$ by p steps, we have

$$\tilde{g}_{j+p} = g_k W_{j+p,k} = g_k W_{pk} W_{jk} \quad \text{by (2.1.5)}$$

$$= g_k \exp(-2\pi ipk/N) W_{jk}, \quad (2.1.13)$$

so that a circular shift of the transform corresponds to multiplication by a linear phase factor.

Finally, we note the important circular convolution and correlation properties. The inverse transform of the product $\tilde{g}_j\tilde{f}_j$ is

$$\begin{aligned}(1/N)\tilde{g}_j\tilde{f}_j W^*_{jp} &= (1/N) g_k W_{jk} f_l W_{jl} W^*_{jp} \\ &= g_k f_l \delta_{k+l,p} \quad \text{by (2.1.6)} \\ &= g_k f_{p-k} \quad (= g_{p-k} f_k), \end{aligned} \quad (2.1.14)$$

so that a circular convolution of the two sequences may be obtained by inverse transforming the products of the transforms of the sequences.

Similarly, a circular correlation may be obtained as

$$(1/N)\tilde{g}_j\tilde{f}_j^* W^*_{jp} = g_k f^*_{k-p} \quad (= g_{k+p} f^*_k). \quad (2.1.15)$$

Note that this implies that a circular autocorrelation may be obtained as the inverse transform of a power spectrum:

$$(1/N)\tilde{g}_j\tilde{g}_j^* W^*_{jp} = g_k g^*_{k-p}. \quad (2.1.16)$$

2.2. Approximation of integral transforms

We now examine the nature of the distortions introduced by representing an integral transform by a finite summation in this way.

We define the integral Fourier transform of a function $g(x)$ by

$$\tilde{g}(t) = \int_{-\infty}^{\infty} g(x) \exp(-2\pi itx)\, dx. \quad (2.2.1)$$

If we construct an "aliased" form of $\tilde{g}(t)$ as

$$\tilde{g}_p(t) = \sum_{n=-\infty}^{\infty} \tilde{g}(t - nT), \quad (2.2.2)$$

we may represent this periodic function by a Fourier *series*

$$\tilde{g}_p(t) = \sum_{n=-\infty}^{\infty} c_n \exp(-2\pi int/T),$$

with coefficients given by

$$c_n = \frac{1}{T}\int_0^T g_p(t)\exp(2\pi int/T)\,dt.$$

Replacing $\tilde{g}_p(t)$ by the summation of (2.2.2) and suitably shifting the variable of integration, we see that

$$c_n = (1/T)\int_{-\infty}^{\infty} \tilde{g}(t)\exp(2\pi int/T)\,dt = (1/T)g(n/T)$$

giving the result that

$$\tilde{g}_p(t) = (1/T)\sum_{n=-\infty}^{\infty} g(n/T)\exp(-2\pi int/T). \qquad (2.2.3)$$

We thus see that the use of an infinite sequence of sample values to represent $g(x)$ results in a transform distorted by overlap between successive terms of (2.2.2) unless $\tilde{g}(t)$ differs from zero only in an interval of less than T.

We now examine the effect of restricting the number of terms in (2.2.3) to N. If we sample $\tilde{g}_p(t)$ at N points in each cycle, with a separation $\Delta t = T/N$, we have

$$\tilde{g}_p(j\Delta t) = (1/T)\sum_{n=-\infty}^{\infty} g(n/T)\exp(-2\pi inj/N). \qquad (2.2.4)$$

Now defining an aliased form of $g(x)$ by

$$g_p(x) = \sum_{m=-\infty}^{\infty} g(x - m/\Delta t),$$

we may reexpress the summation of (2.2.4) in the form

$$\tilde{g}_p(j\Delta t) = (1/T)\sum_{k=0}^{N-1} g_p(k/T)\exp(-2\pi ikj/N). \qquad (2.2.5)$$

Apart from the factor $1/T$, this is identical in form with (2.1.1), the definition of a discrete transform. In general therefore, the use of the discrete representation results in overlap distortion of both the function and its transform. While *one* of these may be of finite extent, avoiding overlap in that plane, it is impossible for *both* to be so (see Section 3.1).

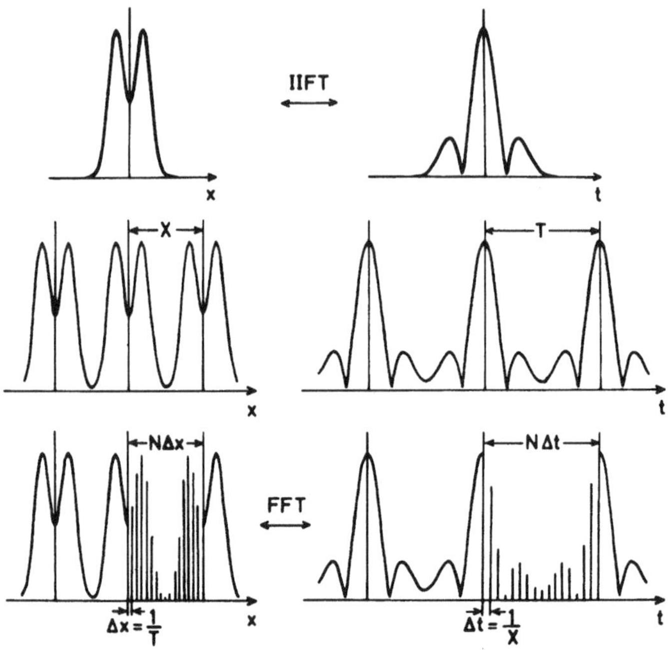

Figure 2.2.1 Sampling and aliasing in one dimension.

Fig. 2.2.1 summarizes the transition from an integral transform pair of functions to a discrete transform pair of sequences. The two functions at the top are integral transforms of each other; the central graphs show the aliased forms, which are *not* transforms of each other; at the bottom, one cycle of each function has been sampled to give a pair of sequences which, apart from the $1/T$ factor in (2.2.5), are discrete transforms of each other. The relations between sampling intervals and aliasing periods are also summarized in the figure.

We note that the number of points N used in the discrete representations is equal to TX in the notation of Fig. 2.2.1. Since the widths of the function/transform pair represented set a lower bound on T and X, we obtain the well-known results that an acceptable representation requires a minimum number of points given by the product of the widths of the function and its transform. The approximation to the infinite integral transform becomes closer as $\Delta x \to 0$ and $\Delta t \to 0$, for example, implying that $N\Delta x \to \infty$.

The representation of periodic functions $g(x)$ in discrete transform processes is analogous but rather simpler. We may express $g(x)$ as a Fourier series

$$g(x) = \sum_{n=-\infty}^{\infty} c_n \exp(2\pi inx/X), \tag{2.2.6}$$

and take sample values

$$g(k\Delta x) = \sum_{n=-\infty}^{\infty} c_n \exp(2\pi ink/N), \tag{2.2.7}$$

if $N\Delta x = X$ as before. An aliased form of the coefficient series c_n is now defined by

$$d_j = \sum_{n=-\infty}^{\infty} c_{j-nN},$$

so that

$$g_k = \sum_{j=0}^{N-1} d_j \exp(2\pi ijk/N), \tag{2.2.8}$$

i.e., $d_j = \tilde{g}_j/N$. Overlap distortion occurs in the Fourier plane only now, and in the special case of *band-limited* periodic functions, where the Fourier series has no more than N terms, the representation is exact. This analogy between "discrete functions" and band-limited periodic functions is exploited further later (Sections 3.3 and 4.5, particularly).

Sampling and aliasing may be conveniently viewed in terms of multiplication by, and convolution with, infinite arrays of delta functions ("comb" functions), the transforms of which are similar arrays with reciprocal spacing. Multiplication in one plane by such an array constitutes sampling and is of course equivalent to convolution in the other plane, i.e., aliasing.

It is appropriate at this point to note the "sampling" theorems about the exact representation of band-limited functions by discrete sample values. In the general case, the theorem follows from (2.2.3), taking $\tilde{g}(t)$ to be nonzero only in a central interval of length less than T, so that over this range, $\tilde{g}_p(t) = \tilde{g}(t)$. Multiplication by a rectangle of width T and inverse transformation yields

$$g(x) = \sum_{n=-\infty}^{\infty} g(n/T)\,\mathrm{sinc}(T(x - n/T)), \tag{2.2.9}$$

with

$$\mathrm{sinc}(x) = \sin(\pi x)/\pi x,$$

establishing that $g(x)$ is completely specified by its values at a sequence of points separated by no more than $1/T$.

In the case of periodic functions, the summation can be reexpressed with finite limits. Only the central terms of (2.2.6) are now nonzero, so that

$$g(x) = \sum_{-m}^{m} c_n \exp(2\pi inx/X). \tag{2.2.10}$$

In terms of sample values of $g(x)$,

$$c_n = (1/N) \sum_{k=0}^{N-1} g(k/T) \exp(-2\pi ink/N),$$

with $N = 2m + 1 = XT$, so that

$$\begin{aligned} g(x) &= (1/N) \sum_{k=0}^{N-1} g(kT) \sum_{n=-m}^{m} \exp(2\pi in(x - k/T)/X) \\ &= \sum_{k=0}^{N-1} g(k/T) s_N((x - k/T)/X), \end{aligned} \tag{2.2.11}$$

with the sampling function

$$s_N(x) = (1/N) \sin(N\pi x)/\sin(\pi x).$$

With $N > 2m + 1$, the range of the summation in (2.2.10) is formally extended, and the result continues to hold except that when N is even

$$s_N(x) = (1/N) \exp(\pm \pi ix) \sin(N\pi x)/\sin(\pi x).$$

With the basic relations between sampling intervals and aliasing periods (fields of view) established, we may briefly note a "reciprocal width" property for discrete Fourier transforms. Given a pair of functions $g(x)$ and $\tilde{g}(t)$ which are an integral transform pair, and defining as measures of their width

$$\sigma_x^2 = (1/E) \int x^2 |g(x)|^2 \, dx, \qquad \sigma_t^2 = (1/E) \int t^2 |\tilde{g}(t)|^2 \, dt,$$

with $E = \int |g(x)|^2\, dx = \int |\tilde{g}(t)|^2\, dt$, we have the well-known inequality

$$\sigma_x \sigma_t \geqslant 1/4\pi, \tag{2.2.12}$$

the equality holding when the functions are Gaussian. We might similarly define the widths of a discrete transform pair g_k and $\tilde{g}_j$ by

$$\sigma_k^2 = (1/E) \sum_{k=-N/2}^{N/2-1} k^2 |g_k|^2, \qquad \sigma_j^2 = (1/NE) \sum_{j=-N/2}^{N/2-1} j^2 |\tilde{g}_j|^2,$$

with $E = g_k g_k^* = \tilde{g}_j \tilde{g}_j^*/N$. It is unfortunately not possible to derive an inequality between σ_k and σ_j by the method used to establish (2.2.12) since this involves differentials of $g(x)$ and $\tilde{g}(t)$. However, it is useful to consider the implication of (2.2.12) for σ_k and σ_j when the sequences approximate the integral transform pair closely. Under these circumstances,

$$\sigma_k = \sigma_x/\Delta x, \qquad \sigma_j = \sigma_t/\Delta t,$$

so that

$$\sigma_k \sigma_j = \sigma_x \sigma_t/(\Delta x \Delta t) \geqslant 1/(4\pi\, \Delta x \Delta t).$$

Now $\Delta x \Delta t = 1/N$ (see Fig. 2.2.1), and we obtain the approximate inequality

$$\sigma_k \sigma_j \geqslant N/4\pi. \tag{2.2.13}$$

As before, the equality will hold most nearly when the sequences approximate Gaussians; as the information content of the sequences rises, $\sigma_k \sigma_j$ will exceed $N/4\pi$ by progressively greater amounts.

2.3. Multidimensional forms

Given a sequence of N values $g_{kmp\cdots}$ parameterized by s subscripts of which the first takes N_1 values, the second N_2, and so on, we define the s-dimensional discrete Fourier transform by

$$\tilde{g}_{jln\cdots} = g_{kmp\cdots}\ {}_1W_{jk}\ {}_2W_{lm}\ {}_3W_{np\cdots}, \tag{2.3.1}$$

with ${}_1W_{jk} = \exp(-2\pi i jk/N_1)$, etc. as previously. The inverse transform follows at once as

$$g_{kmp\cdots} = (1/N)\tilde{g}_{jln}\ {}_1W_{jk}^*\ {}_2W_{lm}^*\ {}_3W_{np\cdots}^*, \tag{2.3.2}$$

with $N = N_1 N_2 N_3 \cdots$.

All the properties established above for one-dimensional transforms have obvious multidimensional analogs, which are listed in the following:

a.

$$\tilde{g}_{jln\cdots}\tilde{h}^*_{jln} = Ng_{jln}h^*_{jln}; \tag{2.3.3}$$

b.

$$\text{if } g_{kmp\cdots} = c, \text{ then } \tilde{g}_{jln\cdots} = c\delta_{j0}\delta_{l0}\delta_{n0}\cdots; \tag{2.3.4}$$

c.

$$\text{if } g_{kmp\cdots} \text{ is real, then } \tilde{g}_{-j-l-n\cdots} = \tilde{g}^*_{jln\cdots}; \tag{2.3.5}$$

d.

$$\text{if } h_{kmp\cdots} = g_{-k-m-p\cdots}, \text{ then } \tilde{h}_{jln\cdots} = \tilde{g}_{-j-l-n\cdots}; \tag{2.3.6}$$

e.

$$\text{if } h_{kmp\cdots} = g^*_{kmp\cdots}, \text{ then } \tilde{h}_{jln\cdots} = \tilde{g}^*_{-j-l-n}; \tag{2.3.7}$$

f. the inverse transform of $\tilde{g}_{j+r,l+s,n+t,\cdots}$ is

$$g_{kmp\cdots}\exp(-2\pi i(rk/N_1 + sm/N_2 + tp/N_3 + \cdots)); \tag{2.3.8}$$

g. the inverse transform of $\tilde{g}_{jln\cdots}\tilde{f}_{jln\cdots}$ is the convolution

$$c_{rst\cdots} = g_{kmp\cdots}f_{r-k,s-m,t-p,\cdots}. \tag{2.3.9}$$

The approximation of multidimensional integral Fourier transforms by finite transforms of sequences of function values sampled on regular lattices also follows the one-dimensional analysis in the preceding. If the sampling lattice is orthogonal, the extension is trivial, and the relations summarized in Fig. 2.2.1 apply to each coordinate independently.

More generally, we may inquire what the finite transform of a sequence which samples a function on an arbitrary lattice represents. (We might, for example, wish to describe the potential distribution in a crystal by sample values on a lattice whose base vectors were parallel to those of the crystal lattice.) The multidimensional infinite integral transform of a function $g(x)$ defined on an s-dimensional space is

$$\tilde{g}(\mathbf{t}) = \int_{-\infty}^{\infty}\int_{-\infty}^{\infty}\cdots g(\mathbf{x})\exp(-2\pi i\mathbf{t}\cdot\mathbf{x})\,d^s\mathbf{x}.$$

If we choose a sampling lattice with base vectors $\mathbf{\Delta x}_1, \mathbf{\Delta x}_2, \cdots$, we may approximate this by

$$\tilde{g}(\mathbf{t}) = D \sum_{k=-\infty}^{\infty} \sum_{m=-\infty}^{\infty} \cdots g_{km\cdots} \exp(-2\pi i \mathbf{t} \cdot (k\mathbf{\Delta x}_1 + m\mathbf{\Delta x}_2 + \cdots)), \quad (2.3.10)$$

where

$$g_{km\cdots} = g(k\mathbf{\Delta x}_1 + m\mathbf{\Delta x}_2 + \cdots),$$

and D is the determinant

$$\begin{vmatrix} \Delta x_{11} & \Delta x_{12} & \cdots \\ \Delta x_{21} & \Delta x_{22} & \cdots \\ \vdots & \vdots & \vdots \end{vmatrix},$$

whose rows are the components of the base vectors $\mathbf{\Delta x}_1, \mathbf{\Delta x}_2, \cdots$. If we divide the cofactors of the elements of each row by D, we obtain a set of vectors $\mathbf{T}_1, \mathbf{T}_2, \cdots$ with the property that

$$\mathbf{T}_n \cdot \mathbf{\Delta x}_m = \delta_{nm}.$$

(For a three-dimensional lattice, these are the base vectors of the familiar reciprocal lattice.) If a sampling lattice is now defined with base vectors

$$\mathbf{\Delta t}_1 = \mathbf{T}_1/N_1, \ \ \mathbf{\Delta t}_2 = \mathbf{T}_2/N_2 \cdots,$$

then the value of (2.3.10) at these lattice points is

$$\begin{aligned} &\tilde{g}(j\mathbf{\Delta t}_1 + l\mathbf{\Delta t}_2 + \cdots) \\ &= D \sum_{k=-\infty}^{\infty} \sum_{m=-\infty}^{\infty} \cdots g_{km\cdots} \exp(-2\pi i(jk/N_1 + lm/N_2 + \cdots)), \end{aligned}$$

and the form of the exponent in this expression is that of (2.3.1). It now becomes clear how the extension to arbitrary multidimensional lattices proceeds; (2.2.2) is replaced, for example, by a summation in which $\tilde{g}(\mathbf{t})$ appears centered on all points of the lattice whose base vectors are $\mathbf{T}_1, \mathbf{T}_2, \cdots$.

Fig. 2.3.1 shows a transition from infinite integral to discrete transforms that is analogous to that of Fig. 2.2.1 for the case of two-dimensional transforms. At the top are shown the reciprocal vectors $\mathbf{T}_1$ and $\mathbf{T}_2$ corresponding to the sampling lattice base vectors $\mathbf{\Delta x}_1$ and $\mathbf{\Delta x}_2$; a pair of infinite integral transforms follow in contour representations; next the aliased forms of these

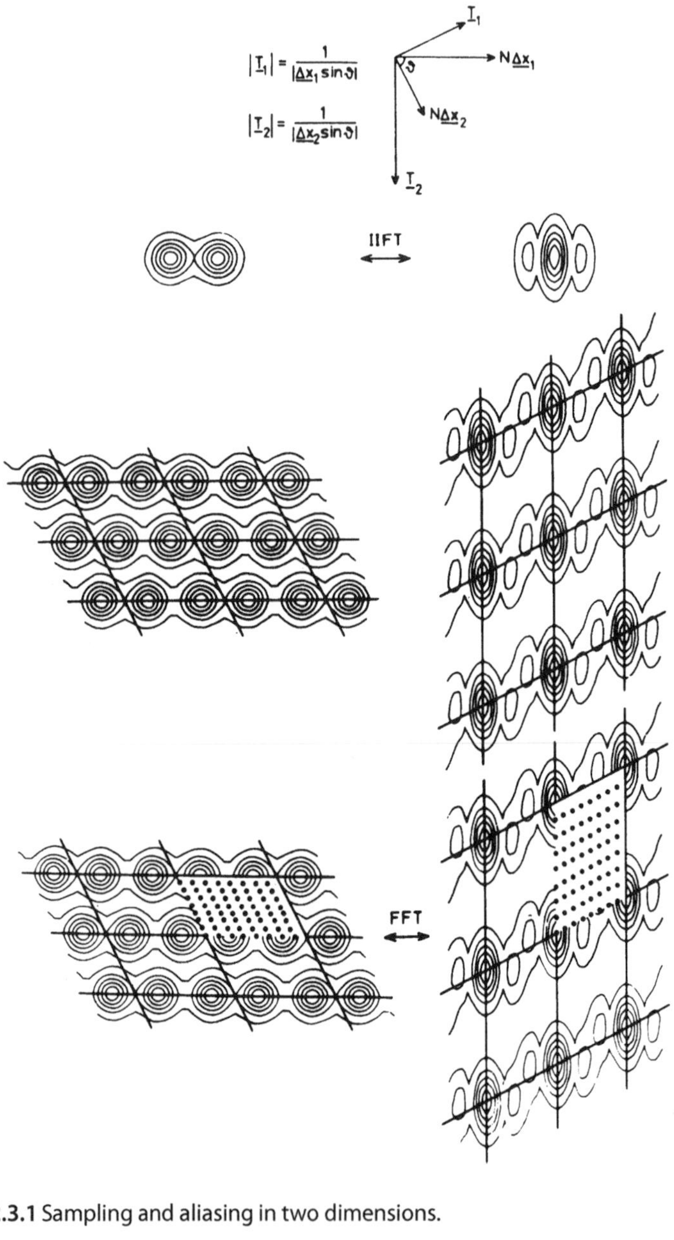

Figure 2.3.1 Sampling and aliasing in two dimensions.

are shown, and finally the sampling lattices are superposed on the aliased forms, the sequences of sample values so obtained being discrete transforms of each other.

CHAPTER THREE

Analytic images☆

W. Owen Saxton
Murray Edwards College, University of Cambridge, Cambridge, United Kingdom
e-mail address: wos1@cam.ac.uk

Contents

3.1. Complex zeros

The fact that electron lenses have a finite bore imposes a number of constraints on the wave function in the image plane, the sampling theorem (2.2.9) being the best-known example; some of the others are collected together in this chapter. In some cases (Sections 3.5–3.7 especially) these constraints can be directly exploited for phase determination from intensity measurements alone, although in others their force depends on a degree of accuracy in the measurements which is not presently attainable. In all cases they are of considerable theoretical interest and serve as a framework for ideas presented subsequently. Constraints of the same type have been previously applied in other fields, particularly particle scattering (e.g., Goldberger et al., 1963), quantum decay (e.g., Khalfin, 1958), optical dispersion, Fourier spectroscopy (e.g., Wolf, 1962), coherent light optics (e.g., Walther, 1963; Arsenault & Lowenthal, 1974), electrical network theory (e.g., Voelcker, 1966; Hofstetter, 1964), and radio astronomy (Bates, 1969). An attempt has been made to present elementary arguments for many of the results in order to make these ideas more widely accessible, and special attention has been given to the cases of periodic, discrete, and two-dimensional images.

☆ Reprinted from Advances in Electronics and Electron Physics, Supplement 10, 1978.

Advances in Imaging and Electron Physics, Volume 214
ISSN 1076-5670
https://doi.org/10.1016/bs.aiep.2020.04.003

Initially, the discussion is one-dimensional, extension to two dimensions being considered from time to time, especially in Section 3.4. This reflects the historical development of the material, the earlier applications being chiefly one-dimensional. These applications were, moreover, to cases where the signal in the Fourier plane vanished over one half of the plane—the one-sided imaging condition discussed in Sections 3.5–3.7. Here we treat this situation as a special instance of the more general results derived earlier in the chapter. The mathematics involved is that of analytic functions of complex variables, applied to Fourier integrals. The close connection between Fourier and Laplace transforms means that some of the references given are to results proved in the context of the latter; however, there is no difficulty in carrying them over to Fourier transform theory.

As outlined in the first chapter, the image plane wave function, under coherent conditions of illumination, is the Fourier transform of that in the diffraction plane

$$g(x) = \int_{-a}^{a} \tilde{g}(t) \exp(2\pi ixt)\, dt, \tag{3.1.1}$$

the integration limits reflecting the finite size of the objective aperture, or the lens bore in its absence. We now regard this as defining a function of a complex variable by replacing x by $z = x + iy$, the function being identifiable with the physical wave function on the real axis. The extension is useful because the function proves to be analytic: the Cauchy–Riemann relations

$$\frac{\partial u}{\partial x} = \frac{\partial v}{\partial y} = -2\pi \int_{-a}^{a} t \exp(-2\pi yt)\, \mathrm{Im}\{\tilde{g}(t) \exp(2\pi ixt)\}\, dt,$$
$$\frac{\partial u}{\partial y} = -\frac{\partial v}{\partial x} = 2\pi \int_{-a}^{a} t \exp(-2\pi yt)\, \mathrm{Re}\{\tilde{g}(t) \exp(2\pi ixt)\}\, dt$$

follow at once from (3.1.1), the integrals existing if that of (3.1.1) does. [Alternatively, see Whittaker & Watson (1950, Section 5.32).] The theory of Fourier integrals is most usefully applied to functions of class L^2, i.e., functions whose squared modulus is Lebesgue integrable. This involves no restriction on physically significant functions defined over a finite range, and we shall occasionally impose other reasonable restrictions for the sake of simplicity. "Delta" functions are not strictly functions at all and are excluded from the results of this section.

The property of analyticity implies immediately that the image plane wave function extends to infinity in both directions, since an analytic function cannot vanish over a finite length of any arc. It also implies that the wave function is determined everywhere by its value over an arbitrarily small region. In practice, exploitation of this property is dependent on arbitrarily accurate measurement of the wave function over this region. A given polynomial, for example, can be closely approximated over a small region by a lower order polynomial whose behavior outside the region will be quite different.

Next the behavior of $g(z)$ as $|z| \to \infty$ is considered. Applying the Schwarz inequality in the form

$$\left|\int g_1 g_2\right|^2 \leqslant \int |g_1|^2 \int |g_2|^2 \tag{3.1.2}$$

to the integral of (3.1.1),

$$|g(z)\exp(-2\pi a|z|)|^2 \leqslant \int_{-a}^{a} |\tilde{g}(t)|^2\,dt \int_{-a}^{a} |\exp(2\pi izt - 2\pi a|z|)|^2\,dt.$$

The first integral being finite, the integrand of the second vanishes as $|z| \to \infty$, except possibly at $t = \pm a$ where it remains finite. Hence

$$\lim_{|z|\to\infty} |g(z)|\exp(-2\pi a|z|) = 0. \tag{3.1.3}$$

On any line parallel to the real axis, $|g(z)|$ vanishes as $x \to \pm\infty$, as is easily seen by applying Parseval's theorem to (3.1.1). The condition (3.1.3) governs the growth of $|g(z)|$ in the y direction and identifies $g(z)$ as a function "of order unity" [Titchmarsh (1939, Section 8.2)]: the condition is in fact unnecessarily strong for the purpose. This is the simplest type of analytic function after polynomials, and a large number of theorems govern their behavior (e.g., Boas, 1954).

Paley and Wiener (1934, Theorem X) show that the condition (3.1.3), together with analyticity and square integrability on the real axis, is in fact sufficient to ensure the vanishing of $\tilde{g}(t)$ outside $(-a, a)$, so that (3.1.3) and (3.1.1) are equivalent. A similar result is established in the following by elementary methods and is sufficient for our present purposes. The result given by Paley and Wiener achieves greater conciseness by the use of theorems of the Phragmén–Lindelöf type to limit the growth of $g(z)$. A further application of (3.1.2) to (3.1.1) easily establishes that

$$\lim_{|y|\to\infty} |g(z)|\exp(-2\pi a|y|) = 0 \tag{3.1.4}$$

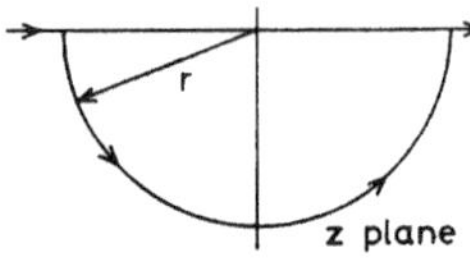

Figure 3.1.1

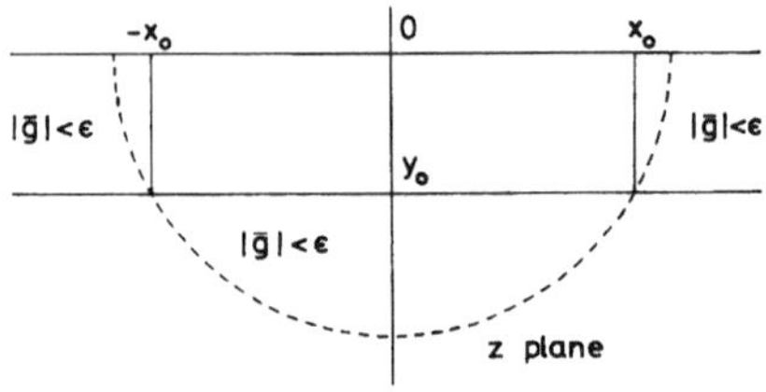

Figure 3.1.2

uniformly with respect to x. This is a stronger condition than (3.1.3) (and includes it, except along the real axis). We now show that this condition, together with the vanishing of $g(z)$ at both ends of any line parallel to the real axis, is sufficient for the vanishing of the inverse transform integral

$$\begin{aligned}\tilde{g}(t) &= \int_{-\infty}^{\infty} g(z)\exp(-2\pi izt)\,dz \\ &= \int_{-\infty}^{\infty} g(z)\exp(-2\pi iza)\exp(-2\pi iz(t-a))\,dz\end{aligned}$$

outside the interval $(-a, a)$. We close a contour over a portion of the real axis with a lower half-plane semicircle as in Fig. 3.1.1 for $t > a$. Jordan's lemma shows that the contribution from the semicircle (and hence from the portion of the real axis) vanishes in the limit $r \to \infty$ provided only that

$$\lim_{|z|\to\infty} |\bar{g}(z)| = |g(z)\exp(-2\pi iaz)| = 0$$

in the lower half-plane. This condition does indeed follow, not quite trivially, from our assumptions; the crucial point is that (3.1.4) is satisfied uniformly with respect to x. For any positive ε, we can therefore choose y_0 such that $|\bar{g}(z)| < \varepsilon$ below $y = y_0$ and then x_0 such that $|\bar{g}(z)| < \varepsilon$ for $|x| > x_0$ and $-y_0 \leqslant y \leqslant 0$, as in Fig. 3.1.2, after which $|\bar{g}(z)|$ is similarly limited for $|z| \geqslant |(x_0^2 + y_0^2)^{1/2}|$ (the broken semicircle). The vanishing of $\tilde{g}(t)$ for $t < -a$ is established by completing the contour in the upper half plane instead.

If we restrict $\tilde{g}(t)$ further, requiring that it take the limiting form

$$\tilde{g}(t) \sim c(t-a)^n \quad \text{as } t \to a \tag{3.1.5}$$

(n being a nonnegative integer), with a similar condition as $t \to -a$, we can derive explicit asymptotic forms for $g(z)$ off the real axis. The point is that as $|y| \to \infty$, the exponential factor in (3.1.1) diverges at a or $-a$ so that integral is governed entirely by the behavior of $\tilde{g}(t)$ near these points. Again the conditions on $\tilde{g}(t)$ are very weak for physically realizable functions, continuity for example being unnecessary except for finite intervals near a and $-a$. In the limit $y \to -\infty$, we now have

$$\begin{aligned} g(z) &= \int_{-a}^{a} \tilde{g}(t) \exp(2\pi i z t)\, dt \\ &\simeq c \exp(2\pi i a z) \int_{-a}^{a} (t-a)^n \exp(2\pi i z(t-a))\, dt \\ &\simeq c \exp(2\pi i a z) \int_{-\infty}^{0} t^n \exp(2\pi i z t)\, dt, \end{aligned}$$

extending the lower limit to infinity, since in the limit $y \to -\infty$ the integrand vanishes except near the upper limit. In this limit then

$$g(z) \simeq (-1)^n c(n!) \exp(2\pi i a z)/(2\pi i z)^{n+1}, \tag{3.1.6}$$

with a similar expression for the limit $y \to \infty$, the exponent and the whole expression having the opposite sign. These limits hold as $r \to \infty$ for any $\theta \neq 0$ or π, but not uniformly with respect to θ. A rigorous formulation of the result in the context of Laplace transforms is given by Doetsch (1950, Volume I, Chapter 14, Section 1). An obvious extension allows the upper and lower bounds of (3.1.1) to have different magnitudes, relaxing (3.1.5) still further.

While $g(z)$ cannot vanish over any finite length of arc, it does vanish at isolated points z_i (its complex zeros). Much of the theory of entire functions of finite order is devoted to the relationship between the asymptotic rate of growth of $g(z)$ and the distribution of its zeros. In fact, apart from two arbitrary constants, $g(z)$ is completely determined by the position of its zeros, and the appropriate expansion, a case of Hadamard's factorization theorem, is now derived. The result is of fundamental importance to subsequent sections. The general analyses of Titchmarsh (1939) and Boas (1954) are combined because they gain considerable clarity in being rewritten for this particular application.

The first step is to limit the asymptotic density of the zeros by means of Jensen's theorem,

$$\int_0^r \frac{n(r')}{r'}\,dr' = \frac{1}{2\pi}\int_0^{2\pi} \ln\left|\frac{g(r\exp i\theta)}{g(0)}\right| d\theta, \tag{3.1.7}$$

where $n(r)$ is the number of zeros of $g(z)$ with modulus less than r. This may be obtained *formally* by radial integration of the familiar expression

$$n(r) = \frac{1}{2\pi i}\oint \frac{g'(z)}{g(z)}\,dz \tag{3.1.8}$$

[see Titchmarsh (1939, Section 3.61)]. In view of (3.1.3), this gives

$$\int_0^r \frac{n(r')}{r'}\,dr' < 2\pi ar$$

in the limit $r \to \infty$. Since $n(r)$ is nondecreasing, we can extract $n(r)$ itself from this inequality:

$$n(r) = n(r)\int_r^{er} \frac{dr'}{r'} \leqslant \int_r^{er} \frac{n(r')}{r'}\,dr' \leqslant \int_0^{er} \frac{n(r')}{r'}\,dr' \leqslant 2\pi ear. \tag{3.1.9}$$

The density of zeros thus falls at least inversely with radius. This is strong enough to make the series

$$\sum_i (1/|z_i|^2)$$

converge absolutely, for if we suppose the zeros ordered so that $|z_i|$ forms a nondecreasing sequence, it gives

$$|z_n| \geqslant n/2\pi ea,$$

so that the series converges absolutely by comparison with $\sum(1/n^2)$.

Second, we construct an analytic function with zeros at the zeros z_i of $g(z)$, much as we might for a polynomial. An additional factor is necessary to ensure that the infinite product converges. We write

$$p(z) = \prod_i (1 - z/z_i)\exp(z/z_i). \tag{3.1.10}$$

The convergence of this is determined by that of the series

$$\sum_i \{\ln(1 - z/z_i) + z/z_i\}.$$

Ignoring the finite number of terms for which $|z/z_i| > \frac{1}{2}$, we expand this to give

$$\sum_i \left\{ -\frac{(z/z_i)^2}{2} - \frac{(z/z_i)^3}{3} - \dots \right\}.$$

The series *within* the brackets converges at least geometrically, its modulus not exceeding $|z/z_i|^2$, and the summation over i converges absolutely by the result in the preceding. Hence (3.1.10) converges *absolutely*. We shall wish to consider infinite subsets of its factors in the subsequent sections.

Finally, the quotient $g(z)/p(z)$ is analytic everywhere and has no zeros. We express it therefore as exp($g(z)$) where $q(z)$ is analytic. Since $g(z)$ is of order unity, the real part of $q(z)$ is bounded by $c|z|$ as $|z| \to \infty$, so that $q(z)$ is a polynomial $c_1 + c_2 z$. This completes the product expansion of $g(z)$:

$$g(z) = c_1 \exp c_2 z \prod_i (1 - z/z_i) \exp(z/z_i). \tag{3.1.11}$$

The constants c_1 and c_2 are easily identified as $g(0)$ and $g'(0)/g(0)$, respectively.

The result is closely related to the more familiar Mittag–Leffler theorem [e.g., Titchmarsh (1939, Section 3.2)] which allows expansion of $g'(z)/g(z)$ in the uniformly convergent series

$$\frac{g'(z)}{g(z)} = \frac{g'(0)}{g(0)} + \sum_i \left\{ \frac{1}{z - z_i} + \frac{1}{z_i} \right\},$$

integration of which leads directly to (3.1.11) if a series of contours tending to infinity can be found on which $g'(z)/g(z)$ is bounded; (3.1.6) gives the limit $\pm 2\pi i a$ for $g'(z)/g(z)$ in the two half planes, but this is not uniform with respect to θ, and it is difficult to establish directly the existence of contours across $\theta = 0$ or π on which $g'(z)/g(z)$ is bounded. Similar difficulties apply to the use of (3.1.8) to establish (3.1.9) directly.

We should mention the *minimum* modulus principle for functions of order unity at this point [e.g., Titchmarsh (1939, Section 8.75)]. This allows us to enclose the zeros of $g(z)$ in circles of arbitrarily small radius, outside which $|g(z)|$ is not *less* than $\exp(-2\pi(a+\varepsilon)|z|)$ for any positive ε in the limit $|z| \to \infty$. This limit on the rate of decay of $g(z)$ parallels the limit (3.1.3) on its rate of growth.

It is easily seen that the product (3.1.11) must always be infinite: for any finite number of factors, the expression gives

$$g(z) = p(z)\exp(cz),$$

$p(z)$ being a polynomial and c a constant. The behavior of this at the ends of the real axis is determined by the real part of c, and in no case can the expression vanish at *both* ends, which is of course required of $g(z)$ by Parseval's theorem. However, Bates (1969) shows that if the Fourier plane wave function itself contains only a finite number of Fourier components so that a finite number of image plane sample values $g_k = g(x_k)$ are adequate to represent the full wave function, only a small number of the zeros lie off the real axis. The image plane function in these circumstances can be expressed in the form

$$g(z) = \sum_{k=-n}^{n} (-1)^k g_k \frac{\sin 2\pi az}{\pi(2az - k)} \tag{3.1.12}$$

[by the sampling theorem (2.2.9)], which may be separated into the two factors

$$\sin 2\pi az / \prod_k (2az - k) \tag{3.1.13}$$

and

$$\prod_k (2az - k) \sum_k (-1)^k g_k / \pi(2az - k), \tag{3.1.14}$$

the former having real zeros only and the latter being a polynomial of degree $2n$ having therefore exactly $2n$ (in general, complex) zeros. Some of the force of this reduction is lost through the difficulty of specifying useful conditions which ensure that the diffraction plane wave function takes the assumed form. The assumption of an object consisting of a regular array of isolated points is adequate only in the absence of aberrations. Section 3.3 however demonstrates that a similar reduction in the number of significant zeros occurs in the important case of periodic objects.

A number of constraints follow at once, from symmetry conditions sometimes applicable on the real axis, on the positions of zeros, all easily established by considering the appropriate continuations of $g(z)$. If $g(x)$ is real or imaginary, the zeros z_i appear in pairs reflected in the real axis; if it is symmetric or antisymmetric, they appear in pairs reflected in the origin, while if it is conjugate symmetric or antisymmetric, they appear in pairs reflected in the imaginary axis. Roman and Marathay (1963) also point out that if $\tilde{g}(t)$ is real and positive, the integrand of (2.1.1) is positive definite for z on the imaginary axis so that $g(z)$ can have no zeros there.

This last result may be extended; if $\tilde{g}(t)$ is real and positive and restricted to the interval $(-a, a)$, the zeros are restricted to lie no nearer the imaginary axis than $1/4a$. We have in this case

$$g(z) = \int_{-a}^{a} \tilde{g}(t) \exp(-2\pi yt) \cos 2\pi xt \, dt - i \int_{-a}^{a} \tilde{g}(t) \exp(-2\pi yt) \sin 2\pi xt \, dt,$$

and both integrals must vanish at a zero of $g(z)$. If $|x| < 1/4a$, however, the integrand of the first is positive definite so that the integral cannot vanish.

Finally, we note that the finite width of the diffraction plane prevents the image from ever being exactly unimodular. The Fourier transform of $|g(x)|^2$ vanishes outside the interval $(-2a, 2a)$ so that $|g(x)|^2$ can be continued off the real axis as $g(z)g^*(z^*)$. If it is unity on the real axis, it will be unity everywhere, conflicting with the requirement that it have an infinite number of zeros.

3.2. Zero flipping

The previous section at first sight represents merely a different characterization of the image plane data without any reduction of their number. The crucial significance of the complex zeros of the analytic continuation of an image wave function lies in a theorem first explicitly stated for optics by Walther (1963). This considers the problem of finding a wave function $g(z)$ consistent with an observed image plane modulus distribution $|g(z)|$ under the constraint that the aperture plane wave function is zero outside the interval $(-a, a)$, i.e., that the Fourier transform $\tilde{g}(t)$ vanishes outside this interval. Walther demonstrates that from any one such wave function all other possible solutions may be obtained by constructing new functions $g(z)$ in which an arbitrary number of the zeros of the known solution are moved to complex conjugate positions. This process is called "zero flipping." The same result is given by Hofstetter (1964) in the context of electrical network theory.

The result follows from a consideration of $|g(x)|^2$. The transform of this being restricted to the interval $(-2a, 2a)$, it may also be continued off the real axis as $g(z)g^*(z^*)$. This has zeros disposed in mirror image pairs about the real axis; *one* of each pair must be present in any solution for $g(z)$ so the only possible manipulations which preserve $|g(x)|^2$ are zero flipping operations. Now flipping a zero z_i involves multiplication by

$$\left(1 - \frac{z}{z_i^*}\right) \Big/ \left(1 - \frac{z}{z_i}\right), \tag{3.2.1}$$

the modulus of which tends to unity as $|z| \to \infty$. Hence the limiting behavior of $g(z)$ is not affected by zero flipping, and $\tilde{g}(t)$ is therefore still limited to the interval $(-a, a)$ by the results of the previous section. The absolute convergence of (3.1.11) allows the flipping of an arbitrary number of zeros. (Note that multiplication by (3.2.1) changes the imaginary part of the constant c_2 in (3.1.11), while the argument of c_1 is arbitrary in any case.)

These results make it possible in principle to construct from an image plane intensity distribution all the wave functions whose aperture plane counterparts are restricted to the region of the aperture. From the continuation of the intensity distribution, the possible locations for the zeros z_i of $g(z)$ are established, and the various solutions for $g(z)$ may then be constructed from (3.1.11) by taking one zero from each pair. In the special case of all the zeros lying on the real axis (e.g., $\sin z/z$) the solution is unique; a priori information may limit the zeros to one half plane (see Section 3.6) also leading to a unique solution. In the general case, the number of solutions is (denumerably) infinite; in the situation discussed by Bates (1969; see Section 3.1) where $\tilde{g}(t)$ is represented by $2n + 1$ Fourier coefficients, there are up to $2n$ zeros which may be flipped, leading to 2^{2n} solutions. A similar limitation applies to periodic images (see Section 3.3).

The number of solutions may thus be formidable, but the fact that the finite width of the aperture plane eliminates all but a set of solutions of measure zero is nevertheless significant. The constraint becomes more powerful in two dimensions rather than less (see Section 3.4).

We may consider briefly how the zeros of a function may be located in practice, given only the function on the real axis. Bates (1969) gives the problem some attention. There are two basic approaches. The first is based on representing $\tilde{g}(t)$ with a finite number of Fourier coefficients, allowing factorization of $g(z)$ as in (3.1.12)–(3.1.14) as the product of a universal function with real zeros only (3.1.13) and a polynomial with coefficients easily determined from the image plane data (3.1.14), after which a standard routine may be used to locate the zeros of the polynomial. The second involves direct calculation of the continuation of the image plane function off the real axis by computing the transform $\tilde{g}(t)$ and using (3.1.1), i.e., by applying an exponential factor to $\tilde{g}(t)$ before inverting the transform to obtain $g(z)$ for a given value of y; the process is then repeated and the sections of $g(z)$ obtained scanned for zeros. This is time consuming, but the polynomial factorization is probably the more limited method since polynomials with more than a handful of terms become rapidly intractable numerically because of rounding errors. Both methods would normally be applied to the intensity $|g(x)|^2$ rather than to the wave function $g(x)$ itself, since the latter is not initially available. This also allows the search for zeros to be restricted to one half plane, which is a significant saving. Once the positions of the zeros have been established, the expansions (3.1.13) and (3.1.14) could be used to construct the various possible complex wave functions.

3.3. Periodic images

The case of periodic images can be derived from the foregoing results by treating the periodic function as the limit of a sequence of aperiodic functions obtained by multiplying the periodic function by a rectangle function of increasing width. However, all the results can be derived directly by elementary methods which are presented in this section.

If $g(x)$ is a periodic function, describing the image wave function formed by a finite-aperture optical system, it may be represented by a finite number of Fourier components

$$g(x) = \sum_{-n}^{n} \tilde{g}_j \exp(2\pi ijx). \tag{3.3.1}$$

This is patently analytic for all finite $z = x + iy$. We continue it off the real axis as $g(z)$ and introduce the transformation $w = \exp(2\pi iz)$ which maps each cycle of the x axis into a circle of unit radius centered on the origin of the w plane, the upper half plane into the region within the circle and the lower half plane into the region outside it. It also maps Fourier series into Laurent series (Titchmarsh, 1939, Section 13.12) so that (3.3.1) becomes the rational function

$$g(z) = h(w) = \sum_{-n}^{n} \tilde{g}_j w^j = w^{-n} \sum_{0}^{2n} \tilde{h}_j w^j, \tag{3.3.2}$$

where $\tilde{h}_j = \tilde{g}_{j-n}$. Apart from the factor w^{-n}, which has no zeros, this is a polynomial of degree $2n$ which has exactly $2n$ zeros, all at finite distances from the origin. Choosing an origin for x such that $h(1) \neq 0$, we can easily recast (2.3.2) as a product over its zeros w_i,

$$h(w) = h(1)w^{-n} \prod_{0}^{2n} \left\{ \frac{w - w_i}{1 - w_i} \right\}. \tag{3.3.3}$$

It is obvious that moving any zero w_i gives another series for $h(w)$ with $2n + 1$ terms (i.e., a function $g(x)$ with the same number of Fourier components); as in the aperiodic case however, the zero manipulations that also preserve $|h(w)|^2$ on the unit circle ($|g(x)|^2$ on the real axis) are severely limited. The continuation of $|h(w)|^2$ off the unit circle is $h(w)h^*(1/w^*)$, which has zeros in pairs w_i and $1/w_i^*$. *One* zero of each pair must occur in any function $h(w)$ of the form (3.3.2) having this squared modulus on the unit circle. It follows that the only permissible adjustment to the zeros w_i of (3.3.3) is to replace an arbitrary number of them by corresponding zeros at $1/w_i^*$, the "image" points in the circle. This zero flipping is achieved by multiplying $h(w)$ by

$$\prod_i \frac{w - 1/w_i^*}{w - w_i} \cdot \frac{1 - w_i}{1 - 1/w_i^*}, \tag{3.3.4}$$

in which the product is extended over any number of zeros w_i of $h(w)$. Flipping zeros with unit modulus is trivial. For subsequent use we re-express the factors of this product in the form

$$a_i(1 + b_i/(w - w_i)), \tag{3.3.5}$$

where $a_i = (1 - w_i)/(1 - 1/w_i^*)$ and $b = w_i - 1/w_i^*$. Note that $|a_i| = |w_i| \neq 1$.

Thus the results established in the previous sections follow in a particularly easy way for band-limited periodic functions. Each zero w_i maps into a set of zeros z_i of $g(z)$, one in each cycle, so that $g(z)$ has exactly $2n$ zeros per cycle, all at finite distances from the real axis. Flipping w_i to $1/w_i^*$ corresponds to flipping the entire set $\{z_i\}$ to the set $\{z_i^*\}$ as illustrated in Fig. 3.3.1.

We note that in the periodic case too, the image plane wave function cannot be unimodular: for such a function, $h(w)h^*(1/w^*)$ would be unity everywhere, conflicting with the requirement imposed by the form of (3.3.2) that it have $4n$ zeros.

The close connection between transforms of discrete data and those of band-limited periodic functions means that the concept of zero flipping

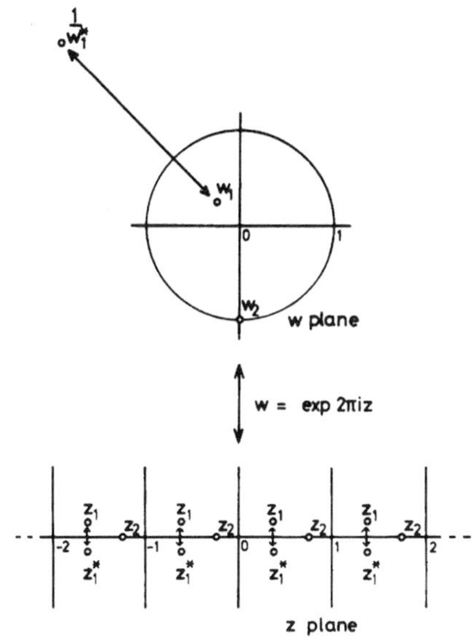

Figure 3.3.1 Zero flipping for periodic functions.

can be introduced successfully even for discrete images, the constraint on the transform width becoming a restriction on the number of nonzero elements of the discrete transform. This section concludes with an outline of this result. The analogy with the earlier treatment is obvious and will not be pointed out in detail.

In the notation of Chapter 2, a discrete image is a set of N values

$$g_k = (1/N)\tilde{g}_j W^*_{jk}, \tag{3.3.6}$$

where $W_{jk} = \exp(-2\pi ijk/N)$, all subscripts being taken modulo N. In general there are N independent values $\tilde{g}_j$, but for our present purposes we shall need to restrict them so that *at least half* are zero. We reexpress (3.3.6) as a unique polynomial in $w = \exp(2\pi ik/N)$, of degree not greater than $N-1$,

$$g_k = (1/N)\tilde{g}_j w^j. \tag{3.3.7}$$

(An infinite number of polynomials of degree $\geqslant N$ could represent g_k.) A similar unique polynomial representation applies to the "intensity" $I_k = |g_k|^2$,

$$I_k = (1/N)\tilde{I}_j w^j, \tag{3.3.8}$$

where, according to (2.1.15), $\tilde{I}_j = \tilde{g}_l \tilde{g}^*_{l-j}/N$. In this way the values g_k are determined by the positions of the zeros w_i of the polynomial (3.3.7). Now if $\tilde{g}_j$ vanishes except for $0 \leqslant j \leqslant m \leqslant (N-1)/2$, (3.3.8) is the ordinary algebraic product of (3.3.7) and

$$(1/N)\tilde{g}^*_j w^{-j},$$

the zeros of which are seen by trial to be $1/w^*_i$ and which equals g^*_k for the appropriate values of w. (This fails if $\tilde{g}_j$ is "wider," since the autocorrelation $\tilde{g}_l \tilde{g}^*_{l-j}$ then overlaps itself circularly.) Under these circumstances, therefore, the zeros of (3.3.8) occur in pairs w_i and $1/w^*_i$, *one* of each pair being present in the representation by means of (3.3.7) of *any* g_k such that $|g_k|^2 = I_k$ and $\tilde{g}_j$ vanishes except for $0 \leqslant j \leqslant m$. Consequently, we find once more that the only operations on g_k preserving I_k and the width of the transform $\tilde{g}_j$ are zero flipping operations on (3.3.7).

3.4. Two-dimensional forms

This section considers how the foregoing concepts may be extended to two-dimensional images, a problem that does not seem to have been investigated rigorously as yet. We are now concerned with functions of two variables $g(x_1, x_2)$ whose two-dimensional Fourier transforms $\tilde{g}(t_1, t_2)$ vanish outside a finite region which we may take to be rectangular without loss of generality:

$$g(x_1, x_2) = \int_{-a}^{a} \int_{-b}^{b} \tilde{g}(t_1, t_2) \exp 2\pi i(x_1 t_1 + x_2 t_2)\, dt_1\, dt_2. \tag{3.4.1}$$

The restrictions imposed on the phase distribution in fact promise to be stronger in two dimensions than in one.

We may reexpress (3.4.1) in the form

$$g(x_1, x_2) = \int_{-a}^{a} \tilde{f}(t_1, x_2) \exp(2\pi i x_1 t_1)\, dt_1, \tag{3.4.2}$$

where

$$\tilde{f}(t_1, x_2) = \int_{-b}^{b} \tilde{g}(t_1, t_2) \exp(2\pi i x_2 t_2)\, dt_2.$$

(We are simply expressing a two-dimensional transform as a product of one-dimensional transforms.) For any fixed value of x_2 therefore, $g(x_1, x_2)$

is the Fourier transform of a function which vanishes outside a finite interval, and the results of the previous sections apply at once; $g(x_1, x_2)$ may be continued off the real axis as $g(z_1, x_2)$ and expressed as a product over its zeros z_i. The only functions with the same modulus for real z_1 as $g(x_1, x_2)$ with transforms limited to the same interval, are those obtained by flipping an arbitrary number of the zeros. The positions of the zeros are now, however, dependent on x_2; they form functions $z_i(x_2)$. Since $g(z_1, z_2)$ is obviously analytic with respect to z_2 as well as z_1 (by interchanging x_1 and x_2 in (3.4.2)), it is reasonable to expect that they will be *continuous* functions, in which case, flipping a given zero involves flipping the entire curve. Instead of squaring the degree of ambiguity, the transition from one to two dimensions or, indeed, to any number of dimensions, may leave it unaltered.

In fact, the functions $z_i(x_2)$ *are* continuous, at least between intersections, as we now demonstrate. A double Taylor series expansion of $g(z_1, x_2)$ about a zero z_i gives

$$g(z_i + \zeta, x_2 + \eta) = g_1\zeta + g_2\eta + (g_{11}\zeta^2 + 2g_{12}\zeta\eta + g_{22}\eta^2)/2 + \cdots \quad (3.4.3)$$

with an obvious notation. To first order in ζ and η, therefore, the zero moves to

$$z_i - g_2\eta/g_1$$

when x_2 is incremented by η, and a limiting process gives

$$\frac{\partial z_i}{\partial x_2} = -\frac{g_2}{g_1}.$$

The derivative is defined, proving $z_i(x_2)$ continuous, unless g_1 vanishes. The vanishing of g_1 as well as of g implies a double zero of g at z_i, i.e., the intersection of two curves $z_i(x_2)$. In such a case, (3.4.3) must be solved to second order, when two zeros emerge at

$$z_i - \{g_{12}\eta \pm (g_{12}^2\eta^2 - g_{11}(g_{22}\eta^2 + 2g_2\eta))^{1/2}\}/g_{11}.$$

Two cases must be considered now. If g_2 vanishes, a limiting process easily gives

$$\frac{\partial z_i}{\partial x_2} = -\frac{\{g_{12} \pm (g_{12}^2 - g_{11}g_{12})^{1/2}\}}{g_{11}},$$

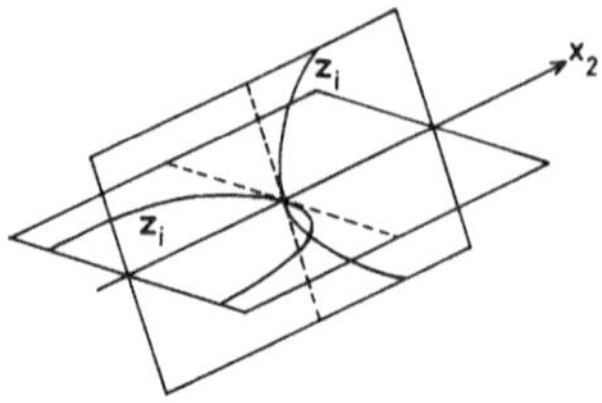

Figure 3.4.1 Zero curves at an intersection.

describing a simple intersection; in the more general case when g_2 does not vanish, however, the behavior is much more complicated. As η diverges from zero, the roots split (infinitely rapidly) in opposite directions, on lines which turn through $\pi/2$ discontinuously as η passes from positive to negative values, as in Fig. 3.4.1, for example. The zeros still form distinct continuous curves that must be flipped as a whole, but it is clear that a more comprehensive discussion of this problem is needed to cover adequately the case of multiple zeros.

The periodic case exhibits similar behavior: we now have

$$g(x_1, x_2) = \sum_{j=-n}^{n} \sum_{l=-m}^{m} \tilde{g}_{jl} \exp 2\pi i(jx_1 + lx_2),$$

or, applying the same transformations as before ($w_1 = \exp 2\pi i z_1$, etc.)

$$g(x_1, x_2) = h(w_1, w_2) = w_1^{-n} w_2^{-m} \sum_{j=0}^{2n} \sum_{l=0}^{2m} \tilde{h}_{jl} w_1^j w_2^l,$$

with $\tilde{h}_{jl} = \tilde{g}_{j-n,l-m}$. For any given value of w_2, this has exactly $2n$ zeros, all at finite distances from the origin, the positions of which are dependent on w_2, forming continuous curves that must be flipped as a whole. Similar difficulties arise in discussing detailed behavior at intersections.

3.5. The one-sided diffraction plane constraint

We now turn to the very much more powerful constraints imposed on an image plane wave function by the assumption that the diffraction plane wave function vanishes not merely outside an interval $(-a, a)$ but also over the interval $(-a, 0)$ comprising one side of the diffraction plane. These have been given considerably more attention in the past than the two-sided

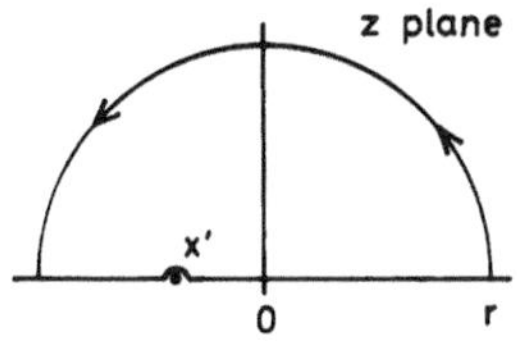

Figure 3.5.1

problem, having applications to all forms of dispersion, coherence theory, and electrical signal theory, as mentioned in Section 3.1.

All the results established so far are of course still valid for one-sided situations; the image plane function (in one dimension) is

$$g(x) = \int_0^a \tilde{g}(t) \exp(2\pi ixt)\, dt, \tag{3.5.1}$$

being an analytic function of $z = x + iy$ for all finite values of z. Provided $\tilde{g}(t)$ is square integrable, it vanishes at both ends of any line parallel to the real axis and exponentially as $y \to \infty$; it diverges to an essential singularity as $y \to -\infty$. In fact a may be infinite, with slightly greater restrictions on $\tilde{g}(t)$ (Whittaker & Watson, 1950, Section 5.32), without affecting the upper half plane behavior, although the lower half plane then has singularities for finite z.

Analyticity in the upper half plane combined with the limiting behavior at infinity is in fact sufficient to ensure the vanishing of the Fourier transform of $g(x)$ for negative values of the argument: the reasoning is similar to that given for the two-sided case (see Fig. 3.1.2) and will not be repeated. A rigorous treatment is given by Titchmarsh (1937, Section 5.4). [See also Paley & Wiener (1934, Theorem VIII).]

Again, zero flipping does not affect the limiting behavior of $g(z)$, and therefore preserves the one-sided character of $\tilde{g}(t)$. In the next section, the possibility is considered of using a priori information to force all the zeros into the lower half plane, giving a unique solution for the phase distribution from $|g(x)|$. In this section, we consider a reciprocal relation between the real and imaginary parts of $g(x)$, valid regardless of the distribution of the zeros.

Since $g(z)$ is analytic, the integral

$$\int \frac{g(z)}{z - x'}\, dz$$

vanishes over the contour in Fig. 3.5.1. In view of the limiting behavior at infinity the only contribution in the limit $r \to \infty$ is from the real axis and the small semicircle about x'. Thus

$$\pi i g(x') = \fint_{-\infty}^{\infty} \frac{g(x)}{x - x'} dx, \qquad (3.5.2)$$

the stroke denoting a Cauchy principal part. Separating the real and imaginary parts of this, writing $g(x) = u(x) + iv(x)$, we obtain the reciprocal relations

$$\begin{aligned} v(x') &= -\frac{1}{\pi} \fint_{-\infty}^{\infty} \frac{u(x)}{x - x'} dx, \\ u(x') &= \frac{1}{\pi} \fint_{-\infty}^{\infty} \frac{v(x')}{x - x'} dx, \end{aligned} \qquad (3.5.3)$$

which are well known under the various names of Hilbert transforms, dispersion relations, and Kramers–Kronig relations [see Jackson (1961)].

The importance of these relations for electron microscopy is that the *scattered* wave in the image plane of a diffraction-limited optical system with one half of the diffraction plane obscured has the properties ascribed above to $g(x)$, and its real part may be approximately observed in bright-field conditions from the image intensity. Taking the image wave function to be $1 + g(x)$, the intensity is

$$1 + 2u(x) + |g(x)|^2, \qquad (3.5.4)$$

allowing direct observation of $u(x)$ for weak scattering conditions [cf. (1.4.1)] and therefore complete solution for the complex wave function [cf. (1.3.7)]. The possibilities of one-sided imaging in electron microscopy were first discussed by Hanszen (1969) [see also Hanszen & Morgenstern (1965)], an optical procedure being described for extracting $v(x)$ for pure weak phase or weak amplitude objects that carried out the operation in (3.5.8) below. An extension to complex objects combining two or more pictures made with different parts of the diffraction plane obscured was proposed by Hoppe et al. (1970) [see also Hoppe (1971a)], with an analysis in the Fourier plane described later in this section. The Hilbert transform formulation of the problem has been put forward recently by Misell et al. (1974) and Misell and Greenaway (1974), iterative application being proposed as a means of extending the method to strong objects [cf. Lannes (1976b) and the next section].

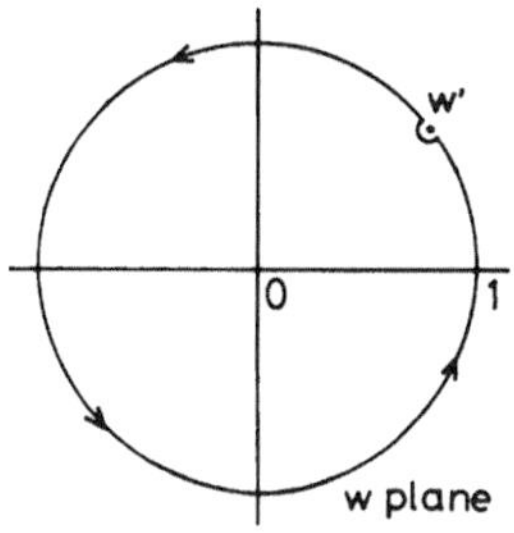

Figure 3.5.2

The form taken by the relations (3.5.3) for periodic objects is now given. We use the formulation of Section 3.3, representing the wave function as in (3.3.2), except that the negative powers vanish, leaving a polynomial

$$h(w) = \sum_{1}^{n} \tilde{h}_j w^j, \tag{3.5.5}$$

where $\tilde{h}_j = \tilde{g}_j$ now rather than $\tilde{g}_{j-n}$. This is now analytic for all finite z. We consider the integral

$$\frac{1}{2\pi} ⨍ h(w) \frac{w + w'}{w - w'} \frac{dw}{w} \tag{3.5.6}$$

around the contour in Fig. 3.5.2, the principal part being taken at $w = w'$. The residue of the integrand vanishes at $w = 0$. Evaluating the integral gives $ih(w')$. Restoring the substitution $w = \exp(2\pi iz)$ and noting that

$$\frac{\exp(2\pi ix) + \exp(2\pi ix')}{\exp(2\pi ix) - \exp(2\pi ix')} = -i \cot \pi (x - x')$$

therefore yields the result

$$ig(x') = ⨍_0^1 g(x) \cot \pi (x - x')\, dx,$$

which we separate, again writing $g(x) = u(x) + iv(x)$, into the relations

$$\begin{aligned} v(x') &= ⨍_0^1 u(x) \cot \pi (x - x')\, dx, \\ u(x') &= ⨍_0^1 v(x) \cot \pi (x - x')\, dx, \end{aligned} \tag{3.5.7}$$

which we may describe as "finite Hilbert transforms." Note that when x is near x', the integrand approximates that of (3.5.3).

A remark is useful at this point on the "background level" of the function $g(x)$, which manifests itself as a delta function at the center of the transform $\tilde{g}(t)$ in the general case or as the zero order coefficient $\tilde{g}_0$ in the periodic case. Inspection shows that any functions defined by the integrals of (3.5.3) or (3.5.7) have a zero background level; the Hilbert transform of a constant vanishes. These relations therefore connect only functions with zero background levels. This does not limit their usefulness seriously because it is usually possible to add a constant to a given function to put it into the required form. It does mean, however, that observation of the real part of such a function will not determine the imaginary part (if any) of the background level; this residual ambiguity will reappear at various points subsequently.

The theory of one-sided imaging has been developed here from the point of view of analytic function theory. Equivalent results can be obtained much more simply however by considering the wave function in the diffraction plane instead. This approach (Saxton, 1974a) reflects the computational methods used in calculating Hilbert transform integrals and is capable of simple extension to more general objects as in the following section.

We take $\tilde{g}(t)$ to vanish for $t < 0$. Separating the real and imaginary parts of $g(x) = u(x) + iv(x)$, this implies that

$$\tilde{v}(t) = i\tilde{u}(t) \quad \text{for } t < 0.$$

If $u(x)$ is given, this at once establishes $\tilde{v}(t)$ over one half of the diffraction plane, its value over the other half following from conjugate symmetry. In general, then,

$$\tilde{v}(t) = -i\,\mathrm{sgn}(t)\tilde{u}(t). \tag{3.5.8}$$

$v(x)$ may thus be determined by Fourier transforming $u(x)$, multiplying by i, reversing the sign over the positive t half plane, and inverting the transform. Note that the presence of a delta function at the origin of $\tilde{u}(t)$ would render (3.5.8) undefined. This is again the background level problem mentioned earlier.

A similar argument applies to the periodic case; there is no advantage here in mapping the real axis on to the unit circle, and we simply write

$$g(x) = u(x) + iv(x) = \sum_{1}^{n} \tilde{g}_j \exp(2\pi ijx) \tag{3.5.9}$$

$$= \sum_{-n}^{n} \tilde{u}_j \exp(2\pi ijx) + i \sum_{-n}^{n} \tilde{v}_j \exp(2\pi ijx). \tag{3.5.10}$$

Comparing coefficients of $\exp(2\pi ijx)$ between (3.5.9) and (3.5.10) we have at once

$$\tilde{v}_j = i\tilde{u}_j \quad \text{for } j < 0,$$

and hence in general

$$\tilde{v}_j = -i\,\mathrm{sgn}(j)\tilde{u}_j, \tag{3.5.11}$$

since $\tilde{u}_j$ and $\tilde{v}_j$ are conjugate symmetric. Both $\tilde{u}_0$ and $\tilde{v}_0$ vanish; a zero order term may be included in (3.5.9) so that they do not, but it is not possible to determine $\tilde{v}_0$ from $u(x)$.

The apparently very different solutions for $v(x)$ represented by (3.5.3) and (3.5.8) may be reconciled by noting that the former is a convolution integral, $u(x)$ being convolved with $1/\pi x$. The convolution can be replaced by multiplication in the Fourier plane as usual. The Fourier transform of $1/\pi x$ exists provided that the integration limits tend to ∞ and $-\infty$ at the same rate,

$$\lim_{r\to\infty} \int_{-r}^{r} \frac{\exp(-2\pi ixt)}{\pi x}\, dx = -i\,\mathrm{sgn}(t), \tag{3.5.12}$$

as is easily verified by completing the contour with a semicircle of radius r in the upper or lower half plane for $t < 0$ and $t > 0$, respectively. The inverse transform integral does not converge but is $(C, 1)$ summable to $1/\pi x$. This demonstrates the equivalence of (3.5.3) and (3.5.8). A similar argument relates (3.5.7) and (3.5.11), the corresponding relations for the periodic case. The former is a convolution of $u(x)$ with $\cot(\pi x)$, and while the Fourier series

$$-\sum_{j\neq 0} i\,\mathrm{sgn}(j) \exp(2\pi ijx)$$

does not converge, it is $(C, 1)$ summable to $\cot(\pi x)$.

A discrete analog of these results is provided by the sequence g_k whose discrete transform $\tilde{g}_j$ vanishes for $j = 0$ (and $j = N/2$ if N is even) as well as over one of the sidebands intervening. We shall have, for example,

$$\tilde{g}_j = g_k W_{jk} = 0 \quad \text{for } j = 0, \quad \text{and} \quad N/2 \leqslant j \leqslant N-1, \tag{3.5.13}$$

so that, separating the real and imaginary parts of $g_k = u_k + iv_k$,

$$\tilde{u}_j + i\tilde{v}_j = 0 \quad \text{for } j \text{ as in (3.5.13)},$$

and therefore

$$\tilde{v}_j = i\tilde{u}_j \quad \text{for } j \text{ as in (3.5.13)},$$

the values of $\tilde{v}_j$ for the rest of the range following from the conjugate symmetry of $\tilde{u}_j$ and $\tilde{v}_j$ as $-i\tilde{u}_j$. Taking these together,

$$\tilde{v}_j = -i\tilde{s}_j\tilde{u}_j, \tag{3.5.14}$$

where

$$\tilde{s}_j = \begin{cases} 1 & \text{for } 1 \leqslant j \leqslant N/2, \\ 0 & \text{for } j = 0 \text{ (and } j = N/2 \text{ if } N \text{ is even)}, \\ -1 & \text{otherwise}, \end{cases}$$

is the equivalent of the sign function occurring in the continuous case.

It is of interest to examine the inverse transform of the multiplier $-i\tilde{s}_j$ of (3.5.14). The multiplication obviously corresponds to circular convolution with $-is_k$. From the definition of $\tilde{s}_j$ we have

$$\begin{aligned} -is_k &= (-i/N)\tilde{s}_j W_{jk}^* \\ &= (i/N)\left\{ \sum_{j=N/2+1}^{N-1} - \sum_{j=1}^{N/2-1} \right\} \exp(2\pi ijk/N), \end{aligned}$$

in which, for convenience only, we have taken N to be even. Evaluating the geometric progressions, we arrive at

$$-is_k = \begin{cases} 2\cot(\pi k/N) & (k \text{ odd}), \\ 0 & (k \text{ even}). \end{cases} \tag{3.5.15}$$

In the limit $k \ll N$, we can approximate the cotangent by the first term of its Laurent series, which gives

$$-is_k = 2/\pi k \quad (k \text{ odd}),$$

the relation to previous results being obvious. The vanishing of the even k values implies, perhaps curiously, the neglect of alternate points in the convolution; this would give half the expected answer were it not for the factor of 2 present in odd k values.

For completeness we mention finally the approach given by Hoppe (1970), which is cast in a slightly different form still. This is restricted to weak scattering situations; $g(x)$ denotes the scattered wave in the image plane normalized so that the unscattered wave is unity. The image plane intensity is then

$$I(x) = 1 + g(x) + g^*(x) + |g|^2$$

[cf. (3.5.4)], the transform of which is

$$\tilde{I}(t) = \delta(t) + \tilde{g}(t) + \tilde{g}^*(-t) + O(|g|^2).$$

If $\tilde{g}(t)$ is one sided, $\tilde{g}^*(-t)$ does not overlap with it, and $\tilde{g}(t)$ is simply equal to $\tilde{I}(t)$ over the open portion of the diffraction plane.

Two examples of computed Hilbert transforms occur in Fig. 3.7.3, described in more detail in Section 3.7. It will be noted that the procedure has some resemblance to differentiation; this is easily understood by reference to (3.5.8). Multiplication by $2\pi it$ would constitute differentiation of $u(x)$, and Hilbert transformation is thus a combination of differentiation and strong low-pass filtering.

3.6. Logarithmic Hilbert transforms

The previous section discussed a relationship between the real and imaginary parts of the image plane wave function valid in any case where one half of the diffraction plane is obscured. Here we consider a relationship between its modulus and phase, valid under an additional restriction: the unscattered wave is to be stronger than the scattered wave everywhere in the image plane—the basic requirement of holography. This is true at least of most unstained biological microscopy. The restriction forces all the complex zeros of the image plane wave function into the lower half plane, allowing a unique solution for the phase distribution. Moreover, a much simpler method is available for calculating the phase than locating the complex zeros as in Section 3.2.

We take $g(x)$ to be the portion of the image plane wave function that was scattered by the object, normalizing so that the unscattered portion is unity: the image plane wave function is thus $1+g(x)$; $g(x)$ has the properties attributed to (3.5.1) in the previous section, in particular satisfying (3.5.2) if half the diffraction plane is obscured. Provided that $|g(x)| < 1$ everywhere, $\ln(1+g(x))$ proves to satisfy the relation (3.5.2) also giving

$$\begin{aligned} \arg(1+g(x')) &= -\frac{1}{\pi} \fint_{-\infty}^{\infty} \frac{\ln|1+g(x)|}{x-x'} dx, \\ \ln(1+g(x')) &= \frac{1}{\pi} \fint_{-\infty}^{\infty} \frac{\arg(1+g(x))}{x-x'} dx. \end{aligned} \tag{3.6.1}$$

There are three distinct approaches to this useful result which are considered in turn. Note that for $|g(x)| \ll 1$, (3.6.1) collapses to (3.5.3).

First, the result can be established by the same method as was used for (3.5.3): $\ln(1+g(z))$ is analytic for all finite z except for logarithmic branch points at the zeros of $1+g(z)$; now $g(z)$ vanishes at infinity throughout the upper half plane and $|g(z)| < 1$ everywhere on the real axis; hence, by the maximum modulus principle, $|g(z)| < 1$ throughout the upper half plane, and $1+g(z)$ has no zeros here. This ensures that $\ln(1+g(z))$ is analytic throughout the upper half plane. We consider

$$\int \frac{\ln(1+g(z))}{z-x'} dz$$

over the contour used to establish (3.5.3). The contribution from the semicircle vanishes at the same rate as before as the radius tends to infinity, giving the analog of (3.5.2),

$$\pi i \ln(1+g(x')) = \fint_{-\infty}^{\infty} \frac{\ln(1+g(x))}{x-x'} dx,$$

from which (3.6.1) follows at once. Note that the condition $|g(x)| < 1$ on the real axis is sufficient but not necessary. What is necessary is the absence of zeros of $g(x)$ in the upper half plane.

An elegant, if circuitous, second approach that has been used to derive dispersion relations in scattering theory (e.g., Jackson, 1961) is based on the solution of an integral equation for a "sectionally analytic" function, analytic in both half planes and vanishing at infinity everywhere but discontinuous across the real axis (Roos, 1969, Chapter 4). Muskhelishvili (1953, Sections 17 and 26) shows that the only such function $f(z)$ whose limiting values from above and below on the real axis $f^+(x)$ and $f^-(x)$ satisfy

$$f^+(x) - f^-(x) = \phi(x) \tag{3.6.2}$$

is

$$f(z') = \frac{1}{2\pi i}\int_{-\infty}^{\infty}\frac{\phi(x)}{x-z'}\,dx, \tag{3.6.3}$$

the integral being taken along the real axis. (This is easily seen to satisfy (3.6.2), the discontinuity being due to the pole at $x = x'$; it is the *only* solution because the *difference* between any two solutions would be analytic everywhere and vanish at infinity, so that by Liouville's theorem it would vanish everywhere.) The integral (3.6.3) only exists if $\phi(x)$ tends to the same limit as x tends to ∞ and $-\infty$ (Muskhelishvili, 1953, Section 43).

The real and imaginary parts of the scattered wave, $g(x) = u(x) + iv(x)$ satisfy the relations (3.5.3), therefore vanishing at both ends of the real axis. We define the sectionally analytic function

$$f(z') = 1 + \frac{1}{\pi}\int_{-\infty}^{\infty}\frac{v(x)}{x-z'}\,dx, \tag{3.6.4}$$

the integral being taken along the real axis; this tends to unity at infinity everywhere. The limiting values on the real axis are

$$\begin{aligned} f^+(x) &= 1 + u(x) + iv(x), \\ f^-(x) &= 1 + u(x) - iv(x), \end{aligned} \tag{3.6.5}$$

to obtain which the contour was completed by semicircles in the lower and upper half planes, respectively, and (3.5.3) was used. From this result,

$$\ln f^+(x) - \ln f^-(x) = 2i \arg(1 + g(x)). \tag{3.6.6}$$

Since $f(z)$ tends to unity at infinity, $\ln f(z)$ vanishes there; under the assumption that $|g(x)| < 1$ everywhere, $\arg(1 + g(x))$ tends to the same limit at both ends of the real axis, and (3.6.6) is thus an integral equation of the type of (3.6.2) with the unique analytic solution

$$\ln f(z') = \frac{1}{\pi}\int_{-\infty}^{\infty}\frac{\arg(1+g(x))}{x-z'}\,dx. \tag{3.6.7}$$

The proof of (3.6.1) is completed by noting that

$$\ln|1+g(x')| = \mathrm{Re}\{\ln f^{+}(x')\}$$

[from (3.6.5)] and using (3.6.7) to evaluate the latter. This gives

$$\ln|1+g(x')| = \frac{1}{\pi}\int_{-\infty}^{\infty}\frac{\arg(1+g(x))}{x-x'}\,dx,$$

which is one part of (3.6.1), the other following by Hilbert inversion.

The method was introduced to optics by Peřina (1972, Section 4.4) who applies it to $g(x)$ itself rather than to $1+g(x)$ for the case when all zeros are restricted to the lower half plane (see Section 3.8). There, however, it appears to fail, for without the term unity in (3.6.4), $f(x)$ vanishes at infinity, and $\ln f(x)$ cannot satisfy the conditions imposed on the solution of (3.6.6)—and, indeed, satisfied by Peřina's solution. There are also problems associated with showing that $\arg g(x)$ tends to the same limit at both ends of the x axis [cf. (3.6.6)]. Burge et al. (1974) rework Peřina's argument, including a constant term. The method appears to offer no advantage over the first method given here, however, in the present application.

The new third method of establishing (3.6.1) appeals to the Fourier transform of $\ln(1+g(x))$. It was established in the previous section that the vanishing of the Fourier transform of a function over the negative half of the range is a necessary and sufficient condition for the real and imaginary parts of the function to be interconnected by Hilbert transforms of the form (3.5.3). It is now shown directly that the Fourier transform of $\ln(1+g(x))$ is indeed one sided, so that (3.6.1) follows at once.

The key is the fact that the convolution of two one-sided functions remains one sided. Thus if $g(x)$ has a one-sided transform, the same is true of $g^2(x)$, and therefore of $g^3(x)$, $g^4(x)$, etc. We might ask in what circumstances the argument extends to infinite series of the form

$$h(x) = \sum_{0}^{\infty} a_n g^n(x).$$

The answer is that the argument holds when the series is uniformly convergent with respect to x, as this permits the term by term integration involved in the Fourier transformation of the individual terms and guarantees the convergence of the resulting series (Titchmarsh, 1939, Section 1.71). It is obviously *sufficient* for uniform convergence that the coefficients a_n be those of a power series with radius of convergence r and that $|g(x)| < r$ everywhere. Thus if the Fourier transform of $g(x)$ is one sided, so is that of $\cos g(x)$ for any $g(x)$, or of $1/(a^2 + g^2(x))^{1/2}$ for $|g(x)| < a$, etc.; the result is far reaching. The case of particular interest is, of course, $\ln(1 + g(x))$, which has a convergent expansion in powers of $g(x)$ provided that $|g(x)| < 1$; its Fourier transform too is therefore one sided, and (3.6.1) follows.

Relations similar to (3.6.1) apply again to the periodic case. The relations

$$\begin{aligned}\arg(1 + g(x)) &= -\fint_0^1 \ln|1 + g(x)| \cot\pi(x - x')\, dx, \\ \ln|1 + g(x)| &= \fint_0^1 \arg(1 + g(x)) \cot\pi(x - x')\, dx\end{aligned} \tag{3.6.8}$$

are easily verified by the first or third methods above, using the treatment used to derive (3.5.7). The Muskhelishvili method can indeed be adapted to the periodic case too, a sectionally analytic function being sought which is discontinuous across the unit circle under the transformation $w = \exp(2\pi iz)$.

It will be noted that the Fourier transform of $\ln(1 + g(x))$ is *not* limited to the interval $(0, a)$ but on the contrary extends to infinity in the positive t direction. In consequence, calculation of a phase distribution by multiplying the transform of $\ln|1 + g(x)|$ by $-i\,\mathrm{sgn}(t)$ will involve rather closer sampling of $g(x)$ to avoid overlap distortion of the transform than is necessary normally (see Section 2.2). In the limit $|g(x)| \to 0$, $\ln(1 + g(x))$ tends to $g(x)$ simply, and the transform, being $\tilde{g}(t)$, shrinks into the interval $(0, a)$. As $|g(x)| \to 1$, however, the number of significant terms in the expansion grows, a progressively larger region of the transform plane contains significant data, and progressively closer sampling of the micrograph is necessary to avoid distortion in the numerical calculation of the Hilbert transform (by *any* method).

Fig. 3.6.1 illustrates this effect in a one-dimensional model problem: (A) shows the phase of a pure phase model object; (B) shows the real and imaginary (continuous and dashed) parts of the image function $g(x)$ formed by an aberration-free one-sided optical system; (C) shows the modulus of $\tilde{g}(t)$; and (D) shows the modulus of the transform of $\ln(1 + g(x))$, similar to (C) since $|g(x)|$ is substantially smaller than unity; (E), (F), and (G) repeat (B), (C), and (D) for an object with a phase distribution similar to that in (A), but scaled so that the maximum phase variation is 2.5 rad rather than 0.5. Now the transform of $\ln(1 + g(x))$ is significantly wider than $\tilde{g}(t)$.

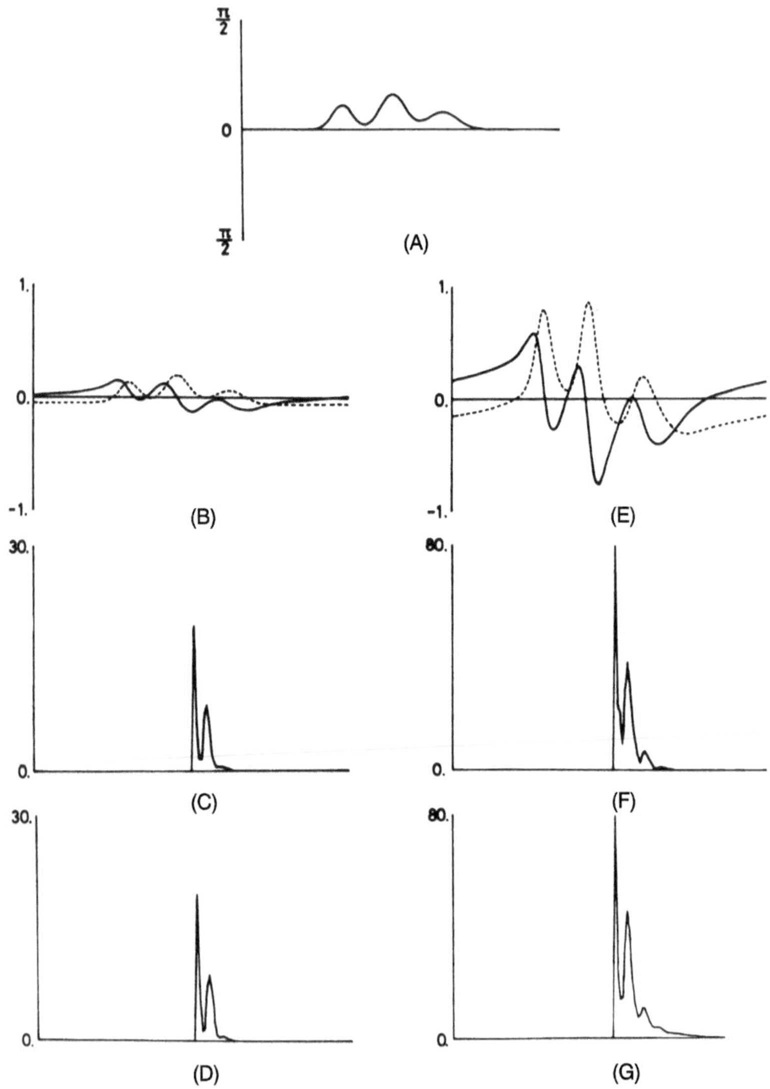

Figure 3.6.1 The relative widths of the transforms of g and $\ln(1+g)$.

This same effect destroys the possibility of a relation analogous to (3.6.1) in the context of discrete images with a one-sided discrete transform. The infinitely wide transform of $\ln(1+g_k)$ ceases to be one sided due to aliasing (circular overlap: see Section 2.2).

The background level problem mentioned in the previous section now takes the form of an unknown phase origin for $1+g(x)$. In combining the wave functions from pairs of complementary half-plane apertured micrographs, it is therefore necessary to establish a common phase origin for the two wave functions by forcing the phase of the unscattered wave (the central point of the transform) to be the same for each.

Some one-dimensional model calculations based on the logarithmic Hilbert transform are given by Misell and Greenaway (1974); a review of the method with some generalizations is given by Burge, Fiddy, et al. (1976).

3.7. The realizability of the one-sided constraint

We consider here the application of the results in the previous sections to two-dimensional situations and how the shapes of physically realizable apertures affect their validity. Again the analysis benefits from the use of the diffraction plane rather than the image plane as the prime object of consideration.

There is no difficulty in extending the basic results presented to two dimensions. As was shown in Section 3.4, each line of a two-dimensional picture is the Fourier transform of a function restricted to a given region (here the interval $(0, a)$) so the Hilbert transform results can be applied line by line to the image. The Fourier plane methods still follow at once if two-dimensional transforms are understood, and indeed a one-dimensional analysis is still straightforward on a line by line basis (see Section 9.3).

In view of the finite width of the primary beam however, it is not possible to use exact half-plane apertures, a small identation being necessary to accommodate the direct beam as in Fig. 3.7.1A. This destroys the Hilbert transform relationship for every line of the picture since the one-dimensional Fourier transform of any given line is no longer exactly one sided. In the weak-scattering case, when we are concerned with the relations (3.5.3) or (3.5.8), it is possible to be precise about the error introduced if we consider the problem in the Fourier plane, as was done in deriving (3.5.8). The crucial requirement for this was that for every point $\mathbf{t}$ at which the wave function $\tilde{u}(\mathbf{t}) + i\tilde{v}(\mathbf{t})$ was to be determined, the image point $-\mathbf{t}$ was obscured by the aperture, allowing the conclusion that $\tilde{v}(-\mathbf{t}) = i\tilde{u}(-\mathbf{t})$. This is true for the aperture of Fig. 3.7.1A except for points within a radius r of the origin. In consequence, $\tilde{g}(\mathbf{t})$ is not determined within this region, and the low frequency components of the reconstruction are inaccurate; the higher frequencies can be exactly determined, however. The use of line-by-line Hilbert transforms will result in some inaccuracy at *all*

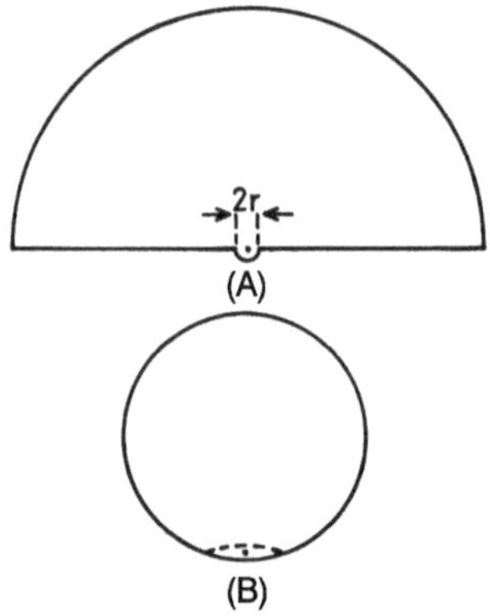

Figure 3.7.1 Two half-plane apertures (see text for details).

frequencies, but it will probably be serious only rarely. Similar arguments can be used to discuss a compromise method in which displaced circular apertures are used as in Fig. 3.7.1B. The wave function $\tilde{g}(\mathbf{t})$ is determined everywhere except within the region around the primary beam bounded by the dashed line. Several such displacements of a circular aperture may be made in turn to build up the information in the diffraction plane (Hoppe, 1971a). Fig. 3.7.2 shows the area within which the wave function may be determined by a four-exposure process of this type; the method suffers from obvious susceptibility to radiation damage, but the portions of the diffraction plane common to two displacements may assist alignment of the four micrographs used.

In the case of the logarithmic relations (3.6.1), whether calculated directly or in Fourier space, the position is more complicated, and the presence of the semi-circle accommodating the primary beam in Fig. 3.7.1A destroys the possibility of exact determination of the data at any point in the diffraction plane. This is again easily seen by considering the third method of establishing the result (3.6.1) in which $\ln(1+g(\mathbf{x}))$ was shown directly to have a one-sided transform. This was based on the fact that if $g(\mathbf{x})$ itself had a one-sided transform, the same was true of any power of $g(\mathbf{x})$, allowing series expansion of $\ln(1+g(\mathbf{x}))$. However, if $\tilde{g}(\mathbf{t})$ is not exactly one sided, as in Fig. 3.7.1A, the transform of $g^2(\mathbf{x})$, being the convolution of $\tilde{g}(\mathbf{t})$ with itself, is not one sided either and indeed spreads twice as far into the "obscured" portion of the diffraction plane. Higher powers of $g(\mathbf{x})$ have transforms that extend still further, and as a result, the range of frequencies over which the Hilbert transform yields meaningless results grows rapidly

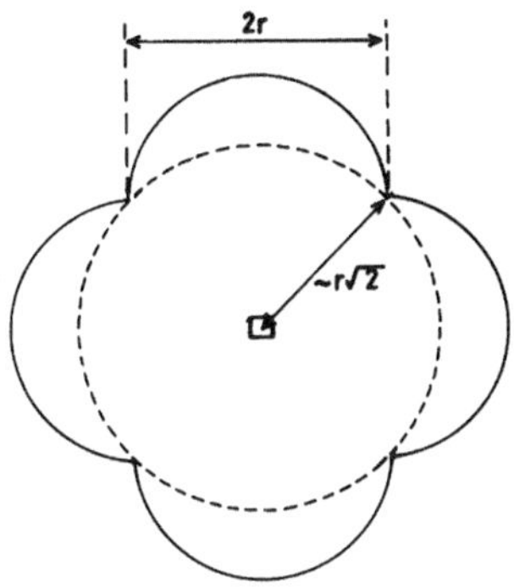

Figure 3.7.2 The area covered by four successive displacements of a circular aperture.

as $|g(\mathbf{x})|$ grows and more terms of the series for $\ln(1 + g(\mathbf{x}))$ become significant. Similar problems apply to the analysis of micrograph sets made with such apertures as those of Fig. 3.7.2.

A related problem concerns the determination of the data on the line common to two complementary apertures of the type shown in Fig. 3.7.1A, for here also points do not occur in mirror pairs inside and outside the aperture: the line is neither inside nor outside. On this line, $\tilde{\nu}(\mathbf{t})$ cannot be established (except perhaps by analytic continuation if the *object* also is assumed to be of finite extent). Unless the wave function has delta functions on it, the problem is not serious in principle because the contribution of a single line to the transform integral only has measure zero; delta functions could arise if the object were periodic and the half-plane aperture were aligned with the crystal lattice, so this arrangement must be avoided in practice. The problem is less easily dismissed for the discrete data used in practice, however, since in this case the data on the common line represent the one-dimensional transform of the projection of the image in rows normal to it, so that the values assigned to the common line points determine the relative background levels of the image rows. If $\tilde{\nu}_{jl}$ cannot be established on the common line, the calculated ν_{km} must remain subject to the addition of an arbitrary imaginary function of the row number. We are forced to conclude that a simple Hilbert transform of a one-sided transform micrograph may be distorted by the common line data and that the only wholly satisfactory answer is to use additional pictures (such as the four-exposure process mentioned above) with a diffraction plane analysis to resolve the problem.

An analogous problem arises of course for the logarithmic case, but it is still more serious in this case. Previously, we have only been prevented from determining the diffraction plane wave function $\tilde{g}(\mathbf{t})$ on the common line, but now we are prevented from determining it exactly *anywhere*. The common line problem only permits the image phase to be determined to within an arbitrary additive function of the row number, i.e., the wave function can be determined only to within a multiplicative unimodular function of the row number, variations in which alter $\tilde{g}(\mathbf{t})$ everywhere and not simply on the central line. Analytic continuation of the transform of $\ln(1+g(\mathbf{x}))$ remains a solution in principle at least, though rather less obviously, if the object is assumed to be of finite extent. The problem disappears if the half-plane aperture is displaced so as to exclude the common line. If $\tilde{g}(\mathbf{t})$ is zero there, it is not hard to see, by the power series expansion method given earlier, that the transform of $\ln(1+g(\mathbf{x}))$ is also zero there, eliminating the common line problem completely. It seems probable that it will frequently be necessary in practice to use more than two pictures to effect a complete phase determination since the use of an exactly complementary pair only will not establish the common line data.

The requirement that $|g(x)| < 1$ everywhere may usefully be examined briefly. This cannot be verified by examining a given micrograph, although the existence of points where $|1+g(x)| > 2$ might *disprove* it occasionally, and must usually be regarded as an a priori assumption. As such, it should be based on known properties of the object, unfortunately it is not possible to express the requirement in terms of useful restrictions on the object. The Schwarz inequality, for example, gives

$$|g(x)|^2 = \left| \int_0^a \tilde{g}(t) \exp(2\pi ixt)\, dt \right|^2 \leqslant a \int_0^a |\tilde{g}(t)|^2\, dt,$$

so that if $g_o(x)$ is scattered wave at the object,

$$|g(x)|^2 \leqslant a \int_{-\infty}^{\infty} |g_o(x)|^2\, dx$$

by Parseval's theorem. If a width X of the object were illuminated, then the rather severe condition

$$|g_o(x)|^2_{\max} \leqslant 1/aX$$

is sufficient to ensure $|g(x)| < 1$, but is obviously unnecessary in most cases. Various looser arguments are useful but not conclusive. If $g_o(x)$ is reasonably smooth and its modulus is less than unity everywhere, the loss of one half of its Fourier transform at the aperture leads to an image $g(x)$ with a modulus typically one half, adequate to ensure $|g(x)| < 1$; a large local value of $g(x)$ implies a broad background in $\tilde{g}(t)$ which in turn implies a large local value somewhere in $g_0(x)$. That object phase shifts much greater than a radian can occur without giving rise to ambiguity in the reconstruction is confirmed by Fig. 3.6.1 where phase shifts as large as 2.5 rad locally do not prevent $|g(x)|$ from being less than unity everywhere. In this connection, again see Misell and Greenaway (1974) who quote the results of model calculations in which nonunique solutions first appeared at image contrast levels of about 70%.

Lest the various considerations above give too pessimistic an air to the prospect of phase determination by these methods, a model problem is presented in Fig. 3.7.3 in which (aberration-free) half-plane apertured micrographs are calculated for a strongly scattering object, and a reconstruction of the object is obtained by the logarithmic Hilbert transformation methods of Section 3.6. The model object appears top left, the amplitude being displayed above the phase; the maximum attenuation of the wave function amplitude by the object is by 30%, and the maximum phase shift is $\pi/2$. Beside this is shown the wave function with the "common line" data (the central column of the transform) set to zero since the data on this line are not determined by the two-micrograph process here modeled. The group bottom left comprises one of the half-plane apertured micrographs (top left, the aperture excluding the "common line" in the transform), its logarithm (bottom left), the actual phase of the image wave function (top right, varying over a full radian), and the phase obtained by Hilbert transforming the micrograph logarithm (bottom right: note the background level shift); corresponding pictures for the complementary micrograph are shown in the group top right, the Hilbert transform now being minus the actual phase. The reconstruction obtained by combining the two half-plane wave functions is shown, once more as amplitude and phase, bottom right, agreeing with the model object (with the common line removed) to better than three significant figures, except for a background phase shift. (The transforms were calculated as described in Section 9.3.)

We shall not consider here the other practical problems peculiar to this method, particularly alignment of the two (or several) component pictures, and the charging up of the apertures used. These are discussed by Downing and Siegel (1973). Note that in certain cases (symmetric object, or pure weak phase or pure weak amplitude object) symmetry conditions in

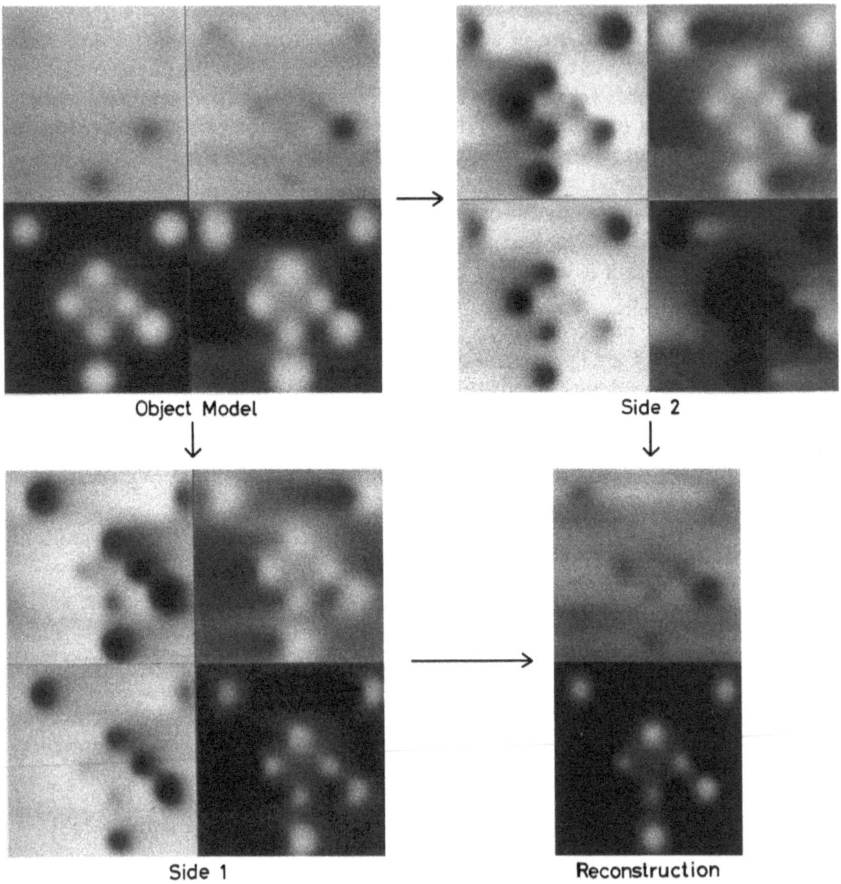

Figure 3.7.3 Logarithmic Hilbert transforms applied to simulated micrographs. The reconstruction bottom right should be compared with the right-hand pair of the block top left (see text for details).

the diffraction plane allow reconstruction from a single half of the diffraction plane, i.e., a single one-sided transform micrograph in the ideal case. An application of the method to uranyl-stained DNA strands is given by Downing and Siegel (1975).

3.8. Logarithmic Hilbert transforms in dark-field conditions

While in Section 3.6 it was shown possible to determine the image phase directly from the observed modulus under bright-field conditions with not too restrictive conditions on the specimen, the same is not true in dark-field conditions when the a priori knowledge of the distribution of the zeros of $g(z)$ between the two half planes is no longer available. There remains a certain value in the use of a logarithmic transform in these circumstances because it enables us to construct at least one solution for the phase, obtaining when all the zeros are in the lower half plane—the so-called "minimal phase." This has been exploited considerably in other fields (see the references in Section 3.1) but almost exclusively in situations where the transform is real as well as one sided, and it appears to be extremely difficult to find a proof of the relationship even in this case. Perina's argument was criticized in Section 3.6, and Papoulis' (1962, Section 10.3) depends on the use of (3.1.6) as a limiting form for $g(z)$ as if it were true uniformly with respect to θ, which is not in general the case.

The basic problem concerns the contribution from the semicircle at infinity to the Cauchy integral used to derive (3.5.3). If $g(z)$ vanishes on this, $\ln g(z)$ becomes infinite, and the contribution does not necessarily vanish. We can limit $\ln|g(z)|$ by the results of Section 3.1 so that it diverges no more rapidly than $|z|$, suggesting the inclusion of an additional factor $1/z^2$ in the Cauchy integral, but it is less easy to limit the phase of $g(z)$. We resort to a rather more subtle argument.

First, the divergence of $\ln|g(z)|$ is limited on the real axis by the condition

$$\int_{-\infty}^{\infty} \frac{|\ln|g(x)||}{1+x^2}\, dx < \infty, \tag{3.8.1}$$

which Paley and Wiener (1934, Theorem XII) show to be necessary and sufficient for a square-integrable $|g(x)|$ to be the modulus of a function with a one-sided transform. Its necessity is not proved here, but its sufficiency is obvious from the construction in the following of a suitable $g(x)$. The function

$$m(x', y') = \frac{1}{\pi} \int_{-\infty}^{\infty} \frac{y' \ln|g(x)|}{(x-x')^2 + y'^2}\, dx = \frac{1}{\pi} \int_{-\infty}^{\infty} \ln|g(x)| \operatorname{Im}\left\{\frac{1}{x-(x'+iy')}\right\} dx \tag{3.8.2}$$

exists everywhere since (3.8.1) converges; its limit on the real axis is $\ln|g(x')|$, as is easily seen if we note that

$$\int_{-\infty}^{\infty} \frac{y'}{(x-x')^2 + y'^2}\, dx = \pi \quad (y' \neq 0), \tag{3.8.3}$$

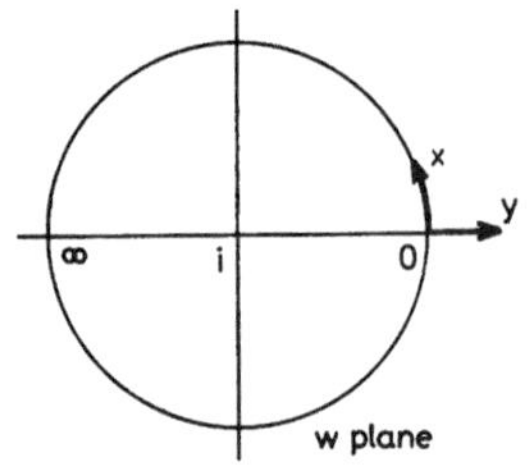

Figure 3.8.1

so that the integrand of (3.8.3) tends to $\pi\delta(x-x')$ as $y' \to 0$. Moreover, it is harmonic since the kernel of the integral is harmonic, and it diverges to $-\infty$ at infinity throughout the upper half plane. Consequently, it is the real part of $\ln g(z')$—this is how Paley and Wiener prove the sufficiency of (3.8.1). We have now to construct the corresponding imaginary part $\phi(x', y')$, the conjugate of $m(x', y')$, which gives the phase of $g(z')$. Formally this is easily effected with the Cauchy–Riemann relations between m and ϕ and gives the unsurprising result

$$\begin{aligned}\phi(x', y') &= -\frac{1}{\pi}\int_{-\infty}^{\infty} \frac{(x-x')\ln|g(x)|}{(x-x')^2 + y'^2}\, dx \\ &= -\frac{1}{\pi}\int_{-\infty}^{\infty} \ln|g(x)|\,\mathrm{Re}\left\{\frac{1}{x-(x'+iy')}\right\} dx, \end{aligned} \tag{3.8.4}$$

with the limit

$$\phi(x') = -\frac{1}{\pi}\,⨍_{-\infty}^{\infty} \frac{\ln|g(x)|}{x-x'}\, dx \tag{3.8.5}$$

on the real axis. Provided that these integrals converge, (3.8.5) is clearly the solution sought. Unfortunately, it is not obvious from (3.8.1) that they do.

The same approach may be employed [Toll (1956); Nevanlinna (1953, Chapter VII, Section 2); Papoulis (1962, Section 10.5)] under the transformation

$$w = \frac{1+iz}{1-iz} \tag{3.8.6}$$

which maps the upper half plane into the unit circle, the real axis into the circle itself, and infinity into the point $w = -1$ (see Fig. 3.8.1). We write $h(w) = g(z)$. $\ln h(w)$ has an essential singularity at -1, diverging to $-\infty$ as

w approaches -1 on or within the circle. However, the integral

$$\int_0^{2\pi} |\ln|h(w)||\,d\theta$$

is finite, being equal to

$$2\int_{-\infty}^{\infty} \frac{|\ln\lfloor g(x)\rfloor|}{1+x^2}\,dx,$$

which is a sufficiently strong condition for $\ln|h(w)|$ to be determined throughout (and on) the unit circle by its boundary value, through a Poisson formula [see, for example, Nevanlinna (1969, Section 11.15)]

$$\ln|h(w')| = \frac{1}{2\pi}\int_0^{2\pi} \ln|h(w)|\,\mathrm{Re}\left\{\frac{w+w'}{w-w'}\right\}d\theta,$$

which is analogous to (3.8.2). The phase of $h(w)$ is the conjugate function

$$\frac{1}{2\pi}\int_0^{2\pi} \ln|h(w)|\,\mathrm{Im}\left\{\frac{w+w'}{w-w'}\right\}d\theta, \qquad (3.8.7)$$

or, restoring the transformation to the z-plane and limiting on to the real axis,

$$\arg g(x') = -\frac{1}{\pi}\int_{-\infty}^{\infty} \frac{\ln\lfloor g(x)\rfloor}{x-x'}\left\{\frac{1+xx'}{1+x^2}\right\}dx, \qquad (3.8.8)$$

which *does* converge. Noting that

$$\frac{1+xx'}{(x-x')(1+x^2)} = \frac{1}{x-x'} - \frac{x}{1+x^2},$$

we can easily relate this to the earlier result (3.8.5) if that converges, the two results differing only by a constant. Papoulis (1962, Section 10.5) gives a result equivalent to (3.8.7) for the case in which $\tilde{g}(t)$ is real, but it is difficult to verify the behavior of his solution near $w = -1$. When $\tilde{g}(t)$ is real, the symmetry of $\lfloor g(x)\rfloor$ allows us to recast (3.8.5) *and* (3.8.8) in the form

$$\arg g(x') = -\frac{2x'}{\pi}\int \frac{\ln\lfloor g(x)\rfloor}{x^2 - x'^2}\,dx, \qquad (3.8.9)$$

which is the form of the result most widely quoted. Practical computation of (3.8.5) and (3.8.8) may be effected conveniently at least in principle by the Fourier methods of Section 3.5, in the latter case *after* the transformation (3.8.6); but in practice the data in both cases are badly conditioned because of the divergence as $x \to \pm\infty$.

A more realistic application of the method is to periodic functions; as in many cases covered earlier, these permit a particularly simple derivation and expression of the result. We start again from the expression

$$h(w) = \sum_{1}^{n} \tilde{h}_j w^j$$

[cf. (2.3.2)], and seek the $h(w)$ for which all the zeros w_i lie outside the unit circle. The zero at the origin is removable by considering $h(w)/w$ instead of w. This is analytic and zero-free throughout the unit circle, and its logarithm is thus analytic there. We can therefore apply the method of (3.5.6) and (3.5.7) directly to $\ln\{h(w)/w\}$; noting that $|h(w)/w| = |h(w)|$ on the unit circle, this leads at once to

$$\arg g(x') = 2\pi x - \frac{1}{\pi}\int_0^1 \ln|g(x)| \cot\pi(x - x')\, dx. \tag{3.8.10}$$

(The argument is easily extended if $\tilde{h}_1$ is zero and so on.) Once more, practical computation of the integral is best performed by the Fourier methods of Section 3.5.

It is repeated that the expression given in this section for the image phase represents only *one* possible solution and that it is necessary to flip its zeros in arbitrary combinations to construct the full set of phase solutions. The results given are easily extended to *two*-sided functions $\tilde{g}(t)$ since these may be made one-sided by a translation which does not affect the image plane modulus.

CHAPTER FOUR

The image and diffraction plane problem: uniqueness☆

W. Owen Saxton
Murray Edwards College, University of Cambridge, Cambridge, United Kingdom
e-mail address: wos1@cam.ac.uk

Contents

4.1. Statement of the problem

The last chapter outlined the surprisingly strong constraints imposed on a wave function by restricting the width of its Fourier transform. In this chapter, we begin to explore the presumably much more powerful constraints obtaining when not merely the extent, but also the modulus, of the transform is known. This idea developed in 1971 in discussion with Dr. R. W. Gerchberg. We might rephrase the problem in the following form:

How far do the modulus of a function and of its Fourier transform determine the phase of the function?

In view of the extensive treatments previously given of the far less promising problem discussed in the preceding chapter, it is surprising that this problem has been given so little attention previously. Reichenbach (1944, Section 20) considers briefly the possibility of determining the quantum mechanical wave function of a particle from the probability distributions of a pair of conjugate observables such as position and momentum but dismisses it after noting some obvious ambiguities of the type mentioned in the next section and goes on to require the time derivative of one of the distributions. A similar problem is treated by Fowle (1964) using

☆ Reprinted from Advances in Electronics and Electron Physics, Supplement 10, 1978.

Advances in Imaging and Electron Physics, Volume 214
ISSN 1076-5670
https://doi.org/10.1016/bs.aiep.2020.04.004

the stationary phase approximation for the transform integrals, and this is examined in detail in Section 4.4. Other work has, however, arisen out of the work presented here and will be referred to at the appropriate points.

The potential value of a solution to the problem will be obvious. Direct phase measurement presents difficulties in numerous wave systems involving electrons and other high energy particles: x-rays, light, microwaves, high frequency sound waves, and electrical circuits. In some cases, holographic techniques have provided a solution, but these are not always viable even where coherent sources exist. In many of these cases, however, the data assumed in the following sections are readily available, and this approach may be useful. X-ray crystallography is at first a very important application for this method; the diffracted wave observed is the (three-dimensional) Fourier transform of the wave at the specimen, and the latter, though not directly observable, is known to have negligible modulus variation, the scattered wave being in quadrature with the incident wave. However, as will be seen in Section 4.3, the low scattering cross sections involved lead to serious and unavoidable ambiguities in the phase derived from the data, so that the present work unfortunately offers no fresh help. The same is not true of many optical and electron microscopical applications, on the other hand, and the hope of a solution to the phase problem by means of this new technique remains of considerable interest.

The nonlinear nature of the problem makes it far from trivial to solve, and no closed form solution to it has yet been found, nor has a complete answer been given to the question of the uniqueness of the solution. This chapter reviews what has been established in this latter direction, while its successor discusses various "direct" and iterative methods for obtaining numerical solutions, and Chapter 6 presents the results of model calculations.

4.2. Data constraints and trivial uniqueness failures

It is not possible to find complex wave functions compatible with *arbitrary* modulus distribution in both planes. A number of constraints must be satisfied by a pair of real square-integrable functions before they can represent the modulus of a function and of its Fourier transform. The most obvious of these is Parseval's theorem which requires that their square integrals be equal [(2.1.8) in the discrete case]. Additionally there are restrictions on their widths. As lens or antenna designers know only too well, it is not possible for a function whose transform is of limited extent to be arbitrarily narrow. Even for "super resolving" designs, the mean square widths of function and transform (as defined in Section 2.2) have a minimum product of $1/4\pi$. Two other theorems are relevant: that due to Paley and Wiener (1934, Theorem XII) which requires that

$$\int_{-\infty}^{\infty} \frac{|\ln |g(x)||}{1+x^2}\, dx < \infty$$

if $\tilde{g}(t)$ is of only *semi*-infinite extent (i.e., vanishing to one side of some given point) thus preventing $|g(x)|$, for example, from being "too small too often"; and that due to Hardy [e.g., Titchmarsh (1937, Section 6.13)], which requires that the transform of a function vanishing at infinity *more* rapidly than a Gaussian of given width must vanish *less* rapidly than a Gaussian of reciprocal width.

Discrete analogs to these conditions, with the exception of Parseval's theorem, are unfortunately not known, though similar conditions must apply in situations where discrete data approximate continuous functions closely. A significant result has however been obtained by Schiske (1974), namely, that whatever *inequalities* might be applicable to discrete modulus distributions, Parseval's theorem is the only *equation* they must satisfy. Schiske expresses the modulus values in terms of the unknown real and imaginary parts of the wave function values, and demonstrates their algebraic independence by explicit calculation of the functional determinant for the resulting set of equations, which is shown to be nonzero in general.

For our purposes, the modulus distributions are available as data and, apart from observational noise, will therefore satisfy any necessary criteria though normalization will be necessary to satisfy Parseval's theorem given the different experiments involved in recording the two intensity distributions. Since our calculations will necessarily be discrete, it would be of interest to know whether an exact solution can exist to the discrete problem posed by the sampled data, with or without observational noise superposed. These remarks, however, represent the only answer so far available, which is at best partial. Practical experience suggests strongly that an exact solution does *not* exist to the problem posed by two arbitrary sets of modulus values satisfying Parseval's theorem and that observational noise normally prevents exact solutions being obtained to real data.

Symmetry properties of the modulus distributions are sometimes reflected in obvious ambiguities in the phase solution. Some such cases are noted here: none of these really poses a serious problem.

First, the absolute phase of the wave cannot be determined since multiplication of a function by a constant phase factor leaves the modulus of its Fourier transform unchanged. However, the absolute phase is physically meaningless since the quantum mechanical wave function for any system may be multiplied by a constant phase factor without changing the probability distribution of any of the observables.

Second, if the intensity distribution in *one* plane is centrosymmetric, the complex conjugate of any given solution for the wave function in the other plane is also a solution [from (2.1.13)].

Third, if the intensity distributions in *both* planes are centrosymmetric, then the complex conjugate or central reflection of a solution for the wave function in either plane is also a solution [(2.1.12) and (2.1.13)]. Similarly, mirror reflection of a wave function solution is possible if the intensity distributions both have mirror symmetry. In connection with these points, we note that centro-symmetry in both intensity distributions at least precludes the corresponding *anti*symmetry of the phase distribution in either plane: antisymmetry of the phase would imply conjugate symmetry in the function, and hence a real transform which would be real *and symmetric*, implying a real symmetric function in contradiction of the initial hypothesis.

Finally, translational ambiguities may arise if the intensity distribution in either plane is constant or has a periodicity higher than that of the wave function itself because a given solution for the wave function could then be shifted laterally without affecting the modulus values in either plane.

More ingenious instances of symmetry giving rise to ambiguity occur in Section 4.4.

4.3. An important failure of uniqueness

Beginning with the least optimistic result established, we discuss first an important class of problem for which the data do not restrict the phase distribution at all for practical purposes. This is the case of images in which the phase variation is small, as it necessarily is for weak objects. As a simple illustration, we may consider the example of a unimodular image whose phase consists of two sinusoidal components:

$$a\cos(2\pi t_1 x) + b\cos(2\pi t_2 x + \phi) \quad (|a|, |b| \ll 1) \tag{4.3.1}$$

The wave function $g(x)$ may be closely approximated by the expansion

$$g(x) = 1 + ia\cos(2\pi t_1 x) + ib\cos(2\pi t_2 x + \phi),$$

the Fourier transform of which is

$$g(t) = \delta(t) + \tfrac{1}{2}ia\{\delta(t-t_1) + \delta(t+t_1)\}$$
$$+ \tfrac{1}{2}ib\{\exp(i\phi)\delta(t-t_2) + \exp(-i\phi)\delta(t+t_2)\},$$

thus consisting of a series of delta functions, the moduli of which are *independent of* ϕ. The intensity data are consequently unable to establish the relative displacement of the two components of the image phase (4.3.1).

More generally, if we write the image plane wave function $g(x)$ in the form

$$g(\mathbf{x}) = 1 + i\phi(\mathbf{x}), \qquad (4.3.2)$$

then a sufficient (though not necessary) condition for the squared modulus of $g(\mathbf{x})$, i.e., the image intensity, to differ only negligibly from unity, is that $\phi(\mathbf{x})$ is real and $|\phi(\mathbf{x})| \ll 1$. The diffraction plane intensity data fix the modulus $|\tilde{\phi}(\mathbf{t})|$ of the Fourier transform of $\phi(\mathbf{x})$. Let us assign the phase of $\tilde{\phi}(\mathbf{t})$ *at random* over the half plane to one side of the origin; provided that we choose the phase for the other half plane so that $\tilde{\phi}(\mathbf{t})$ is conjugate symmetric, i.e.,

$$\arg\tilde{\phi}(-\mathbf{t}) = -\arg\tilde{\phi}(\mathbf{t}),$$

then the corresponding $\phi(\mathbf{x})$ will be real. Moreover, if the scattering is weak enough, the diffraction plane integral

$$\int |\tilde{\phi}(\mathbf{t})|\, d^2\mathbf{t} \qquad (4.3.3)$$

will be small compared with unity, and this integral is larger than $|\phi(\mathbf{x})|$, so that $|\phi(\mathbf{x})|$ will also be small regardless of the diffraction plane phase distribution; indeed, it will remain small for most phase distributions even when (4.3.3) is less small. An almost random choice of diffraction plane phase thus leads to an image intensity showing negligible contrast, and we conclude that the image phase is not determined by the data.

Far reaching as this conclusion is, it is possible to generalize the ambiguity still further so as to embrace the case of an image wave function which, although still involving only small phase variations, exhibits a large degree of modulus variation as well. We now have a wave function

$$g(\mathbf{x}) = m(\mathbf{x}) + i\phi(\mathbf{x}), \qquad (4.3.4)$$

where $m(\mathbf{x})$ and $\phi(\mathbf{x})$ are real, and $|\phi(\mathbf{x})| \ll m(\mathbf{x})$. The image intensity will differ only negligibly from $m^2(\mathbf{x})$, thus being independent of $\phi(\mathbf{x})$. Provided the real part of the scattered wave is not too large, we may take $m(\mathbf{x})$ to be positive and therefore observable. The corresponding diffraction plane wave is

$$\tilde{g}(\mathbf{t}) = \tilde{m}(\mathbf{t}) + i\tilde{\phi}(\mathbf{t}),$$

where $\tilde{m}(\mathbf{t})$ and $\tilde{\phi}(\mathbf{t})$ will be conjugate symmetric so that

$$\tilde{g}(-\mathbf{t}) = \tilde{m}^*(\mathbf{t}) + i\tilde{\phi}^*(\mathbf{t}).$$

Denoting $\tilde{g}(-\mathbf{t})$ by g_-, the diffraction plane intensity data then allow us to write the equation pair

$$|\tilde{m} + i\tilde{\phi}| = |\tilde{g}|, \qquad |\tilde{m} - i\tilde{\phi}| = |\tilde{g}_-| \tag{4.3.5}$$

to be solved for $\tilde{\phi}$ over one half of the diffraction plane, the value over the other half following from the symmetry of $\tilde{\phi}$. Now the first equation restricts $i\tilde{\phi}$ to a circle of radius $|\tilde{g}|$ about $-\tilde{m}$, and the second restricts it to a circle of radius $\tilde{g}_-$ about $\tilde{m}$; consequently *two* solutions for $\tilde{\phi}$ exist in general at any point in the diffraction plane, one for each of the two intersections of the circles (see Fig. 4.3.1). Occasionally (e.g., when $\tilde{\phi} = 0$), the intersections will coincide and there will be no ambiguity; on the other hand, when $\tilde{m}$ vanishes, the circles become coincident, and *any* $\tilde{\phi}$ with the appropriate modulus is permissible; if $\tilde{m}$ vanishes everywhere, this is the case presented previously. If continuity arguments can be applied in the diffraction plane, for example, because the object is also assumed to be of finite extent, the ambiguity in the solution for $\tilde{\phi}$ may be greatly reduced since it will be possible to follow one solution from point to point over extended regions of the diffraction plane; in the case of periodic or discrete images, no such arguments will be possible. It is not of course obvious that all solutions will satisfy the condition $|\phi(\mathbf{x})| \ll m(\mathbf{x})$ everywhere in the image plane, but most will certainly do so unless $m(\mathbf{x})$ is "small often." The overall ambiguity in $\tilde{\phi}$ (and hence the image phase) will be enormous.

These arguments thus demonstrate that, regardless of the presence of modulus variation in an image, if the phase variation is small it is not possible to determine the phase from the diffraction plane intensity data. This conclusion applies to bright-field electron microscopy of thin unstained biological materials for example. However, it does not continue to hold when

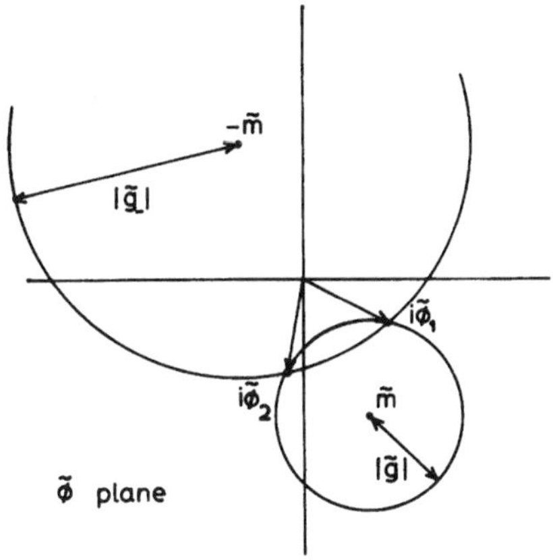

Figure 4.3.1 Uniqueness failure for weak objects: graphical solution of (4.3.5).

the assumptions used here are no longer applicable. It loses its force for images of objects with strong phase components, such as crystals of moderate thickness; it fails if the object is of finite extent, since $m(\mathbf{x})$ must then decay to zero at the outside of the image: moreover, the uniqueness failure depends on our assumed *experimental* failure to measure the image intensity sufficiently accurately to detect the terms of second order in $\phi(\mathbf{x})$ so that we have not established a *theoretical* failure of uniqueness. In the subsequent sections, we examine the general problem, first for aperiodic images with continuous transforms and second for the special but important case of periodic and discrete images.

4.4. The continuous aperiodic problem

The relation noted in Section 4.2 between symmetry and ambiguity has also been stressed by Schiske (1974) who presents for the one-dimensional case a number of ingeniously constructed wave functions such that the intensity data in the two planes permit an arbitrary degree of phase ambiguity in the solution. The simplest of these will suffice here to illustrate their general nature. The functions are built recursively from a base function $\tilde{g}_0(t)$, restricted to the interval $(-\frac{3}{2}, \frac{3}{2})$, by means of the recursion formula

$$\tilde{g}_{n+1}(t) = \tilde{g}_n(t) + q_n(\tilde{g}_n(t - 3^n) + \tilde{g}_n(t + 3^n)), \tag{4.4.1}$$

in which q_n is an arbitrary complex constant. This describes the repeated addition of multiples of itself to either side of $\tilde{g}(t)$. With this restriction on the width of $\tilde{g}_0(t)$, the added portions do not overlap with the original, and the modulus is thus independent of the phase of the factors q_n. The Fourier transform of the relation (4.4.1) gives a similar recursion formula for $g_n(x)$, namely

$$g_{n+1}(x) = g_n(x)(1 + 2q_n \cos(2\pi 3^n x)), \tag{4.4.2}$$

with a squared modulus depending on $\mathrm{Re}(q_n)$ and $|q_n|^2$ only; the intensity data in the two planes thus permit a twofold ambiguity in the imaginary part of every q_n. Applying (4.4.1) sufficiently often thus constructs a function with an arbitrarily large number of solutions for the phase. While Schiske is able to relax the restriction on $\tilde{g}_0(t)$ substantially, his constructions are all of the same character and are unlikely to present serious problems in practice.

A further example with a twofold solution is given by Grünbaum (1974). If the symmetric and antisymmetric parts $g_s(x)$ and $g_a(x)$ of a *real* function $g(x)$ are nonzero only over disjoint internals, then the function

$$f(x) = g_s(x) - g_a(x)$$

clearly has the same modulus as $g(x)$. The squared modulus of the transform $\tilde{g}(t)$ is

$$|\tilde{g}(t)|^2 = |\tilde{g}_s(t)|^2 + |\tilde{g}_a(t)|^2 + 2\,\mathrm{Re}\{\tilde{g}_s(t)\tilde{g}_a^*(t)\},$$

while

$$|\tilde{f}(t)|^2 = |\tilde{g}_s(t)|^2 + |\tilde{g}_a(t)|^2 - 2\,\mathrm{Re}\{\tilde{g}_s(t)\tilde{g}_a^*(t)\}.$$

With these assumptions, however, $\tilde{g}_s(t)$ is real and $\tilde{g}_a(t)$ imaginary; the product $\tilde{g}_s(t)\tilde{g}_a^*(t)$ has no real part, and the transform moduli $|\tilde{g}(t)|$ and $|\tilde{f}(t)|$ are identical. Accordingly the modulus data allow a sign change in $g(x)$ over the intervals for which $g_a(x)$ is nonzero; again, this situation is unlikely to be of real practical interest since $g(x)$ is assumed real.

A result due to Fowle (1964), which may be established by the use of the stationary phase approximation applied to the Fourier integral, now deserves comment. We separate $g(x)$ into modulus and phase factors, writing

$$g(x) = |g(x)| \exp i\phi(x).$$

The stationary phase approximation is useful only when $\phi''(x)$ has the same sign everywhere, so that $\phi'(x)$ is monotonic. This is completely unrealistic for wave functions in the electron microscope, but conclusions based on the use of the approximation are nevertheless of general interest and are directly relevant to such fields as electrical signal theory. Under these circumstances, the rapid oscillations of the phase factor in the Fourier transform integral

$$\tilde{g}(t) = \int_{-\infty}^{\infty} |g(x)| \exp(i\phi(x) - 2\pi ixt)\, dx \tag{4.4.3}$$

mean that the integral is dominated by the contribution from the neighborhood of the single point $x_s(t)$ for which

$$\phi'(x_s) = 2\pi t, \tag{4.4.4}$$

i.e., the phase is stationary. Expansion of the phase in the form

$$\phi(x) = \phi(x_s) + 2\pi t(x - x_s) + \tfrac{1}{2}\phi''(x_s)(x - x_s)^2 + \cdots$$

thus allows approximation of the integral (4.4.3) by the form

$$\tilde{g}(t) = \exp\{i\phi(x_s) - 2\pi ix_s t\} \int_{-\infty}^{\infty} |g(x)| \exp\{\tfrac{1}{2}i\phi''(x_s)(x - x_s)^2\}\, dx,$$

which gives the stationary phase approximation to (4.4.3),

$$\tilde{g}(t) = |g(x_s)| \left\{ \frac{2\pi}{|\phi''(x_s)|} \right\}^{1/2} \exp i \left\{ \phi(x_s) - 2\pi x_s t \pm \frac{\pi}{4} \right\} \tag{4.4.5}$$

($+$ for $\phi''(x_s) > 0$ and $-$ for $\phi''(x_s) < 0$; the same distinction applies below), with a squared modulus

$$|\tilde{g}(t)|^2 = 2\pi |g(x_s)|^2 / |\phi''(x_s)|, \tag{4.4.6}$$

x_s depending on t according to (4.4.4), of course. The approximation becomes closer as ϕ'' increases.

Fowle is concerned with finding a phase distribution *approximately* consistent with *arbitrary* modulus distributions in the two planes. His argument, however, also implies that if the problem is known a priori to be such that the stationary phase approximation (4.4.5) for the Fourier transform is close, then there exist exactly two solutions for ϕ consistent with the intensity data in both planes, one with ϕ'' everywhere positive and one with ϕ'' negative. This will follow if t can be shown to be a unique function of x_s

since it will then be possible to integrate (4.4.4) with respect to x_s to obtain ϕ; this indeed proves to be the case. From (4.4.4) and (4.4.6),

$$\frac{dt}{dx_s} = \frac{\phi''(x_s)}{2\pi} = \pm\frac{|g(x_s)|^2}{|\tilde{g}(t)|^2}. \tag{4.4.7}$$

In view of the constant sign of dt/dx_s, the only doubt about a one-to-one correspondence between t and x_s arises from the constant introduced by integrating (4.4.7). If we define the integrals

$$X(x_s) = \int_{-\infty}^{x_s} |g(x)|^2\, dx, \qquad T(t) = \int_{-\infty}^{t} |\tilde{g}(t')|^2\, dt',$$

then the formal solution of (4.4.7) is

$$x_s = X^{-1}\{\pm T(t) + c\}.$$

T and X both lie in the range 0 to $X(\infty)$ ($= T(\infty)$ by Parseval's theorem) and the inverse functions X^{-1} and T^{-1} exist only for arguments in this range; however, x_s must exist for all t if the stationary phase approximation applies, and the possible solutions are thus

$$x_s = X^{-1}\{T(t)\}$$

for $\phi'' > 0$, and

$$x_s = X^{-1}\{T(\infty) - T(t)\}$$

for $\phi'' < 0$, which may be rephrased as

$$t = T^{-1}\{X(x_s)\} \quad \text{and} \quad t = T^{-1}\{X(\infty) - X(x_s)\}. \tag{4.4.8}$$

These allow immediate integration of (4.4.4) to give solutions for ϕ determined up to an arbitrary additive constant.

Fowle shows, by means of a number of convincing examples, that for a given $g(x)$, the solution phase $\phi(x)$ obtained by this method is progressively more accurate as $|\tilde{g}(t)|$ becomes wider. This is probably because as $|\tilde{g}(t)|$ becomes wider, the value at any given point drops (on average) so as to maintain the Parseval integral, and consequently $|g(x)|^2/|\tilde{g}(t)|^2$ increases; from (4.4.7) therefore, ϕ'' increases and with it the validity of the stationary phase approximation. As a result, we can state that if the product of the widths of the intensity distributions in the two planes is *large* it will be possible to construct two approximate solutions for the phase distribution by this method. The resulting wave function will of course be quite unrealistic for electron microscopical purposes, but a numerical procedure might well find one of these solutions, unless suitably constrained, instead of the true wave function.

Fowle's argument does not extend easily into two dimensions unfortunately; ϕ'' is replaced by its two-dimensional analog $\{\phi_{11}\phi_{22} - \phi_{12}^2\}$, leading to the equation

$$\frac{\partial t_1}{\partial x_1}\frac{\partial t_2}{\partial x_2} - \left(\frac{\partial t_1}{\partial x_2}\right)^2 = \pm|g(\mathbf{x})|^2/|\tilde{g}(\mathbf{t})|^2$$

in place of (4.4.7), the subscript s having been dropped. This is difficult to integrate.

We turn next to an important result established by the methods of Chapter 3, supposing the width of the diffraction plane to be restricted by an aperture and using the result established in Section 3.2 that every image plane wave function with the same modulus as a given wave function and with a transform restricted to the same interval can be constructed simply by flipping arbitrary sets of the complex zeros associated with the given wave function. We may ask what the effect is on the modulus distribution *within* the diffraction plane aperture of flipping a given zero of the image; the circumstances under which it remains unchanged are very severely restricted.

We take $g_0(z)$ to be the original wave function (the image plane coordinate x is continued off the real axis as z) and construct $g_1(z)$ by multiplication by $(z - z_i^*)/(z - z_i)$ according to (3.2.1). We define a function

$$f(z) = g_0(z)/(z - z_i) = g_1(z)/(z - z_i^*). \tag{4.4.9}$$

First we show that its transform $\tilde{f}(t)$ vanishes outside the diffraction plane aperture, like $\tilde{g}_0(t)$. Noting that

$$(z - z_i^*)/(z - z_i) = 1 + (z_i - z_i^*)/(z - z_i),$$

we can write $g_1(z)$ in the form

$$g_1(z) = g_0(z) + (z_i - z_i^*)f(z).$$

Since $\tilde{g}_1(t)$ and $\tilde{g}_0(t)$ are both restricted to the aperture region, the same must be true of $\tilde{f}(t)$. In terms of this function, the Fourier transform of

$g_0(z)$ is

$$\tilde{g}_0(t) = \int f(z)(z - z_i)\exp(-2\pi izt)\, dz = (i\tilde{f}'(t)/2\pi) - z_i\tilde{f}(t) \qquad (4.4.10)$$

(it is easy to show that $\tilde{f}(t)$ not merely exists but is differentiable as well), while the transform of $g_1(z)$ is

$$\tilde{g}_1(t) = (i\tilde{f}'(t)/2\pi) - z_i^*\tilde{f}(t). \qquad (4.4.11)$$

Now if (4.4.10) and (4.4.11) are to have the same modulus, the phases of $\tilde{f}'$ and $\tilde{f}$ must differ by $\pi/2$: equality of the moduli requires that $i\tilde{f}'/2\pi$ lies equidistant from $z_i\tilde{f}$ and $z_i^*\tilde{f}$, i.e., on the line

$$\arg(t) = \arg\tilde{f}(t)$$

(see Fig. 4.4.1). Consequently,

$$\tilde{f}'(t) = ip(t)\tilde{f}(t)$$

with $p(t)$ *real* and integrating this gives

$$\tilde{f}(t) = \tilde{f}(0)\exp\left\{i\int_0^1 p(s)\, ds\right\}. \qquad (4.4.12)$$

This has a constant modulus $|\tilde{f}(0)|$. We have established therefore that a zero z_i of a wave function $g_0(z)$ may be flipped without affecting the diffraction plane modulus if and only if the transform of $f(z) = g_0(z)/(z - z_i)$ has a constant modulus throughout the diffraction plane aperture. This is obviously a rare occurrence and is in fact sometimes impossible.

In the situation treated by Bates (1969) for example (see Section 3.1), where the diffraction plane wave function is itself composed of a finite number of Fourier components, it is certainly impossible to flip any zero in isolation. $\tilde{f}(t)$ is restricted to the same region as $\tilde{g}_0(t)$, so the same sampling points determine $f(z)$ as $g_0(z)$; however, the zero sample values that allow the truncation of the sum (3.1.12) remain zero on division by $(z - z_i)$, and $\tilde{f}(t)$ therefore contains the same finite number of Fourier components as $\tilde{g}_0(t)$. Finally, it was proved in Section 2.3 that no function representable by a finite number of Fourier components is unimodular so that $\tilde{f}(t)$ is *not* unimodular, and no zeros can be flipped individually.

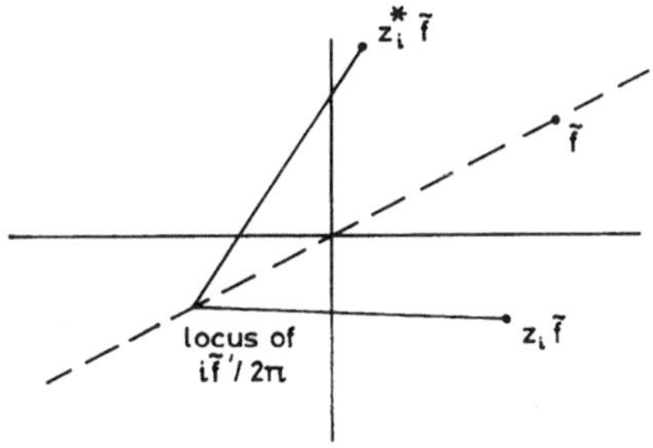

Figure 4.4.1

We might expect to be able to use these results to construct functions $g(z)$ whose phases are not uniquely determined by $|g(z)|$ and $|\tilde{g}(t)|$ by choosing some unimodular $\tilde{f}(t)$ and multiplying its inverse transform $f(z)$ by a factor $(z - z_i)$. The process is not straightforward, however, because the multiplication may destroy the square integrability of $f(z)$ so that the $g(z)$ constructed may not have a Fourier transform; this happens, for example, for $\tilde{f}(t) = 1$, $\exp(ikt)$ or $\exp(ik^2t^2)$ within the diffraction plane aperture. In fact this can be proved always to be the case. The square integrability of the constructed $g(z)$ would imply that of $-2\pi izf(z)$, and hence that of its Fourier transform, $\tilde{f}'(t)$; however, since $\tilde{f}(t)$ is unimodular within the aperture and zero outside it, it must be discontinuous at the edge and cannot therefore have a square-integrable derivative.

The argument can probably be extended into two dimensions along the lines of Section 3.4. Its serious shortcoming is its inability to treat the simultaneous flipping of *sets* of zeros which introduce too many terms into (4.4.10) and (4.4.11). That symmetry in $g_0(z)$ sometimes allows the flipping of sets of zeros without effect on the diffraction plane modulus is obvious from the examples constructed by Schiske described at the beginning of the section. Replacing q_n by q_n^* in (4.4.2) flips *all* the zeros of the factor $(1 + 2q_n\cos(2\pi 3^n x))$, since it replaces the expression by its complex conjugate; there are two zeros in each cycle of the function in the x direction.

The foregoing represents the somewhat weak conclusions so far possible for cases where no further assumptions are made about the wave function than that the width of the diffraction plane aperture is finite. An important fact has recently been established by Huiser et al. (1976), however at least for the one-dimensional case, on the assumption that the object also is of finite extent so that the diffraction plane wave function is analytic and therefore capable of representation by a power series expansion. They show that if the diffraction plane modulus is nonzero at the extremes of the aperture and takes different values at the two sides, the solution for the phase is unique (apart, of course, from the trivial additive constant); moreover, they show that if the modulus takes the same value at both extremes, the ambiguity is only twofold.

After this rather more encouraging result, we turn to the case of periodic and discrete wave functions, where a reasonably satisfactory statement is possible of the circumstances in which a unique solution may be expected.

4.5. The periodic and discrete problems

The examples of phase ambiguities given in the previous section are easily adapted to periodic functions or discrete sequences, so it is clear that sufficient symmetry will still be capable of leading to ambiguous solutions in these cases too. However, these granted, the general case is very promising.

Periodic objects can be effectively treated by applying the rational function mapping used in Section 3.3, expressing the wave function in the form given by (3.3.2). The diffraction plane modulus is the modulus of the coefficients $\tilde{h}_j$. Recalling (3.3.5), we express the zero flipping as a multiplication of $h(w)$ by a factor

$$a_i(1 + b_i/(w - w_i)).$$

The second term inside the bracket is the most significant and is considered first. We assume $|w_i| < 1$, the other cases being merely the same set of flippings in reverse. With this assumption, $1/(w - w_i)$ may be expanded on the unit circle to give

$$\frac{1}{w - w_i} = \frac{1}{w} + \frac{w_i}{w^2} + \frac{w_i^2}{w^3} + \cdots$$

or

$$\frac{1}{w - w_i} = \sum_{-\infty}^{\infty} \tilde{f}_j w^j,$$

with

$$\tilde{f}_j = w_i^{-j-1} \quad \text{for } j < 0 \text{ otherwise } 0. \tag{4.5.1}$$

Thus the product $h(w)/(w - w_i)$ becomes

$$w^{-n}\sum_{j=0}^{2n}\sum_{l=-\infty}^{\infty}\tilde{h}_j\tilde{f}_l w^{j+l} = w^{-n}\sum_{j=0}^{2n}\sum_{l=-\infty}^{\infty}\tilde{h}_j\tilde{f}_{l-j} w^l. \tag{4.5.2}$$

This is equivalent to a convolution in the diffraction plane. Using (4.5.1) it reduces to

$$w^{-n}\sum_{0}^{2n-1}\tilde{c}_l w^l, \tag{4.5.3}$$

where

$$\tilde{c}_l = w_i^{-l-1}\sum_{l+1}^{2n}\tilde{h}_j w_i^j. \tag{4.5.4}$$

(For $l \geqslant 2n$, all the $\tilde{f}$ components in (4.5.2) vanish; for $l < 0$, we obtain

$$\tilde{c}_l = w^{-l-1}\sum_{0}^{2n}\tilde{h}_j w_i^j$$

which vanishes since w_i is a zero of $h(w)$—cf. (3.3.2).) The polynomial factor in (4.5.3) is one degree lower than that in (4.5.2), corresponding to the removal of one root by the multiplication by $1/(w - w_i)$.

The wave function with a flipped zero thus becomes

$$h'(w) = w^{-n}\sum_{0}^{2n}\tilde{h}'_j w^j, \tag{4.5.5}$$

with coefficients

$$\tilde{h}'_j = a_i(\tilde{h}_j + b_i\tilde{c}_j), \tag{4.5.6}$$

where we define $\tilde{c}_{2n} = 0$ for convenience. If the modulus of any coefficient $\tilde{h}'_j$ differs from that of the corresponding $\tilde{h}_j$, we shall be able to argue that flipping the zero w_i, although preserving the finite width of the diffraction plane wave function, has altered the modulus distribution within the nonvanishing portion, so that the data eliminate phase ambiguity associated with this particular zero. But this is true for the coefficient $\tilde{h}'_{2n} = a_i\tilde{h}_{2n}$, since $\tilde{c}_{2n}$ vanishes, and $|a_i| = |w_i|$, as noted in connection with (3.3.5), being here less than one. Moreover, flipping any number of zeros for which $|w_i| < 1$

reduces the modulus of $\tilde{h}_{2n}$ by $\prod_i |w_i|$, therefore being ruled out. The permissible operations are thus flipping only sets of zeros w_i, some outside and some inside the unit circle, such that $\prod_i |w_i| = 1$. The existence of such a set is obviously a coincidence. The same result is achieved by considering the only other coefficient $\tilde{h}'_j$ which is easily evaluated, namely $\tilde{h}'_0$. From (4.5.4), $\tilde{c}_0$ is easily shown to be $-\tilde{h}_0/w_i$ so that (4.5.6) becomes

$$\tilde{h}'_0 = a_i(1 - b_i/w_i)]\tilde{h}_0,$$

and recalling that $b_i = w_i - 1/w_i^*$, we have

$$\tilde{h}'_0 = \frac{a_i}{|w_i|^2}\tilde{h}_0,$$

so that $|\tilde{h}'_0| = |\tilde{h}_0|/|w_i|$. If a set of zeros w_i exists for which $\prod_i |w_i| = 1$, it becomes necessary to consider the other terms of (4.5.6): these have so far proved intractable.

It may be helpful to express two simple cases of residual phase ambiguity in the formalism used here. First, if a diffraction pattern is symmetric, a given image plane wave function cannot be distinguished from its complex conjugate. This is equivalent to the statement that in such a case flipping *all* the zeros leaves $|\tilde{h}_j|$ unaltered. The continuation $h(w)$ is replaced by $h^*(1/w^*)$ when the complex conjugate is taken, and this has zeros at $1/w_i^*$. From (3.3.2), it may be expressed in the form

$$h^*(1/w^*) = w^n \sum_0^{2n} \tilde{h}_j^* w^{-j} = w^{-n} \sum_0^{2n} \tilde{h}_{2n-j}^* w^j,$$

so that if $|\tilde{h}_{2n-j}| = |\tilde{h}_j|$, the coefficients of $h^*(1/w^*)$ and $h(w)$ are indistinguishable. It is obvious that $\prod_i |w_i| = 1$, for the product $\prod_i w_i$ is of course $\tilde{h}_0/\tilde{h}_{2n}$, which is here unity. Similarly, if the intensity distribution in *both* planes is symmetric, a given wave function $g(x)$ cannot be distinguished from its central reflection $g(-x)$. The continuation is now $h(1/w)$, with zeros $1/w_i$, the expansion of which is

$$h(1/w) = w^n \sum_0^{2n} \tilde{h}_j w^{-j} = w^{-n} \sum_0^{2n} \tilde{h}_{2n-j} w^j,$$

allowing the same conclusion as before. The set $\{1/w_i\}$ is included in the set $\{w_i, 1/w_i^*\}$ in this case because the symmetry of $|h(w)|^2$ causes the zeros of $h(w)h^*(1/w^*)$ to occur in *fours* at w_i, $1/w_i^*$, w_i^* and $1/w_i$.

We have established, then, that in the case of periodic objects, the phase distribution will normally be uniquely determined by the modulus data and that ambiguity is exceptional though it may arise if the modulus data have sufficient symmetry properties.

The close analogy between discrete functions and band-limited periodic functions noted in Section 2.2 has already been exploited in Section 3.3 to extend the scope of the zero flipping theory to embrace the discrete case too. It now permits us to show that the conditions under which a unique set of phase values is determined for a sequence or complex numbers by the modulus values of the sequence and of its discrete Fourier transform are exactly analogous to those established for the periodic object problem. The analog to the restricted width of the diffraction plane for continuous functions was shown in Section 3.3 to be a requirement that the transform $\tilde{g}_j$ is nonzero only over a consecutive range of j covering not more than half the total of N sample points (which we can take to be $0 \leqslant j < (N-1)/2$ without loss of generality). Under these circumstances, a given g_k can be expressed as a polynomial in $w = \exp(2\pi ik/N)$ of degree less than $N/2$,

$$g_k = (l/N)\tilde{g}_j w^j, \tag{4.5.7}$$

and the only operations on g_k which preserve $|g_k|$ and the range of j over which $\tilde{g}_j$ is zero are zero-flipping operations on the zeros w_i of this polynomial. Most of the existing proof can now be reworked directly; the flipping is achieved by the same multiplication which we again break up as in (3.3.5). The effect on the $\tilde{g}_j$ is identical in spite of the fact that the convolution (4.5.2) becomes circular; this can be established directly, but it is simpler to regard g_k as a set of sample values of the periodic function $g(z)$, thus identifying the $\tilde{g}_j$ with the Fourier components of the latter. In the same way as before therefore, the effect of flipping a set of zeros $\{w_i\}$ of (4.5.7) is to multiply the modulus of the highest numbered $\tilde{g}_j$, and divide that of the lowest by $\prod_i |w_i|$. "In general" the discrete problem can thus also be expected to have a unique solution if the width of $\tilde{g}_j$ is restricted in this way. In an indirect way this result bears on the aperiodic problem of the previous section too since it is possible to choose sampling intervals such that a discrete transform pair closely approximate an aperiodic continuous transform pair (see Section 2.2).

Schiske (1974) has tackled the uniqueness problem in the discrete case by algebraic methods, using the Galois theory of extension fields. The approach does not unfortunately offer a practical method of extracting the phase distribution, but his results are nevertheless reassuring. In the general case *without* any restriction on the number of nonzero diffraction plane sample values, he shows that the existence of *one* solution (the true phase) implies the existence of at least one further (distinct) solution, though it is not possible to establish the total number of solutions in this way. *With* the restriction on the diffraction plane width, he demonstrates that intensity values permitting more than one phase solution are the exception rather than the rule.

These general conclusions are incidentally confirmed by the computer searches of Section 6.4, which, although disclosing as many as seven solutions to one problem with no restriction on the range of j for which $\tilde{g}_j$ was nonzero, failed to find any alternative solutions when the range was restricted as in this section.

4.6. Summary

All numerical applications of this method are strictly to the discrete problem, and perhaps the most important result demonstrated in the preceding is therefore that with a "finite width" condition on the discrete transform, the phase solution is unique "in general." The method *can* thus do what we set out to achieve, namely establish phase distributions outside the range of validity of the weak-scattering approximation.

The words "in general" cover occasional ambiguities due to symmetry properties of the wave function—the minor ones of Section 4.2 and the more interesting ones of Section 4.4. The presupposition that the intensities in both planes can be measured with arbitrary accuracy leads to difficulties in many practical cases, compounded by the fact that the "finite width" condition in the diffraction plane requires that the image plane data be sampled at twice the usual rate; in particular, the method cannot be applied when the phase variation in one plane is weak (which includes the x-ray case unfortunately) and is likely to suffer from ill-conditioning of the data near this limit. On the other hand, the bright-field weak object case in electron microscopy is easily handled by conventional methods (see Chapter 1), and the accuracy requirements for the present approach should be substantially relaxed in the situations for which it has been developed, namely dark-field microscopy or bright-field microscopy of strong objects.

CHAPTER FIVE

The image and diffraction plane problem: numerical methods☆

W. Owen Saxton
Murray Edwards College, University of Cambridge, Cambridge, United Kingdom
e-mail address: wos1@cam.ac.uk

Contents

5.1. Direct methods

The numerical methods presented in this chapter are formulated in one dimension rather than in two for simplicity, extension to two-dimensional methods being considered from time to time as necessary. They are also naturally formulated in terms of discrete sequences of numbers interconnected by discrete Fourier transforms rather than in terms of continuous functions. The relationship between the two has however been extensively discussed in previous chapters. This section is concerned with a case where the representation of the latter by the former is exact, namely that of band-limited periodic functions with the additional assumption that the transform values are known a priori to be zero over at least half their range. This is the case appropriate to microscopy of periodic objects with finite apertures, provided the image is sampled at twice the standard frequency, and the wave function solution was shown in the last chapter to be unique "in general." A simple noniterative solution for the wave function is possible from the modulus data in these circumstances, outlined below, which is capable of systematic generation of all wave functions consistent

☆ Reprinted from Advances in Electronics and Electron Physics, Supplement 10, 1978.

Advances in Imaging and Electron Physics, Volume 214
ISSN 1076-5670
https://doi.org/10.1016/bs.aiep.2020.04.005

with the data if the latter are known *exactly* but which we shall unfortunately find to be very unreliable in the presence of even small levels of noise and to be limited by computer time requirements to very small sets of data.

We shall denote the N image wave function values by g_k, and the transform values by $\tilde{g}_j$, the subscripts running from $-N/2$ through zero to $N/2-1$ (rounding down if N is odd). We shall assume the range of j for which $\tilde{g}_j$ is nonzero to be evenly disposed about the origin so that the extreme nonzero values are $\tilde{g}_n$ and $\tilde{g}_{-n}$. The extension necessary to remove this assumption is straightforward and will not be given here. Our assumed data are $|g_k|$ and $|\tilde{g}_j|$; from $|g_k|$ we can obtain the autocorrelation

$$\tilde{I}_j = \sum_j \tilde{g}_l \tilde{g}^*_{l-j} \tag{5.1.1}$$

as the discrete transform of $|g_k|^2$ [see (2.1.16)]. Summation is not assumed in this section over repeated subscripts and is indicated in the normal way where necessary. The assumed restriction on the nonzero values of $\tilde{g}_j$ prevents any circular overlap in calculating $\tilde{I}_j$.

The solution works from the outside of the transform range to the center. $\tilde{I}_j$ being conjugate symmetric, we are interested only in the positive j range. The outermost nonzero value is

$$\tilde{I}_{2n} = \tilde{g}_n \tilde{g}^*_{-n}, \tag{5.1.2}$$

involving only two of the $\tilde{g}_j$ values. We fix a phase origin by assigning an arbitrary phase to $\tilde{g}_n$; this done, $\tilde{g}_{-n}$ follows at once from (5.1.2). We move then to the next nonzero correlation value, namely,

$$\tilde{I}_{2n-1} = \tilde{g}_n \tilde{g}^*_{-(n-1)} + \tilde{g}_{n-1} \tilde{g}^*_{-n}, \tag{5.1.3}$$

which introduces two more unknown $\tilde{g}_j$ values. Now $|\tilde{g}_j|$ being known, the moduli of the two terms in (5.1.3) are known; it will therefore be obvious geometrically that a twofold choice is in general possible for these terms, and so for the pair $\tilde{g}_{n-1}$ and $\tilde{g}_{-(n-1)}$, since $\tilde{g}_n$ and $\tilde{g}_{-n}$ are known. Exceptionally, a single solution only will be permitted, but we must expect two in general. Accordingly we begin the construction of a tree of solutions from this initial branching, taking each solution in turn and proceeding to $\tilde{I}_{2n-2}$ which introduces two more unknowns $\tilde{g}_{n-2}$ and $\tilde{g}_{-(n-2)}$ in an equation with the same form as (5.1.3). Besides the possibility of further branching, we now encounter for the first time the possibility that the moduli of the two terms involving the unknowns allow no solution at all, in which case we have to strike off this branch and revert to exploring the alternative solution to the previous equation.

In this way we work in and out until reaching at some point the equation involving $\widetilde{I}_n$:

$$\widetilde{I}_n = \tilde{g}_n\tilde{g}_0^* + \tilde{g}_0\tilde{g}_{-n}^* + \text{other known terms.} \tag{5.1.4}$$

This introduces $\tilde{g}_0$ for the first time and determines the last unknown $\tilde{g}_j$ value, thus providing a possible complete solution. However, we have only used half of the $\widetilde{I}_j$ values so far, so it is to be expected that many branches will extend as far as this point. Whenever it is reached, we must calculate from the $\tilde{g}_j$ values then current the expected correlation values for $0 \leqslant j < n$ and compare these with the observed values (the calculation proceeding by way of the corresponding g_k of course). From the results of the previous chapter, we expect that only one branch will meet this final test in general, but it would naturally be prudent to apply it to all branches yielding a complete set of $\tilde{g}_j$ values in case exceptional multiple solutions exist to the particular data being tackled.

In order to keep the foregoing explanation of this approach reasonably simple, an exceptional source of difficulty has been so far overlooked. If for some j, the condition

$$\tilde{g}_n\tilde{g}_{-j}^* + \tilde{g}_j\tilde{g}_{-n}^* = 0 \tag{5.1.5}$$

on the wave function values holds, then the tree generation breaks down at the equation introducing $\tilde{g}_j$ and $\tilde{g}_{-j}$ since this will reduce to (5.1.5) and leave the phase of the two terms arbitrary, provided only that they differ by π. As this *is* an exceptional case, it will not be labored here, but an indication of the necessary action may be useful. If we assign a phase to either term at random, we introduce an unknown phase error into $\tilde{g}_j$, and the opposite error into $\tilde{g}_{-j}$; examination of the subsequent equations easily reveals that *all* subsequently found $\tilde{g}_j$ values for positive j share the same error and for negative j share the opposite error. When (5.1.4) is encountered, $\tilde{g}_0$ and $\tilde{g}_0^*$ are treated as independent unknowns, the former deriving from the positive j series and the latter from the negative j series; the unknown phase error introduced earlier is then easily identified by requiring that the solutions for $\tilde{g}_0$ and $\tilde{g}_0^*$ be complex conjugates of each other. If we are unfortunate enough to encounter a problem for which (5.1.5) holds for several values of j, we have to treat $\tilde{g}_j$ and $\tilde{g}_j^*$ as independent for a wider range of j values, and pursue the tree below the $\widetilde{I}_n$ equation.

Two-dimensional extension of the method is straightforward; instead of working inwards from the ends of the one-dimensional transform, we work inward along the rows of the two-dimensional transform, beginning with the extreme nonzero rows and progressing to the next when all elements in these have been established.

At this point, however, the method's drawbacks must be pointed out. These are two, both of them serious. First, the method is extremely noise sensitive. Because of the sequential nature of the solution, errors are magnified rapidly as they propagate inward. The crucial outside data values are normally the least accurately known so that the solution tree rapidly comes to bear little resemblance to that for ideal data. Admittedly, there are additional data (the innermost $\tilde{I}_j$ values) permitting cross checking, but no simple recipe exists for correcting the $\tilde{g}_j$ values in the light of these. Second, the simplicity of the solution may have disguised the long computing time required to explore all branches; this grows exponentially with the number of data points, and although reasonable for 64 points or less, it might easily extend to a million years for only 256 points (even in one dimension)!

Dallas (1975), in discussing the method, does not appear to be aware of its limitations. Although a variant will appear in Chapter 7, we will pursue it no further but will turn instead to faster iterative solutions which promise to be more noise tolerant.

5.2. Steepest descent methods

This section reviews several ways in which the problem can be approached through minimizing the residues of nonlinear equation sets (the residue of an equation being the amount by which its two sides are in imbalance for a given estimate of the solution). Being based on progressive global minimization these methods are much more noise tolerant than the sequential solution of the preceding section.

For completeness, the form given by Gerchberg and Saxton (1971) is outlined first, though this is not subsequently pursued. The present discrete formulation allows a much more rapid derivation of the equations involved. We denote the N wave function values in the image and diffraction plane by g_k and $\tilde{g}_j$ as usual, separating the modulus and phase factors of $\tilde{g}_j$ to give

$$\tilde{g}_j = \tilde{h}_j \tilde{\phi}_j \quad \left(\sum\right),$$

where $|\tilde{\phi}_j| = 1$, and $\tilde{h}_j$ is observable and real. The image plane wave function samples follow, by Fourier transformation, as

$$g_k = N h_m \phi_{k-m} \tag{5.2.1}$$

[see (2.1.14)]. The intensity at a given image point thus becomes

$$I_k = N^2 h_m h_p^* \phi_{k-m} \phi_{k-p}^* \quad \left(\sum_k\right). \tag{5.2.2}$$

(Note that the summation convention of Section 2.1 is again in force except where explicitly stated otherwise.) This comprises a set of N equations in $2N$ unknowns (the real and imaginary parts of ϕ_k): the imaginary part of the set vanishes for any ϕ_k. Additional equations can be adduced expressing the fact that the Fourier transform of ϕ_k is unimodular:

$$\tilde{\phi}_j \tilde{\phi}_j^* = 1 \quad \left(\not\sum\right),$$

so that from (2.1.16),

$$\phi_p \phi_{p-k}^* = \delta_{k0}. \tag{5.2.3}$$

This set is conjugate symmetric in k for any ϕ_k; moreover, the imaginary parts of the equations for $k = 0$ and $N/2$ vanish for all ϕ_k. The set thus provides exactly N further equations, and we regard (5.2.2) and (5.2.3) as comprising a set of $2N$ quadratic equations to be solved for the $2N$ unknown real and imaginary parts of ϕ_k. Gerchberg and Saxton (1971) demonstrate their solution for a small model problem using a standard "library" routine.

The quadratic form of the equations unfortunately offers no further information on the uniqueness of the solution; even a set of two quadratics (conic sections) can have any number of solutions from zero to four. Nor does it offer a specially rapid method of residue minimization compared with transcendental forms. Since these allow us to halve the number of equations, we proceed to examine three such forms.

Taking our unknowns to be the N phases θ_k of the image wave function values g_k, we might inverse transform the diffraction plane intensity $\widetilde{D}_j$ to give the set of N equations

$$D_k = N g_p g_{p-k}^*$$

[by (2.1.16)]. Solution would amount to minimizing the (complex) residues

$$f_k(\theta_p) = N g_p g^*_{p-k} - D_k \tag{5.2.4}$$

for $0 \leqslant k \leqslant N/2$ (the remaining values being related by symmetry). Equally, we might consider the transform of this set, namely, the N real residues

$$f_j(\theta_k) = |\tilde{g}_j|^2 - \widetilde{D}_j \tag{5.2.5}$$

with $0 \leqslant j \leqslant N-1$, which measures diffracted intensity defect or simply a set measuring modulus defect

$$f_j(\theta_k) = |\tilde{g}_j| - \tilde{d}_j, \tag{5.2.6}$$

where $d_j = \sqrt{\widetilde{D}_j}$ is the diffraction plane wave function modulus.

We thus face the common problem of tackling a set of nonlinear equations with the general form

$$f_i(x_j) = 0$$

or at least minimizing f_i according to some suitable criterion. The core of conventional approaches to this is the expansion of f_i about x_j to give

$$\Delta f_i \simeq \frac{\partial f_i}{\partial x_j} \Delta x_j \tag{5.2.7}$$

(the derivatives being evaluated at x_j). Provided x_j is reasonably near the root, we can choose $\Delta f_i = -f_i(x_j)$ and estimate our distance Δx_j from the root by solving the *linear* equation set (5.2.7). The neglect of the nonlinear terms is accounted for by iteration of the process. The method is the multidimensional analog of the one-dimensional Newton–Raphson method and is referred to subsequently, in view of the linear equation solution involved, as the matrix inversion method. The derivative matrix may in the last resort be calculated by computing the set f_i repeatedly, incrementing each of the x_j in turn, but in the present case the derivatives can be evaluated analytically and expressions are now given appropriate to each of the equation sets (5.2.2)–(5.2.6).

For the set (5.2.2), for which the residues are

$$f_k = N^2 h_m h_p \phi_{k-m} \phi^*_{k-p} - I_k \quad \left(\sum_k\right),$$

the derivatives we require are those with respect to the real and imaginary parts of ϕ_p, i.e., $\mathrm{Re}\{\partial f_k/\partial\phi_p\}$ and $-2\,\mathrm{Im}\{\partial f_k/\partial\phi_p\}$, according to (5.A.1.4) and (5.A.1.5). These are, respectively,

$$2N\,\mathrm{Re}\{h_{k-p}g_k^*\} \quad \Big(\sum_k\Big)$$

and

$$-2N\,\mathrm{Im}\{h_{k-p}g_k^*\} \quad \Big(\sum_k\Big),$$

using (5.2.1), while for the complementary set (5.2.3), the residues are

$$f_k = \phi_p\phi_{p-k}^* - \delta_{k0},$$

and the derivatives are

$$\phi_{p-k}^* + \phi_{p+k} \quad \text{and} \quad i(\phi_{p-k}^* - \phi_{p+k}).$$

Of greater interest, in view of the smaller number of equations involved, is the set (5.2.4) for which the required derivatives, according to (5.A.1.3), are

$$\frac{\partial f_k}{\partial\theta_p} = Ni(g_pg_{p-k}^* - g_p^*g_{p+k}) \quad \Big(\sum\Big). \tag{5.2.8}$$

Like the derivatives above these are rapidly evaluated, but it should be noted that calculation of the residues themselves involves convolution or correlation sums which are of course most rapidly obtained through two Fourier transformations (see Sections 2.1 and 9.3).

Only one transformation is necessary to obtain the residues for the intensity defect set (5.2.5). These can be reexpressed in terms of g_k as

$$f_j = g_kg_m^*W_{jk}W_{jm}^* - \tilde{D}_j \quad \Big(\sum_j\Big),$$

allowing the derivatives

$$\frac{\partial f_j}{\partial\theta_p} = 2\,\mathrm{Im}\{g_p^*\tilde{g}_jW_{pj}^*\} \quad \Big(\sum\Big) \tag{5.2.9}$$

to be obtained easily from (5.A.1.6). Since $\tilde{g}_j$ is calculated in the course of evaluating f_j, these are also rapidly calculated. The derivatives for the modulus defect set (5.2.6) are found to be

$$\frac{\partial f_j}{\partial \theta_p} = \frac{1}{|\tilde{g}_j|} \operatorname{Im}\{g_p^* \tilde{g}_j W_{pj}^*\} \quad \left(\sum\right) \tag{5.2.10}$$

and require a similar computational effort in calculation.

It should be noted that in each of the nonlinear equation sets, one of the equations is trivial or dependent on the others [e.g., the equation for $k=0$ in (5.2.4) is trivial, while $\sum_j f_j = 0$ for (5.2.5) because of Parseval's theorem (2.1.7)]. This reflects the fact that the data cannot determine *absolute* phase (Section 4.2), and accordingly it is sets of $N-1$ rather than N linear equations (5.2.7) that actually result.

This conventional method has actually been applied to the set (5.2.4), for which the derivatives are given in (5.2.8). The choice of this particular set was accidental, the method in fact being developed from a quite different point of view. Its rapid convergence properties—all the forms so far discussed exhibit the second order convergence of the one-dimensional Newton–Raphson process—were exploited in a search for multiple exact solutions to particular problems in conjunction with a theoretical uniqueness study (see Sections 4.5 and 6.4). However, even allowing for the numerous variants in use with improved speed and stability, all methods based on (5.2.7) are seriously limited for the present purposes by their computing requirements and their instability when far from the root or when an exact root does not exist (e.g., because of observational noise). The storage necessary is proportional to N^2 (for the element of the derivative matrix) and the time to N^3 (for the solution of the linear equation set) so that N is limited to a few hundred on even the largest computers. Accordingly, we turn instead to a variant exhibiting first order convergence only but requiring storage and time proportional to N only and showing greater reliability in convergence when far from the root. A variant will subsequently be presented to meet the special problem arising near the root when no exact root exists.

It is obviously necessary and sufficient for the vanishing of a set of residues f_k that the single function

$$E = f_k f_k^* \tag{5.2.11}$$

should vanish. The problem is thus reduced to locating the zero(s) of a single positive function of N independent variables. A typical numerical

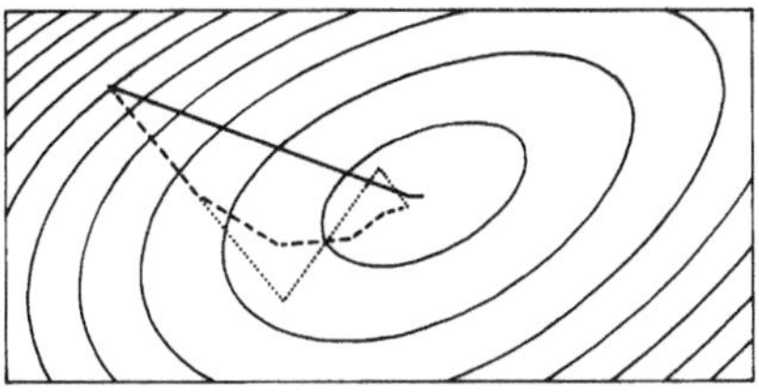

Figure 5.2.1 The convergence pattern of three iterations. Solid: the matrix inversion method. Broken: the steepest descent method. Dotted: the steepest descent method with double step.

technique for this purpose involves making successive steps in the direction of steepest descent of the surface E, and we shall refer to this method subsequently as the "steepest descent" method, though this name is sometimes applied to the matrix inversion method of (5.2.7). Specifically, given an estimate x_k of the root, the gradient $E_k = \nabla_k E$ is calculated, and a correction

$$\Delta x_k = -\frac{E}{E_j E_j} E_k \qquad (5.2.12)$$

is added to the estimate x_k, the process being repeated until it converges. Fig. 5.2.1 compares the successive steps taken by the matrix inversion method (solid) and the steepest descent method (broken) from a common starting point for the two-variable problem,

$$f_1 = x_2 - \sin(x_1)/2, \qquad f_2 = x_1 - \exp(-2x_2),$$

against a background of a contour map of the sum square residue $E = f_1^2 + f_2^2$. (The contour levels are evenly spaced from zero, but the central one is at a height of one *quarter* the interval between the others.) The second step of the matrix inversion method reaches a point that matches the position of the root to four significant figures while five steps of the steepest descent method leave it further from the root than a single step of the matrix inversion method. A substantial improvement can be made at no extra cost by *doubling* the correction (5.2.12). The root is in fact a minimum of E, since E is always positive, and a given section through it is thus in general quadratic. On such a section, it is easy to see that the usual steepest descent method takes steps half as large as it should. The dotted route in Fig. 5.2.1 shows the progress of the steepest descent method with this modification. Obviously, it should be thus modified only near a root;

even so, three double-length steps achieve, in the example of the figure, more than five normal steps.

In view of the elimination of matrix inversion, even five steps of the steepest descent method are much faster to execute than one of the matrix inversion method. The only factor not yet considered is evaluation of the gradient E_k, and this proves to be reasonably easily calculated. Again, analytic expressions are well worth obtaining, and we consider these now.

The method may be conveniently applied to the intensity defect set (5.2.5); we require

$$E_k = \frac{\partial E}{\partial \theta_k} = 2f_j \frac{\partial f_j}{\partial \theta_k},$$

and substituting the expression (5.2.9) for $\partial f_j / \partial \theta_k$, we obtain

$$E_k = 4\,\mathrm{Im}\{g_k^* f_j \tilde{g}_j W_{kj}^*\} \quad \left(\sum_k\right), \tag{5.2.13}$$

involving the inverse Fourier transform of $f_j \tilde{g}_j$. Two transformations are therefore required to evaluate f_j and E_k from a given set θ_k. For comparison, two transformations are necessary to evaluate the residues alone for the set (5.2.4) which involves an autocorrelation, while the gradients E_k involve a further transformation.

In the same way, the gradient E_k for the modulus defect set (5.2.6) is found to be

$$E_k = 2\,\mathrm{Im}\left\{g_k^* \frac{f_j \tilde{g}_j}{|\tilde{g}_j|} W_{kj}^*\right\} \quad \left(\sum_k\right)$$

from (5.2.10), or, replacing f_j by (5.2.6), when the first inverse transformation simply yields g_k,

$$E_k = -2\,\mathrm{Im}\left\{g_k^* \frac{d_j \tilde{g}_j}{|\tilde{g}_j|} W_{kj}^*\right\} \quad \left(\sum_k\right). \tag{5.2.14}$$

It is in fact this last form, based on the set (5.2.6), that is pursued most thoroughly subsequently. This is partly because the same definition of a sum square residue function E appears as a natural error criterion also for the quite different approach presented in the next section, but also because iterations based on (5.2.13), i.e., (5.2.5), appear to converge much more slowly in practice than those based on (5.2.14), although the reasons for

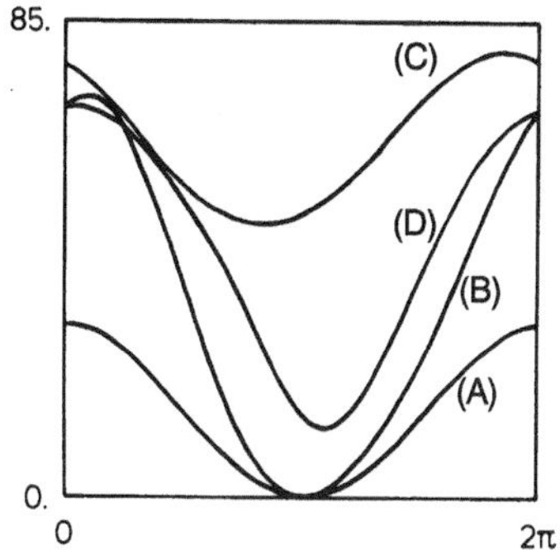

Figure 5.2.2 The sum square residue E as a function of individual image phases (see text for details).

this are not clear. Computationally the two formulations make virtually identical demands.

The behavior of the function E, defined in this final way, is illustrated in Fig. 5.2.2 which shows the variation of E as various individual phases traverse the full 2π range. The image plane wave function here was defined on 64 points, at which the phases and moduli were taken at random from uniform distributions between 0 and 1. Curves (A) and (B), which exhibit zeros, are sections through the root, i.e., all the phases, except the one being varied over the 2π range, are correct; in curve (C), one additional phase is displaced by $\pi/2$ from the root; in curve (D), all the phases are displaced from the root by amounts taken at random from a uniform distribution between 0 and 0.6 rad. The smooth character of the surface is reassuring, but it is obvious that there may be many local minima in the surface besides the root(s).

Although the steepest descent iteration outlined in the foregoing largely overcomes the problems of storage and time requirements associated with the matrix inversion method, it continues to fail catastrophically when no exact root to the equations $f_j = 0$ exist, for in this situation E has a minimum but no actual zero, and the step Δx_k of (5.2.12) becomes infinite at such a minimum since the gradient E_k vanishes. It is therefore necessary to switch from a zero-seeking iteration to a minimum-seeking one as the neighborhood of the root is approached. We now consider how this can be done.

It is necessary and sufficient for E to be extremal that the function

$$B = E_k E_k \tag{5.2.15}$$

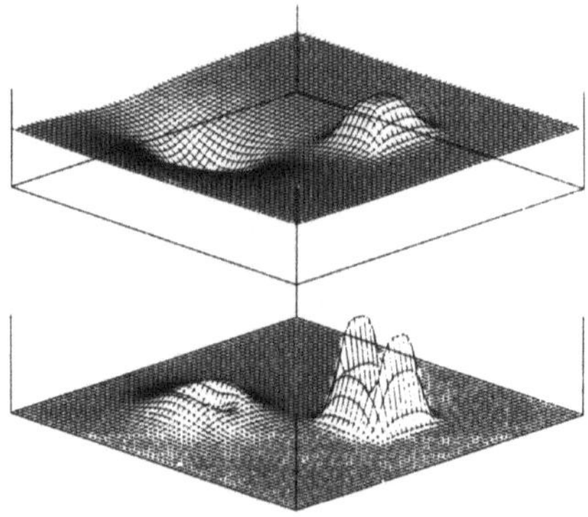

Figure 5.2.3 Above: the function E for a two-variable problem; below: the corresponding function B.

vanish. Treating B as a function of the N independent variables (the unknown phases θ_p), we can apply the steepest descent method of (5.2.12) again, this time to find the zero(s) of B, this function also being everywhere positive, but there being no longer any possibility of catastrophic divergence in the neighborhood of a near-zero of E, since B is guaranteed to have a zero at the extremum of E. On the other hand, B has also many zeros which are not zeros or even minima of E (saddle points or maxima of E, for example). This is illustrated in Fig. 5.2.3 which shows in perspective a function E of two variables, and the corresponding B. This also makes it clear that the region around a zero of E within which the direction of steepest descent on B is *toward* rather than away from the root is much smaller than the region within which a steepest descent method applied to E would converge. The obvious practical course is to reach the neighborhood of a minimum of E with the original method given earlier, doubling the step length when E is sufficiently small, and switching to a steepest descent of B when the first process shows signs of instability.

Again it is well worth obtaining an analytic expression for $B_p = \nabla_p B$, as its direct numerical calculation requires N^2 Fourier transforms! In fact four transforms only are necessary. The gradients are

$$B_p = \frac{\partial B}{\partial \theta_p} = 2E_k \frac{\partial E_k}{\partial \theta_p}. \qquad (5.2.16)$$

Using the expression given in (5.2.14) for E_k, and (5.A.1.6), we have

$$B_p = -4E_k \,\mathrm{Im}\left\{ \frac{g_p^*}{i} \frac{\partial}{\partial g_p^*} \left([g_k^* \tilde{g}_j W_{jk}^* - g_k \tilde{g}_j^* W_{jk}] \frac{\tilde{d}_j}{|\tilde{g}_j|} \right) \right\} \quad \left(\sum_p \right)$$

$$= 4E_k \operatorname{Re}\left\{ g_p^* \left([\tilde{g}_j W_{jp}^* \delta_{pk} - g_k W_{jp}^* W_{jk}] \frac{\tilde{d}_j}{|\tilde{g}_j|} - [g_k^* \tilde{g}_j W_{jk}^* - g_k \tilde{g}_j^* W_{jk}] \frac{\tilde{d}_j \tilde{g}_j}{2|\tilde{g}_j|^3} W_{jp}^* \right) \right\} \quad \left(\sum_p \right),$$

in which the result has been used that

$$\frac{\partial}{\partial g_p^*} \tilde{g}_j^* = \frac{\partial}{\partial g_p^*} g_l^* W_{jl}^* = \delta_{pl} W_{jl}^* = W_{jp}^*.$$

We may take E_k within the "Re" operator since it is real; combining the second terms from the two inner brackets gives

$$B_p = 4 \operatorname{Re}\left\{ g_p^* \left(E_p \frac{\tilde{g}_j \tilde{d}_j}{|\tilde{g}_j|} W_{jp}^* - \frac{1}{2} \left[E_k g_k W_{jk} + E_k g_k^* \frac{\tilde{g}_j^2}{|\tilde{g}_j|^2} W_{jk}^* \right] \frac{\tilde{d}_j}{|\tilde{g}_j|} W_{jp}^* \right) \right\} \quad \left(\sum_p \right). \tag{5.2.17}$$

In each term, the j-summation constitutes a Fourier transformation. After this, the second and third terms involve sums of the convolution type, so that it is obviously advantageous to use Fourier transformation to evaluate them. Accordingly we introduce the transform of $E_k g_k$:

$$E_k g_k = (1/N) \tilde{b}_l W_{lk}^*.$$

Substituting this in the second and third terms and denoting $\tilde{g}_j \tilde{d}_j / |\tilde{g}_j|$ by $\tilde{a}_j / N$, we have

$$B_p = 4 \operatorname{Re}\left\{ g_p^* \left(E_p a_p - \frac{1}{2N} \left[\tilde{b}_l W_{lk}^* W_{jk} + \tilde{b}_l^* W_{lk} \frac{\tilde{g}_j^2}{|\tilde{g}_j|^2} W_{jk}^* \right] \frac{\tilde{d}_j}{|\tilde{g}_j|} W_{jp}^* \right) \right\} \quad \left(\sum_k \right).$$

The k-summation yields factors of $N\delta_{jl}$ (2.1.2), giving finally,

$$B_p = 4 \operatorname{Re}\left\{ g_p^* \left(E_p a_p - \frac{1}{2} \left[\tilde{b}_j + \tilde{b}_j^* \frac{\tilde{g}_j^2}{|\tilde{g}_j|^2} \right] \frac{\tilde{d}_j}{|\tilde{g}_j|} W_{jp}^* \right) \right\} \quad \left(\sum_p \right). \tag{5.2.18}$$

This expression allows the calculation, with only four transformations, of the residues f_j the gradient E_k and the gradient B_p. A single transformation yields f_j and allows us to construct $\tilde{a}_j$; an inverse transformation then gives E_k (from (5.2.14)) and the first term of B_p (5.2.18), besides enabling us to construct b_k from $g_k E_k$; a third transform then leads to $\tilde{b}_j$ from which the inverse transformand may be calculated for the last transform which provides the remaining terms of B_p (5.2.18).

Not the least convenient feature of the use of discrete Fourier transforms to calculate (5.2.18) is the fact that if two-dimensional transforms are used in place of one-dimensional transforms, the whole analysis extends directly to two-dimensional problems. Appendix 5.A.2 contains a short Fortran program which performs the necessary calculations in this case, the routine FOUR2 being used to perform the transformations in situ. Precautions must be taken against the possibility of any $\tilde{g}_j$ vanishing. Mathematically this leaves (5.2.10) and various subsequent expressions undefined; numerically it causes overflow unless circumvented as in the program. A case where this actually occurs for *every* $\tilde{g}_j$ except $\tilde{g}_0$ is in the problem of a pure phase image ($I_k = 1$) with the initial estimate $\theta_k = 0$, which gives $\tilde{g}_j = N\delta_{j0}$. For residues defined as in (5.2.5) instead of (5.2.6) (the intensity rather than modulus defect form), the corresponding case gives a vanishing gradient E_p (5.2.13), and indeed a vanishing gradient B_p (the expression is not given here, but is similar to (5.2.18)). The estimate is clearly extremely in error. Avoidance of the singularity is thus a convenient feature of the intensity defect form, but is by no means sufficient to outweigh its poor convergence rate.

Some results from the steepest descent method are presented in the next chapter, all based on the modulus defect form which this section has examined in most detail. There are indeed problems of instability, but the method is certainly sufficiently modest in its requirements of storage and time to be viable with large numbers of data points.

A short digression is made here because the mathematics involved draws heavily on the development above of the gradients of B; it considers the confidence levels to which the phases are determined by the data available (discussed again in Section 6.1).

We take the curvature of the surface $E(\theta_p)$ to measure the accuracy with which the phases are determined. Measuring θ_p from a zero (or minimum), we may write the value of E on a line in the direction of a given phase θ_p as

$$E = E_0 + \frac{1}{2}\frac{\partial^2 E}{\partial \theta_p^2}\theta_p^2 \cdots = E_0 + \frac{1}{2}E_{pp}\theta_p^2 \cdots .$$

Choosing an arbitrary change ΔE in E (say 0.1% of the wave function energy) as significant, the extreme values of θ_p to be tolerated are $\pm(2\Delta E/E_{pp})^{1/2}$. We require only a convenient way of calculating E_{pp}. An expression for this is now given; it relates closely to the expression above for B_p (5.2.18), a similar amount of computation being needed to evaluate it. We require $\partial E_p/\partial \theta_p$; comparing this with (5.2.16), the only difference is a factor of $2E_k$ and a summation over k. We can thus pick up the calculation at (5.2.17), dropping a factor $2E_k$ and replacing k by p throughout:

$$\begin{aligned}\frac{\partial^2 E}{\partial \theta_p^2} &= 2\,\mathrm{Re}\left\{g_p^*\left(\frac{\tilde{g}_j \tilde{d}_j}{|\tilde{g}_j|} W_{jp}^* - \frac{1}{2}\left[g_p W_{jp} + g_p^* \frac{\tilde{g}_j^2}{|\tilde{g}_j|^2} W_{jp}^*\right] \frac{\tilde{d}_j}{|\tilde{g}_j|} W_{jp}^*\right)\right\} \quad \Big(\sum_p\Big) \\ &= 2\,\mathrm{Re}\left\{g_p^*\left(a_p - \frac{1}{2} g_p \frac{\tilde{d}_j}{|\tilde{g}_j|} + g_p^* \frac{\tilde{g}_j^2 \tilde{d}_j}{|\tilde{g}_j|^3} W_{j,2p}^*\right)\right\} \quad \Big(\sum_p\Big), \qquad (5.2.19)\end{aligned}$$

writing as $\tilde{g}_j \tilde{d}_j/|\tilde{g}_j|$ as $N\tilde{a}_j$ as before. Four Fourier transforms are necessary to compute the expression. One obvious remark may be made about the form of the expression: the factor g_p^* occurring throughout means that the expression will be small for low-modulus points and that the phase of these points will in consequence be least accurately determined, as we might have expected.

5.3. The iterative transform method

The problem is now approached from a completely new direction, with the development of an algorithm based on repeated Fourier transformation of the wave function estimate with modulus corrections being made symmetrically in the two data planes. It is completely stable, being incapable of diverging, though capable of locking, and is flexible enough to find other important applications (see Chapter 7) besides the present one. Before the final version of the algorithm is presented, an attempt is made to indicate the line of thought leading to its development.

It is useful, first of all, to consider a simple iteration for extracting the square root of a (possibly complex) number y. From a given estimate x_i lying to one side of the root, a further (equally inaccurate) estimate lying to the other side is obtained from y/x_i; averaging these two produces an estimate x_{i+1} with considerably reduced error. That is, the iteration is defined by

$$x_{i+1} = (x_i + y/x_i)/2.$$

The convergence is, in fact, second order. Now one of the ways in which the phase determination problem can be cast is that of finding a function

g of given modulus with a given autocorrelation function D (see (5.2.4)). If we start with a given estimate of g, we might attempt to follow the square root algorithm above and construct a new estimate, equally inaccurate but erring in the opposite direction, by "decorrelating" D with g (i.e., by Fourier transforming g and D to give $\tilde{g}$ and $\tilde{D}$, and then inverse transforming $\tilde{D}/\tilde{g}^*$), and then averaging these two estimates in some way, correcting the modulus of the new estimate to the value given before beginning the process again.

Various forms of average are possible, and most appear to result in convergent iterations. The relative weight given to the original and decorrelated estimate in the averaging process also appears relatively unimportant, and convergence still occurs if the decorrelated estimate is taken on its own.

Thus far the two data planes are not symmetrically treated. We might on the other hand impose the modulus corrections on the transformed functions (while performing the decorrelation) as well as on the image plane estimates. This procedure also converges quite reliably in practice. If we consider the precise effect of the adjustments in the transform plane, though, a significant result emerges. Having calculated $\tilde{g}$, we take $\tilde{D}/\tilde{g}^*$, which has the *same phase* as $\tilde{g}$, and then correct its modulus to $\tilde{D}$. In other words, the adjustment in the transform plane amounts to correcting the *modulus* of the estimate to the observed value while preserving its *phase*. In view of the fact just noted that the relative weights given to the original and decorrelated estimates in the averaging process do not greatly affect convergence, we might take simply the decorrelated estimate as our new starting point; this procedure too converges.

This particular iteration is of considerable significance, and is the subject of the rest of this section. It has been described previously by Gerchberg and Saxton (1972, 1973a). The steps involved are restated and clarified in the block diagram of Fig. 5.3.1. Starting with any available estimate of the image plane wave function g, we compute the corresponding diffraction plane wave function $\tilde{g}$ by a Fourier transformation and correct the modulus of this function to the observed diffraction plane modulus $\sqrt{\tilde{D}}$ while preserving its phase. By inverse transformation we compute a new estimate of the image plane wave function, which will in general differ in modulus and phase from the original estimate. In this plane too we correct the modulus to the observed value, preserving the phase. This completes one cycle of the iteration, and the sequence is now repeated until successive estimates of the image plane wave function cease to differ significantly. The process may

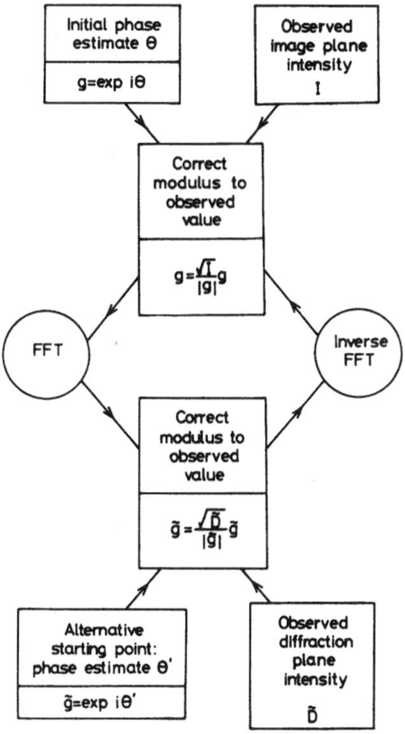

Figure 5.3.1 The iterative transform algorithm.

clearly be commenced in either plane with equal success. Appendix 5.A.3 contains a short Fortran program implementing the algorithm.

A quantitative criterion is required for the agreement between the observed intensity distributions and the wave function estimate at any stage; this provides a basis for deciding when to halt the iteration and also a mathematical parameter in terms of which the iteration behavior can be discussed subsequently. A suitable measure is a normalized form of the sum square residue function E associated in the last section with (5.2.6). Given a wave function estimate in either plane (the modulus of which matches the observed modulus), we Fourier transform it to obtain the corresponding estimate in the other plane and then sum the squared difference between *its* modulus and the observed modulus in that plane. An appropriate normalizing factor is the summed (observed) intensity in that plane. Thus if $\widetilde{D}_j$ is the observed diffraction plane intensity, the error in a given estimate g_k of the image plane wave function is given by

$$E = \frac{\sum_j (|\tilde{g}_j| - \sqrt{\tilde{D}_j})^2}{\sum_j \tilde{D}_j}. \tag{5.3.1}$$

This measure, subsequently referred to as the "sum square error" (or SSE) in an estimate, and frequently quoted as a percentage, measures the Hilbert space norm between the observed and estimated transform moduli and vanishes if and only if the estimate is completely consistent with the observed moduli in both planes. It is of course equally applicable to estimates in either plane.

We may sum up the convergence properties of the algorithm in the statement that the action of the algorithm cannot increase the SSE in an estimate. This implies great stability in the iterative process, though at the same time the fact that a stronger statement is not possible means that convergence may be slow. The statement is now proved by a method due to Gerchberg (Gerchberg & Saxton, 1972, 1973a). The proof is stated in discrete terms as usual, but the extension to the continuous case is here particularly obvious and straightforward.

Fig. 5.3.2 shows, in Argand diagram form, the wave function estimate for an arbitrary point in either plane at a series of stages of the iteration. At each stage, the circle shown has a radius equal to the observed wave function modulus for the point in question. The first stage (top left) shows a given estimate g_k with the observed modulus, but an incorrect phase. On transformation, this yields an estimate $\tilde{g}_j$ in the other plane, the modulus of which differs from the observed modulus (unless g_k is already a solution). Correction of the modulus in this plane is accomplished by the addition of $\tilde{c}_j$ which is everywhere parallel or antiparallel to $\tilde{g}_j$. Referring to (5.3.1), we note that the SSE in the initial estimate g_k is the summed squared modulus of $\tilde{c}_j$,

$$E_1 = \sum_j |\tilde{c}_j|^2, \tag{5.3.2}$$

if the problem is so scaled that the normalization factor in the denominator of (5.3.1), which is constant throughout the iteration, is unity. The corrected estimate $\tilde{g}_j + \tilde{c}_j$ is now restored by inverse transformation to the original plane where the modulus is again corrected, as shown in the third stage of the figure, by addition of d_k. This provides a new estimate for comparison with g_k. The SSE in this is obtained by a further transformation, to the last stage shown in the figure, where addition of $\tilde{e}_j$ performs the

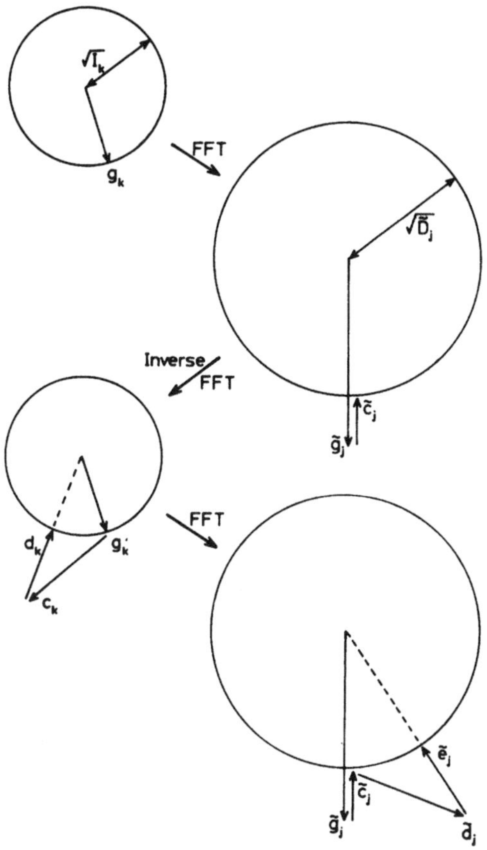

Figure 5.3.2 The convergence of the iterative transform algorithm.

modulus correction, as

$$E_2 = \sum_j |\tilde{e}_j|^2. \tag{5.3.3}$$

Considering the fourth stage shown, it follows from the triangle inequality,

$$|x + y| \leqslant |x| + |y|,$$

that at each point in this plane,

$$|\tilde{e}_j| \leqslant |\tilde{d}_j|,$$

and hence that

$$E_2 \leqslant \sum_j |\tilde{d}_j|^2.$$

By Parseval's theorem (2.1.7), however

$$N\sum_{k}|d_k|^2=\sum_{j}|\tilde{d}_j|^2,$$

so that

$$E_2\leqslant N\sum_{k}|d_k|^2.$$

Applying the triangle inequality now to the third stage in the figure, we have

$$|d_k|\leqslant|c_k|, \tag{5.3.4}$$

so that

$$E_2\leqslant N\sum_{k}|c_k|^2.$$

By a further application of Parseval's theorem, however,

$$N\sum_{k}|c_k|^2=\sum_{j}|\tilde{c}_j|^2=E_1, \tag{5.3.5}$$

leading to the result that

$$E_2\leqslant E_1, \tag{5.3.6}$$

which proves the statement made earlier.

We note that while this result compares the wave function estimate at the first and third stages of the algorithm (before and after one complete cycle) it also follows from the argument above that the wave function estimate "improves" at every single stage. The SSE in the estimate at the second stage, for example, is

$$N\sum_{k}|d_k|^2,$$

which, by (5.3.4) and (5.3.5), cannot exceed E_1. We may thus choose if we wish, to regard each cycle of the algorithm as presented in Fig. 5.3.2 as two cycles of a simpler iteration, each cycle of which involves Fourier transformation followed by modulus correction.

It should be stressed that the convergence proof presented here depends simply on the linearity and unitarity (Parseval's theorem) of the Fourier transform, discrete or continuous; the algorithm presented will therefore converge when *any* linear unitary transformation replaces Fourier transformation. This point is pursued in Chapter 7, which explores further applications of the algorithm as presented and some useful modified forms.

The foregoing discussion makes it clear that this method of solution is completely stable against catastrophic divergence, no matter how far from the root the initial phase estimate may be. This fact allows us to use, as a satisfactory initial estimate in the absence of any real prior information as to the phase distribution, a function whose modulus matches the observed modulus and whose phase is generated by taking successive values from a random number generator covering the range $-\pi$ to π, though a phase distribution reflecting any expected properties of the solution, even if only in the size of the maximum phase shifts present, would always be preferable to one containing no information at all. Certain simpler initial phase estimates are quite unsatisfactory however, in particular cases, and must be avoided. For example, the iteration fails if commenced in a given plane with all phases equal when the intensity distribution in the other plane is centrosymmetric. Such an initial estimate is a *real* function (on removal of a constant phase factor) and its Fourier transform is thus conjugate symmetric. In view of the symmetry of the intensity distribution, the modulus correction in the Fourier plane preserves the conjugate symmetry so that the new estimate in the original plane remains real. The modulus correction in this plane then restores the original function; the iteration "locks," making no change in the wave function estimate.

The "locked" condition, more generally, occurs as the limiting case of (5.3.6) when the equality holds. This occurs when c_k has the same phase as g_k everywhere so that the correction made in the third stage of Fig. 5.3.2 leads directly back to the first stage, and no progress at all is made by the algorithm. A locked iteration will clearly propagate indefinitely both forward *and backward* in time so that a locked iteration cannot result from an initially converging one. However, an initially converging iteration may approach the locked condition asymptotically, and this behavior is indeed not infrequently observed, the sum square error tending to a finite nonzero value. The condition is also invariably approached as an iteration approaches a solution, the rate of convergence dropping progressively in a characteristic manner. The first few iterations are the most useful, and the rate of correction of the phase estimate dwindles subsequently.

We can express the condition for a locked iteration formally in terms of the wave function estimate and intensity distributions. We require that c_k has the same phases as g_k everywhere, i.e., that $c_k g_k^*$ is real or that the correlation

$$\tilde{c}_l \tilde{g}_{l-j}^*$$

is conjugate symmetric (2.1.9). The correction $\tilde{c}_j$ is

$$\tilde{c}_j = \tilde{g}_j(\sqrt{\tilde{D}_j} - |\tilde{g}_j|)/|\tilde{g}_j|,$$

so the condition becomes that the correlation

$$\tilde{g}_l \tilde{g}_{l-j}^*(\sqrt{\tilde{D}_l} - |\tilde{g}_l|)/|\tilde{g}_l| \tag{5.3.7}$$

is conjugate symmetric. An obvious case in which this is satisfied is that of a conjugate symmetric estimate $\tilde{g}_j$ and a symmetry intensity distribution $\tilde{D}_j$. In this case, both $\tilde{c}_j$ and $\tilde{g}_j$ are conjugate symmetric, and their correlation has the same property. This is the case mentioned earlier, for if $\tilde{g}_j$ is conjugate symmetric, then g_k is real, and its phase is constant everywhere.

We turn now to an alternative approach to the convergence properties of the iteration which permits us to explore mathematically the prospects for over- or under-correction of the estimates as a means of improving convergence. It is limited to estimates which are already close to the solution and uses a different measure of the error in an estimate which will not recur after this section.

If the image plane wave function is f_k, we may write a close approximation to it as

$$g_k = f_k + \varepsilon_k \quad (|\varepsilon_k| \ll |f_k|),$$

the error E_1 in g_k being defined as $\sum_k |\varepsilon_k|^2$. An obvious weakness of the criterion is its sensitivity to multiplication of g_k by a constant phase factor, but it is sufficient for our present purpose. The application of the algorithm to this results initially in the transform

$$\tilde{g}_j = \tilde{f}_j + \tilde{\varepsilon}_j \quad (|\tilde{\varepsilon}_j| \ll |\tilde{f}_j|),$$

after which the modulus correction is performed by replacing $\tilde{g}_j$ by $\tilde{g}_j|\tilde{f}_j|/|\tilde{g}_j|$, giving a new estimate

$$\tilde{g}_j' = (\tilde{f}_j + \tilde{\varepsilon}_j)|\tilde{f}_j|/|\tilde{f}_j + \tilde{\varepsilon}_j| \tag{5.3.8}$$

$$
\begin{aligned}
&= \tilde{f}_j \left[\frac{1 + \tilde{\varepsilon}_j / \tilde{f}_j}{1 + \tilde{\varepsilon}_j^* / \tilde{f}_j^*} \right]^{1/2} \\
&= \tilde{f}_j + \frac{\tilde{\varepsilon}_j}{2} - \frac{1}{2} \frac{\tilde{f}_j}{\tilde{f}_j^*} \tilde{\varepsilon}_j^* + O(|\varepsilon|^2) \\
&= \tilde{f}_j + \frac{\tilde{\varepsilon}_j}{2} - \frac{\tilde{\delta}_j}{2},
\end{aligned} \tag{5.3.9}
$$

writing $\tilde{\delta}_j = \tilde{f}_j \tilde{\varepsilon}_j^* / \tilde{f}_j^*$ and neglecting powers of ε greater than unity. The algorithm continues by inverting the transform and evaluating the new error E_2. The latter may as easily be evaluated in this plane, however, because of Parseval's theorem (2.1.8), and is

$$
\frac{1}{N} \sum_j \left| \frac{\tilde{\varepsilon}_j}{2} - \frac{\tilde{\delta}_j}{2} \right|^2 .
$$

The modulus of $\tilde{\delta}_j$ is the same as that of $\tilde{\varepsilon}_j$, however, and consequently

$$
|\tfrac{1}{2}\tilde{\varepsilon}_j - \tfrac{1}{2}\tilde{\delta}_j| \leqslant |\tilde{\varepsilon}_j|.
$$

Thus,

$$
E_2 \leqslant (1/N) \sum_j |\tilde{\varepsilon}_j|^2 = E_1 .
$$

This treatment will now be applied to examine two obvious adjustments to the basic algorithm, the first involving over- or under-correction of the moduli and the second taking as a new estimate a linear combination of the original estimate and that obtained by modulus correction.

The effect of over- or under-correction of the moduli is to replace (5.3.8) by

$$
\tilde{g}_j' = (\tilde{f}_j + \tilde{\varepsilon}_j) \left[\frac{|\tilde{f}_j|}{|\tilde{f}_j + \tilde{\varepsilon}_j|} \right]^{\alpha},
$$

α being greater or less than unity according to whether the estimate g_k is to be over- or under-corrected. The expansion of (5.3.9) now becomes

$$
\tilde{g}_j' = \tilde{f}_j + (1 - \tfrac{1}{2}\alpha)\tilde{\varepsilon}_j - \tfrac{1}{2}\alpha\tilde{\delta}_j,
$$

with an error

$$
E_2 = \frac{1}{N} \sum_j |(1 - \tfrac{1}{2}\alpha)\tilde{\varepsilon}_j - \tfrac{1}{2}\alpha\tilde{\delta}_j|^2 .
$$

As before, $|\tilde{\delta}_j| = |\tilde{\varepsilon}_j|$, and so by the triangle inequality, for $\alpha < 2$

$$|(1 - \tfrac{1}{2}\alpha)\tilde{\varepsilon}_j - \tfrac{1}{2}\alpha\tilde{\delta}_j| \leqslant (1 - \tfrac{1}{2}\alpha)|\tilde{\varepsilon}_j| + \tfrac{1}{2}\alpha|\tilde{\varepsilon}_j| = |\tilde{\varepsilon}_j|,$$

and $E_2 \leqslant E_1$. The cancellation of α in this last expression shows that there is no reason to over- or under-correct the moduli, since this will not affect the rate of convergence. (The factors in this last expression involving α should in fact be enclosed in modulus signs: if $\alpha > 2$, this becomes significant and allows $E_2 > E_1$.)

If we adopt as a new estimate a linear combination of the original estimate and that obtained by modulus correction, we have, for (5.3.9),

$$\tilde{g}'_j = \lambda(\tilde{f}_j + \tilde{\varepsilon}_j) + \mu(\tilde{f}_j + \tfrac{1}{2}\tilde{\varepsilon}_j - \tfrac{1}{2}\tilde{\delta}_j).$$

We choose $\lambda + \mu = 1$ to balance the zero order terms and so obtain an error E_2 given by

$$E_2 = \frac{1}{N}\sum_j |\lambda\tilde{\varepsilon}_j + \tfrac{1}{2}(1-\lambda)\tilde{\varepsilon}_j - \tfrac{1}{2}(1-\lambda)\tilde{\delta}_j|^2.$$

With $0 \leqslant \lambda \leqslant 1$, we have by the triangle inequality

$$|\lambda\tilde{\varepsilon}_j + \tfrac{1}{2}(1-\lambda)\tilde{\varepsilon}_j - \tfrac{1}{2}(1-\lambda)\tilde{\delta}_j| \leqslant \lambda|\tilde{\varepsilon}_j| + \tfrac{1}{2}(1-\lambda)|\tilde{\varepsilon}_j| + \tfrac{1}{2}(1-\lambda)|\tilde{\varepsilon}_j| = |\tilde{\varepsilon}_j|,$$

giving once more $E_2 \leqslant E_1$, so that this approach too is unpromising. (With λ outside the range 0 to 1, we may have $E_2 > E_1$.)

Both of these disappointing results appear to be borne out in practice. The only device which has had any success in improving the convergence of the process is the accentuation of the detail in the observed modulus distribution. The algorithm is allowed to run for a number of cycles on a synthetic problem made up by exaggerating the intensity variations observed in the two planes (preserving, of course, the Parseval integral), after which the degree of exaggeration is relaxed somewhat and a further few cycles of the algorithm executed. The problem being solved thus gradually approaches the actual problem posed: the hope is that the accentuation of the intensity variations will force the phase corrections more positively in the right direction.

A convenient formula is needed to define an accentuated modulus distribution with a single parameter to vary the degree of accentuation. Denoting the original modulus by m, and its mean by μ, one such suitable

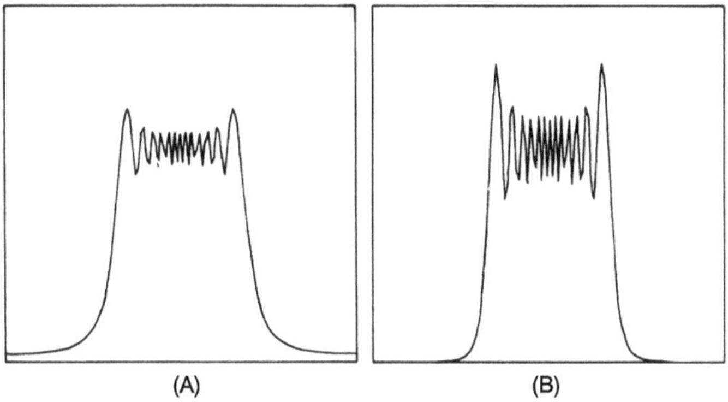

Figure 5.3.3 A modulus distribution (A) before and (B) after accentuation.

function is

$$m' = m(1 + \lambda(m - \mu)), \tag{5.3.10}$$

in which λ is the parameter required. The distribution is unchanged for $\lambda = 0$ and is accentuated progressively until $\lambda = 1/\mu$, when it is squared. For higher values of λ, the curve has a negative portion, and $1/\mu$ is thus the largest useful value for λ. This problem does not affect the curve for $m > \mu$, and a separate larger value of λ may be used here. Fig. 5.3.3 shows a modulus distribution (that of the transform of a quadratic phase factor with five cycles of phase to either side of the origin) in its original form (A) and accentuated (B) with $\lambda = 1/\mu$. Chapter 6 gives some results from this acceleration technique: it is somewhat erratic.

We have now completed the presentation of the basic numerical methods devised for the extraction of phase distributions from image and diffraction plane intensity data; the performance of these is examined in the next chapter which gives some results obtained by these methods in one- and two-dimensional model problems, thus providing an "experimental" assessment of this approach to phase determination.

Appendix 5.A

5.A.1 Derivatives of functions of complex variables

Some expressions are established in the following for certain derivatives of a function $f(z_i, z_i^*)$ of several complex variables. Writing $z = x + iy = r\exp(i\theta)$ with x, y, r and θ real, and treating z_i and z_i^* as formally independent, we have

$$\frac{\partial f}{\partial x_i} = \frac{\partial f}{\partial z_i}\frac{\partial z_i}{\partial x_i} + \frac{\partial f}{\partial z_i^*}\frac{\partial z_i^*}{\partial x_i} = f_{z_i} + f_{z_i^*}. \tag{5.A.1.1}$$

Similarly,

$$\frac{\partial f}{\partial y_i} = if_{z_i} - if_{z_i^*}. \tag{5.A.1.2}$$

Also,

$$\frac{\partial f}{\partial \theta_i} = \frac{\partial f}{\partial z_i}\frac{\partial z_i}{\partial \theta_i} + \frac{\partial f}{\partial z_i^*}\frac{\partial z_i}{\partial \theta_i} = iz_i f_{z_i} - iz_i^* f_{z_i^*}. \tag{5.A.1.3}$$

If $f(z_i, z_i^*)$ is *real* the derivatives must also be real; from (5.A.1.1) then,

$$f_{z_i} + f_{z_i^*} = f_{z_i^*} + f_{z_i^*}^*,$$

while (5.A.1.2) gives

$$f_{z_i} - f_{z_i^*} = -f_{z_i}^* + f_{z_i^*}^*.$$

Subtracting these, we see that

$$f_{z_i^*} = f_{z_i}^*$$

in this case. The preceding expressions now simplify considerably:

$$\frac{\partial f}{\partial x_i} = 2\,\mathrm{Re}\{f_{z_i}\}, \tag{5.A.1.4}$$

$$\frac{\partial f}{\partial y_i} = -2\,\mathrm{Im}\{f_{z_i}\} = 2\,\mathrm{Im}\{f_{z_i^*}\}, \tag{5.A.1.5}$$

$$\frac{\partial f}{\partial \theta_i} = -2\,\mathrm{Im}\{z_i f_{z_i}\} = 2\,\mathrm{Im}\{z_i^* f_{z_i^*}\}. \tag{5.A.1.6}$$

5.A.2 Sum square residue gradient calculation

```
      SUBROUTINE GRAD2D(IC,N,PH,IM,D,F,FSQ,/GRADE/,/ESQ/,
     >/GRADB/,/BSQ/)
C
C  CALCULATES RESIDUES, GRADIENTS AND HIGHER ORDER GRADIENTS
C  FOR DIFFRACTION PLANE MODULUS FIT FOR GIVEN IMAGE PHASES
C     IC     - CONTROL VALUE (1,2 OR 3 - SEE COMMENTS)
```

```
C     N(2)  - DIMENSIONS OF PICTURE ARRAYS (POWERS OF 2)
C     PH    - IMAGE PHASES
C     IM    - IMAGE PLANE MODULUS
C     D     - DIFFRACTION PLANE MODULUS
C     F     - RESIDUES
C     FSQ   - SUM SQUARE RESIDUE (E)
C     GRADE - GRADIENT OF E
C     ESQ   - SUM SQUARE E-GRADIENT (B)
C     GRADB - GRADIENT OF B
C     BSQ   - SUM SQUARE B-GRADIENT
C
      COMPLEX G(4096),GT(4096),B(4096),Z
C  PRESENT DIMENSIONS LIMIT N(1)*N(2): INCREASE AS NECESSARY
C
      REAL PH(1),F(1),GRADE(1),GRADB(1),IM(1),D(1)
      INTEGER N(2),P
      MN=N(1)*N(2)
      DO 1 P=1,MN
      G(P)=CMPLX(COS(PH(P)),SIN(PH(P)))*IM(P)
1     GT(P)=G(P)
      CALL FOUR2(GT,N,2,-1,1)
      FSQ=0.
      DO 2 J=1,MN
      T=CABS(GT(J))
      IF (IC.EQ.1) GOTO 6
C  TRAP VANISHING ELEMENTS
      Z=0.
      IF (T.NE.0.) Z=GT(J)*D(J)/T
      B(J)=Z
6     F(J)=T-D(J)
2     FSQ=FSQ+F(J)*F(J)
C
C  RETURN IF RESIDUES ONLY REQUIRED
      IF (IC.EQ.1) RETURN
      CALL FOUR2(B,N,2,1,1)
      ESQ=0.
      DO 3 P=1,MN
      Z=CONJG(G(P))*B(P)
      DL=-2.*AIMAG(Z)
```

```
      GRADE(P)=DL
      ESQ=ESQ+DL*DL
      IF (IC.GE.3) GRADB(P)=4.*REAL(Z)*DL
3     B(P)=G(P)*DL
C
C  RETURN UNLESS B-GRADIENTS REQUIRED
      IF (IC.EQ.2) RETURN
      CALL FOUR2(B,N,2,-1,1)
      DO 4 J=1,MN
      T=CABS(GT(J))
C  TRAP VANISHING ELEMENTS
      Z=0.
      IF (T.NE.0.) Z=(B(J)+CONJG(B(J))*GT(J)*GT(J)/(T*T))*D(J)/T
4     B(J)=Z
      CALL FOUR2(B,N,2,1,1)
      BSQ=0.
      DO 5 P=1,MN
      GRADB(P)=GRADB(P)-2.*REAL(CONJG(G(P))*B(P))
5     BSQ=BSQ+GRADB(P)*GRADB(P)
      RETURN
      END
```

5.A.3 The image and diffraction plane algorithm

```
      SUBROUTINE GSALG(PSI,M,N,IMAGE,BFP,NPRINT,ITMAX,SSQERR,ACC)
C
C  ITERATIVE CALCULATION OF COMPLEX IMAGE WAVEFUNCTION COMPATIBLE
C  WITH MODULUS DATA IN IMAGE AND DIFFRACTION PLANES.
C    PSI(M,N) - COMPLEX ARRAY CONTAINING INITIAL IMAGE ESTIMATE
C               (M AND N ARE POWERS OF 2)
C    IMAGE    - IMAGE PLANE MODULUS
C    BFP      - DIFFRACTION PLANE MODULUS
C    NPRINT   - NUMBER OF CYCLES BETWEEN ERROR PRINTOUT
C    ITMAX    - MAXIMUM NUMBER OF ITERATIONS TO BE PERFORMED
C    SSQERR   - SUM SQUARE ERROR IN IMAGE PLANE ESTIMATE
C    ACC      - REQUIRED ACCURACY
C
C  THE ITERATION QUITS IF :
C    1. SSQERR < ACC
C    2. ITMAX CYCLES HAVE BEEN PERFORMED
```

```
C     3. THE CONVERGENCE GETS VERY SLOW
C
      COMPLEX PSI
      REAL IMAGE
      DIMENSION PSI(1),IMAGE(1),BFP(1),NN(2)
      MN=M*N
      NN(1)=M
      NN(2)=N
C
C   CHECK NORMALISATION OF MODULUS DATA AND
C   FORCE MODULUS OF SUPPLIED PSI
      E1=0.
      E2=0.
      DO 1 I=1,MN
      E1=E1+IMAGE(I)*IMAGE(I)
1     E2=E2+BFP(I)*BFP(I)
      T1=SQRT(E1/E2)
      DO 2 I=1,MN
      S=CABS(PSI(I))
      PSI(I)=PSI(I)*IMAGE(I)/S
2     BFP(I)=BFP(I)*T1
C
C   INITIALISE ITERATION
      RMN=MN
      RMN=1./SQRT(RMN)
      RNORM=100./E1
      TR=1E50
      TC=.003+NPRINT
      NC=1
      NIT=NPRINT
      WRITE (6,14)
C
C   COMPUTE ERROR AND CORRECT MODULUS IN DIFFRACTION PLANE
5     CALL FOUR2(PSI,NN,2,-1,1)
      T1=0.
      DO 3 I=1,MN
      S=CABS(PSI(I))
      P=S*RMN-BFP(I)
      T1=T1+P*P
```

```
3     IF (S.NE.0.) PSI(I)=PSI(I)*BFP(I)/S
      T1=T1*RNORM
C
C  REPEAT CORRECTION IN IMAGE PLANE
      CALL FOUR2(PSI,NN,2,1,1)
      DO 4 I=1,MN
      S=CABS(PSI(I))
4     IF (S.NE.0.) PSI(I)=PSI(I)*IMAGE(I)/S
C
C  CHECK EXIT CRITERIA AND LOOP
      IF (NIT.EQ.NPRINT) GOTO 6
      NIT=NIT+1
      GOTO 7
6     WRITE (6,10) NC,T1
      NIT=1
7     IF (T1.GT.ACC) GOTO 9
      WRITE (6,11)
      SSQERR=T1
      RETURN
9     IF (NC.LT.ITMAX) GOTO 20
      WRITE (6,12) ITMAX
      SSQERR=T1
      RETURN
20    IF (TR-T1.GT.TR*TC) GOTO 8
      WRITE (6,13)
      RETURN
      SQERR=T1
8     TR=T1
      NC=NC+1
      GOTO 5
10    FORMAT (I8,10X,F13.3,'%')
11    FORMAT (/' REQUESTED ACCURACY ACHIEVED'//)
12    FORMAT (/16,' ITERATIONS NOW PERFORMED'//)
13    FORMAT (/' CONVERGENCE SLOW - ITERATION TERMINATED'//)
14    FORMAT ('1NUMBER  OF          SSQ ERROR IN'
     >/' ITERATIONS          FOURIER PLANE'//)
      END
```

CHAPTER SIX

The image and diffraction plane problem: computational trials☆

W. Owen Saxton
Murray Edwards College, University of Cambridge, Cambridge, United Kingdom
e-mail address: wos1@cam.ac.uk

Contents

6.1. Real problems and mathematical models

This chapter presents a series of model calculations using the methods outlined in the preceding chapter, thus providing an evaluation of their performance. The data defining the model problems solved are all computer generated rather than real data, and therefore we first consider how the former approximate the latter, and so how far the results presented form a fair assessment of the method.

There are certain problems associated with recording the intensity distributions in the two Fourier planes (image and diffraction planes); these derive from the requirement that the diffraction pattern corresponds to the area of the image that is sampled, from the dynamic range of the diffraction pattern, the noise level in the electron beam, and the need to establish the relative alignment of the two intensity distributions. The first of these is irrelevant to periodic specimens; provided only that a large number of unit cells are illuminated, the maximum intensities of the diffraction spots represent sample values of the transform of a single cell of the image, which is thus the natural unit for computation, the discrete Fourier transformation representing an *exact* interrelation between the data in the two planes if an objective aperture is present (see Section 2.2). However in the aperiodic case, only the object area used in the reconstruction can be illuminated; the former is severely restricted by computer space requirements,

☆ Reprinted from Advances in Electronics and Electron Physics, Supplement 10, 1978.

ISSN 1076-5670
https://doi.org/10.1016/bs.aiep.2020.04.006

and it is extremely difficult to illuminate so small an area. Second, although a diffraction pattern is at least independent of the objective lens spherical aberration, its dynamic range may be so wide that several have to be recorded with different electron exposures; this too is possible only for periodic specimens for which a diffraction pattern can be obtained with a relatively low electron dose and for which data may be easily combined from several different preparations. The accuracy of the recorded intensities is ultimately limited by the total electron dose which the specimen will tolerate and the desired resolution level, by way of electron shot noise (see Section 1.6); the distributions are also distorted to some extent by the superposition of the inelastically scattered electrons (see Sections 1.2 and 1.3) which contribute a diffuse background throughout the image (rather less diffuse at low resolution) and distort the diffraction pattern chiefly in the low-angle region. Finite beam coherence and energy spread both degrade the image in some degree also, though presenting no serious problem in the diffraction plane. Again, it is possible to reduce the noise level in images of periodic specimens greatly by averaging the images of many unit cells, while additional micrographs taken at the optimum *inelastic* defocus permit some estimate to be made of the inelastic contribution to the image (see Section 1.5); it is impossible to separate the inelastic and elastic contributions to the diffraction pattern however without an energy filter on the microscope. Finally, there is usually an alignment problem only for aperiodic specimens where there is no array of spots to define a diffraction plane lattice which may be compared with the image plane lattice; it is the orientational alignment which is problematical, a lateral misalignment only introducing a linear phase factor in one plane. Taken together, these four considerations rule out any attempt to apply the method to aperiodic specimens in the context of electron microscopy at least, and while the *general* uniqueness problem was nevertheless also investigated in Chapter 4, it is principally with *periodic* specimens that we are subsequently concerned.

The sampling interval for the diffraction plane (designated "bfp" for "back focal plane" in the figures) is fixed by the spacing of the spots, but the image plane sampling interval is at our disposal; decreasing it involves using more data points, i.e., including more diffraction spots. From the point of view of the results available on the theoretical uniqueness of the solution (Section 4.5), it is desirable that at least half the diffraction plane sample values are zero. The frequent use of contrast-free images in the subsequent examples conflicts with this since a unimodular wave function cannot have a finite number of nonzero Fourier components (see Section 3.3); these examples are certainly heavily oversampled nevertheless, and useful solutions are still obtained.

Since aberration correction is straightforward if the wave function is once determined, no aberrations have been included in generating the data moduli for the examples. Accordingly, except where a diffraction limit has been imposed, the in-focus image of a pure phase object under bright-field conditions is free of contrast, and the corresponding modulus (unity everywhere) has been used as one of the two sets of data in several examples without its being included in the figures. In such cases it is also possible to consider the iteration as being applied between the *object* and the diffraction plane.

Where problems are presented in both bright- and dark-field conditions, the interchange is made simply by suppressing the zero order transform value so that the central stop dark-field mode is implied. The background intensity remaining between "atoms" in the dark-field images, although perhaps surprising at first, is a feature of periodic specimens and not an artifact of the numerical process.

Both one- and two-dimensional examples are given, the former allowing a more accurate visual assessment of the quality of the solutions but the latter being of course the realistic case. Computational requirements for the former are negligible; for the latter, about 15 seconds of processor time are needed on an IBM 370/165. In view of the relatively unimportant ambiguities described in Section 4.2, adjustments have sometimes been made before display, where permissible, in the solutions obtained; this applies particularly where the image is free of contrast and the phase distribution is free to move relative to it, cross correlation being used to remove translational (and sometimes sign) differences between actual and obtained phase distributions. Background level differences have not usually been removed, and this sometimes results in brightness differences in the figures for the two-dimensional examples which are in fact not significant; the contrast scales in these vary considerably, many distributions being shown at higher than actual contrast level for clarity, while the zero order diffraction spot has always been suppressed in diffraction pattern displays.

With each result, the corresponding value of the sum square error (SSE) defined in (5.3.1) is given. Useful as this is as a mathematical convergence parameter, it gives little idea of how closely observed modulus values are actually reproduced by the solution or how accurately the phases are established so a brief consideration of these points is useful. The general treatment was touched on in Section 5.2 [particularly (5.2.19)], but a simple assessment is possible at least for low contrast images near the solution. First, an rms fractional defect f in the modulus estimate for a unimodular image with N sample points implies an SSE of Nf^2/N of f^2, so that an SSE of 0.1%, for example, implies that the modulus values are reproduced with an rms error no better than 3.2%, and it is not surprising that a lower SSE is frequently needed in bright-field problems to give a satisfactory phase solution. Second, if the phase error in such a wave function is fairly small, we can denote the difference between the actual and estimated wave function sample values by ϕ_k, with $|\phi_k|$ very nearly equal to the phase error at the point k. If we suppose the iteration to be in the final stage of slow convergence to the root, then the transform $\tilde{\phi}_j$ has everywhere almost the same phase as the diffraction plane wave function (cf. Section 5.3), and the SSE in the image wave function estimate is simply $\sum|\tilde{\phi}_j|^2/N^2$. By Parseval's theorem (2.1.8) however, $\sum|\tilde{\phi}_j|^2 = N\sum|\tilde{\phi}_k|^2$, so that if we denote the rms value of ϕ_k by ϕ, we have an SSE of ϕ^2: an SSE of 0.1% thus implies an rms phase error of 0.032 rad.

Considering these expressions from another point of view, a required phase accuracy of ϕ implies reducing the SSE to ϕ^2, which in turn implies that the fractional error in the modulus data must be rather less than ϕ; we have thus linked the error in our data and solution in a convenient and useful way. A phase error of 0.1 rad might be satisfactory for strong phase objects, and would imply the reasonably easily attained figure of 5% in the modulus values, i.e., 10% in the intensities. Whether the 1% accuracy necessary to permit an error no greater than 0.01 rad will be possible is an open question.

6.2. The steepest descent method

This section considers the performance of the most promising of the methods described in Section 5.2, namely the steepest descent method in which the iteration seeks directly to minimize an error criterion E (an unnormalized form of the SSE) which is the sum of the squares of the residues defined by (5.2.6), a given phase estimate being corrected by the addition of the expression (5.2.12), the derivatives of E (given in (5.2.14)) being computed by means of the subroutine listed in Section 5.A.2. Except when the SSE exceeds 5%, double length steps are used (see Fig. 5.2.1); various criteria are used to halt the iteration, including a limit on the processor time used (about 2 sec on an IBM 370/165) or the number of iteration cycles, the achievement of a "sufficiently low" SSE, or a pronounced *growth* in the

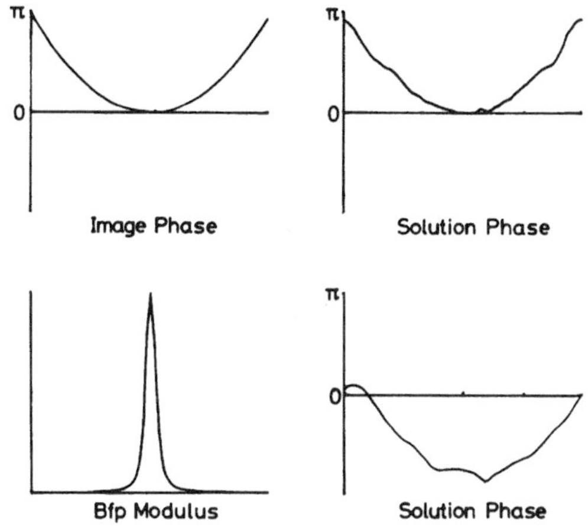

Figure 6.2.1 Two solutions to a bright-field problem. SSEs: above 0.10%, below 0.08%.

SSE between iterations (indicating instability). In some cases, when the SSE drops below 1%, the descent is made on the surface $B = |\nabla E|^2$ instead of the E surface [see (5.2.15): the derivatives of B, given in (5.2.18), may also be computed with the subroutine in Appendix 5.A.2], always with a double length step. Sixty-four sample points were used for the one-dimensional models, and a 32 by 32 array for the two-dimensional one.

The starting point for the iteration is an "estimate" of the phase values usually generated by adding to the true phase values numbers taken from a random number generator with a uniform distribution of variable width. The variability of the distribution width provides a simple control on how far from the solution a given trial starts. When the distributions are in the range $-\pi$ to π, the estimate is independent of the true phase; sometimes, however, the iteration is successful only if the range is in some degree restricted, and in such cases we may conclude that the particular problem under discussion could not be solved in this way without a reasonable first approximation to the solution. An initial estimate of zero everywhere should be avoided for the reason given near the end of Section 5.2.

In the first model (Fig. 6.2.1) as in many subsequent ones, the image is supposed contrast free, corresponding to a pure phase object, and the image modulus is not included in the figure. The actual image phase grows quadratically from the center to a maximum of π. The upper solution was obtained from an initial phase up to $\pm\pi/4$ from the true phase; the SSE fell by a factor of three during each of the four cycles needed for solution. The lower solution was obtained from a random initial estimate in nine cycles;

Table 6.2.1 SSE decay for the steepest descents method.

Iteration number	SSE (%)	*B* (unnormalized)	Iteration number	SSE (%)	*B* (unnormalized)
0	148.70	526744	5	6.52	45168
1	69.50	252063	6	1.71	11318
2	28.18	76921	7	0.49	2087
3	14.76	42957	8	0.20	409
4	10.00	23215	9	0.08	178

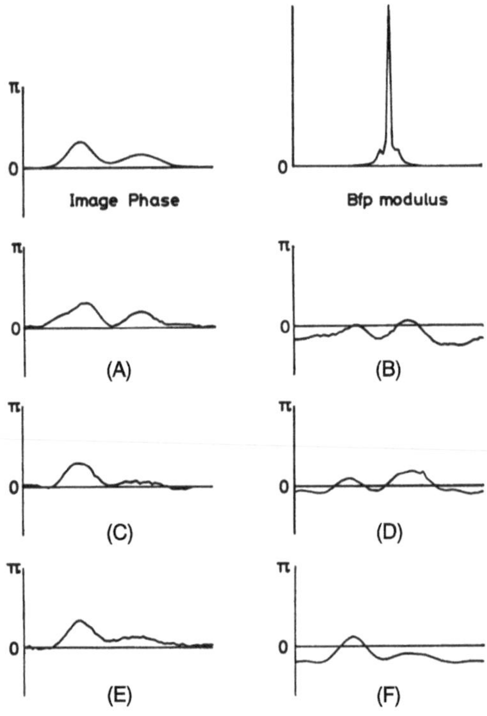

Figure 6.2.2 Six solutions to a bright-field problem. SSEs: (A) 0.09%, (B) 0.10%, (C) 0.07%, (D) 0.17%, (E) 0.08%, (F) 1.00%.

the general convergence behavior of the method is illustrated by the SSE decay for this case, listed in Table 6.2.1 together with the corresponding B values. The final SSE values could clearly have been improved by further iteration. The descents were exclusively on the E surface in both cases.

Fig. 6.2.2 considers a problem in which the phase distribution consists of a pair of Gaussians with a maximum value of 1 rad. In the figure, (A) and (B) are solutions obtained from starting points up to ± 1 rad and $\pm\pi/2$, re-

Table 6.2.2 SSE decay for the steepest descents method on *B*.

Iteration number	SSE (%)	*B* (unnormalized)	Iteration number	SSE (%)	*B* (unnormalized)
0	26.072	236 786.7	20	1.732	8 348.7
1	6.771	52 150.8	21[a]	1.003	8.8
2	2.311	5 886.5	22	1.003	1.9
3	5.613	49 014.6	23	1.003	4.1
4	1.704	7 318.5	24	1.002	1.3
5	2.592	17 500.7	⋮	⋮	⋮
⋮	⋮	⋮	30	1.002	1.0
19	2.395	15 864.4			

[a] The first set of values after the switch to the *B* surface.

spectively, from the true phase, both in three cycles; (C) and (D) are those obtained from random initial estimates, the phases for (C) being restricted to the range $-\pi/2$ to $\pi/2$, in 5 and 16 cycles, respectively. Note that (B) and (D) show reversed phase distributions; this corresponds to conjugation in the diffraction plane (cf. Section 4.2), the conjugation having been removed before the display. In (E), the solution is based on data in which 10% "noise" has been superposed on the diffraction plane modulus data (by the use of a random multiplier distributed uniformly in 0.9 to 1.1), the moduli being renormalized to satisfy Parseval's theorem; the starting point was ± 1 rad from the true phase, and four cycles were necessary. Finally, (F) shows a case where the descent was made on the B rather than the E surface; the possibility of an exact solution was destroyed by increasing the diffraction plane modulus values by 10% so that Parseval's theorem was *not* satisfied. Table 6.2.2 lists the values of the SSE and of B as the iteration proceeds from a starting point ± 1 rad from the true phase, the first set of values after the switch to the B surface being indicated by footnote; in a single cycle, the value of B is reduced by a factor of 1000, and the minimum is found within limits apparently imposed only by rounding errors in the gradient calculation. Used far from the root, however, the B surface descent proceeds rather slowly, and it should clearly not be resorted to until the E surface method shows signs of instability. At slightly higher SSE values than those in the figure, a subsidiary peak is often present in the "solution" outside the smaller of the two genuine peaks.

Fig. 6.2.3 considers the problem of Fig. 6.2.2 translated into dark-field conditions. Here the method did not succeed in reaching a low SSE from a completely random initial phase estimate. The solution (A) was obtained in

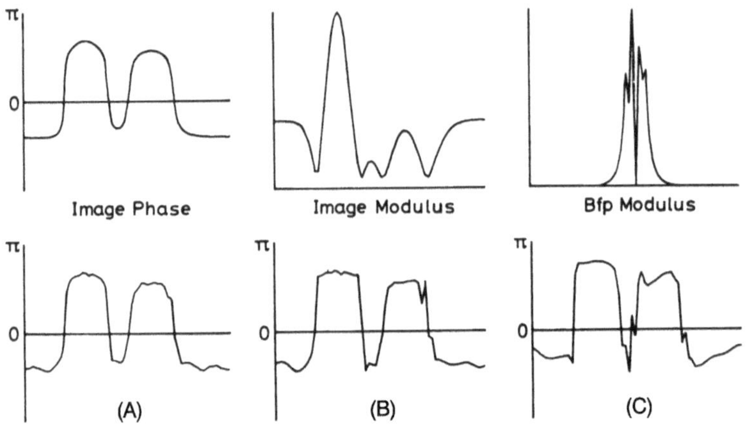

Figure 6.2.3 The problem of Fig. 6.2.2 in dark-field conditions. SSEs: (A) 0.16%, (B) 0.35%, (C) 0.56%.

four cycles from a starting point ±0.25 rad from the true phase; (B) started ±1 rad from the true phase, E dropping to 0.9% in 11 cycles, after which the descent was switched to the B surface and E only dropped to 0.35% over the next 80 cycles; similarly (C), starling ±2 rad from the true phase, reached an SSE of 0.58% in 25 cycles but remained static for about 30 cycles when switched to the B surface.

Fig. 6.2.4 gives a solution to a two-dimensional model problem similar to that of Fig. 6.2.2. The image is again taken to be contrastless, and the phase has a peak value of 1 rad. The solution shown results from five cycles applied to an initial estimate ±1 rad from the true phase; iterations commenced with a random initial phase did not achieve a low SSE.

With Fig. 6.2.5 we come to the first problem where the diffraction plane wave function is sharply restricted in spatial extent: only at five points out of the total of 64 are the sample values nonzero. The problem was generated by choosing a modulus and phase at random for each of these points and then computing the corresponding image wave function. In view of the smaller number of unknown phases in the diffraction plane than in the image, the numerical process was reversed for this problem (and for the two remaining problems in this section) the phase in the *diffraction* plane being adjusted to minimize the sum square difference between true and predicted *image* moduli. There is no advantage in terms of labor; mathematically the analysis remains identical apart from the interchange of transform directions, and the same subroutine was used for the computation. It was hoped that the descent might become more stable in view of the reduction of its

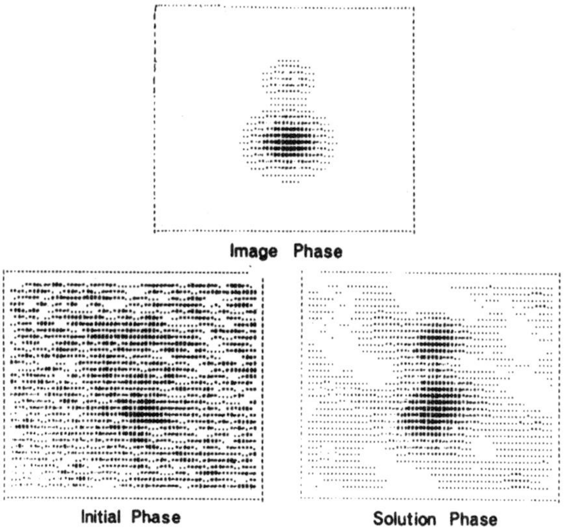

Figure 6.2.4 A two-dimensional bright-field problem. SSE: 0.018%.

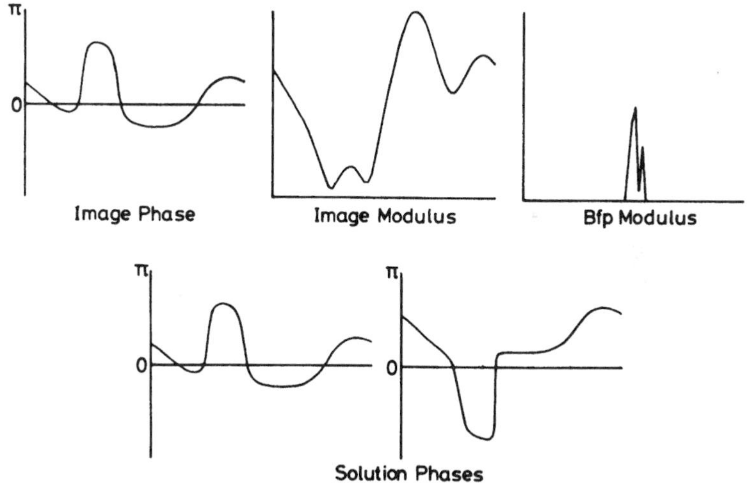

Figure 6.2.5 Two solutions to a problem with a narrow diffraction plane. SSEs: left 0.005%, right 0.14%.

dimensionality; while it is sufficiently reliable to find solutions from completely random initial phase estimates, it is debatable whether there is any improvement over the earlier form however. The figure shows two solutions, obtained in less than 20 cycles from completely random starting points; the first, virtually indistinguishable from the true phase, is the most

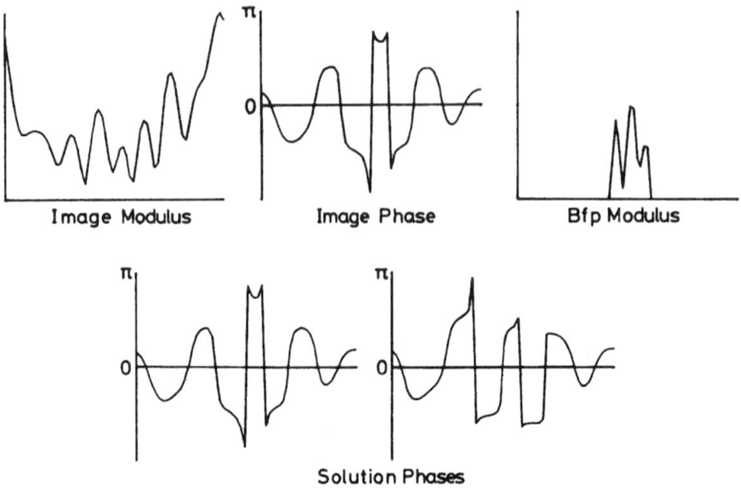

Figure 6.2.6 Two solutions to a problem with a narrow diffraction plane. SSEs: left 0.16%, right 5.1%.

common, but the second one occurs occasionally, and it is worth examining it in detail. It differs from the true phase chiefly in the replacement of the central hump by a dip; however, across the bottom of the dip, where the image modulus is large, the phase is shifted by about 2π from the true phase, which is of course trivial, while at its edges where the solution phase values are less satisfactory, the modulus is very small. This pattern is observed quite frequently, phase solutions drifting by 2π from the true value across low modulus regions, and is obviously not a serious problem. No other problems of uniqueness have been encountered for problems with a restricted back focal plane width except when the image phase variation is weak. This confirms the predictions of Section 4.5. The rather high quality of the solutions obtained to this particular problem reflects the size of the diffraction plane modulus values as well as their restricted number, as it implies much better conditioned data.

Fig. 6.2.6 parallels Fig. 6.2.5 but with 11 nonzero diffraction plane points instead of 5. Again the phase solutions are always excellent near the ends of the axis where the modulus is large, but they may drift by 2π across the low modulus points. Two solutions are shown, both obtained in 20 cycles from completely random starting points, one effectively perfect and one with some drifting present. The latter has the extremely high SSE of 5%, yet only for the small central maximum is the phase solution less than satisfactory; this again reflects the improved performance possible

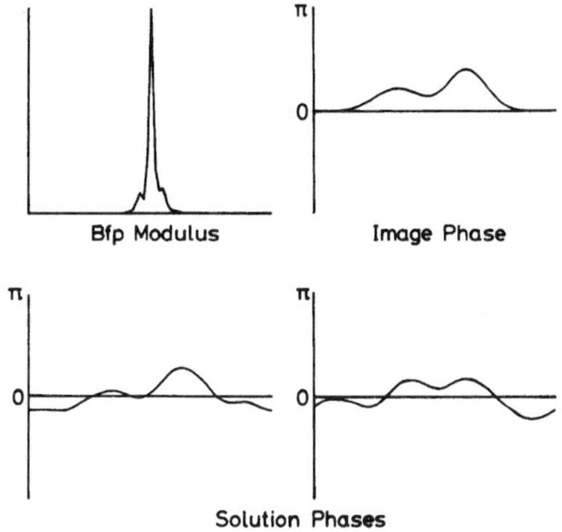

Figure 6.2.7 Two solutions to a bright-field problem with data planes interchanged. SSEs: left 0.46%, right 0.57%.

if the diffraction plane sample values are all of the same order of magnitude.

Finally, Fig. 6.2.7 shows the method with the planes reversed applied to an image similar to that of Fig. 6.2.2. Its performance is much the same as in Fig. 6.2.2. Two solutions are shown, each after 20 cycles from a completely random starting point; the final SSE is rather poor for both, and the second shows a subsidiary peak, again as was noted in connection with Fig. 6.2.2.

In summary, it can be said that the steepest descents method is viable with realistic numbers of data points, though decreasingly stable when started progressively further from the solution. It can be successful for relatively high information cases without any initial phase estimate, provided that the back focal plane modulus values are sufficiently large. No serious uniqueness problems have been encountered in the examples investigated.

6.3. The iterative Fourier transform method

This section presents a number of model problems solved by the method described in Section 5.3, in which a wave function estimate is repeatedly transformed between the two planes, and the modulus corrected in each, while the phase is preserved. The examples here confirm the stability of the process, most examples being solved from completely random initial phase estimates, though this is achieved at some cost in terms of con-

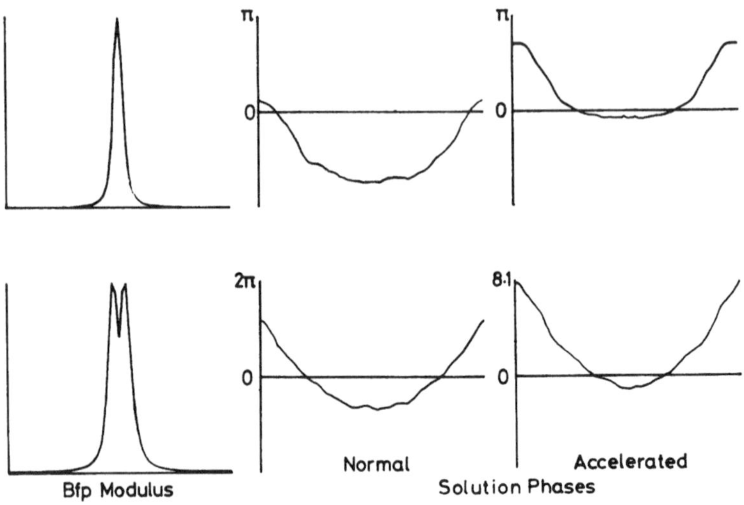

Figure 6.3.1 Solutions from normal and accelerated iterations for a bright-field problem. SSEs: above center 0.003%, right 0.00003%, below center 0.004%, right 0.002%.

vergence rates. As pointed out in Section 5.3, it is the first few cycles that are the most useful, and 20 cycles will usually establish all but the finer details of the solution; however, 50 to 100 cycles may be necessary to resolve these also. The computation involved in a cycle of this iteration is the same as that required for a cycle of the steepest descent method, so the figures can be compared directly. Appendix 5.A.3 lists a subroutine suitable for performing the computation.

The first problem tackled here, as for the steepest descents method, is that of an image without contrast, the phase of which varies quadratically as in Fig. 6.2.1; here, however, more than one value of the maximum phase shift has been explored. Fig. 6.3.1 shows the back focal plane modulus (top left) and an image phase solution (top center) for the case where the peak phase shift is π, as previously, obtained in about 100 cycles. Table 6.3.1 lists the SSE decay for this case to illustrate the general pattern of convergence of the method. The lower row shows a similar pair for a problem in which the maximum phase shift was 2π instead. Above this value solutions are only occasionally obtained, the SSE decaying only slowly to an asymptote other than zero. No serious uniqueness problems have been encountered at least. The most difficult feature of the phase distributions to reproduce accurately is the discontinuity in the gradient at the outside of the image, and although the phenomenon disappears as the SSE is forced

Table 6.3.1 SSE decay for the iterative transform method.

Iteration number	SSE (%)	Iteration number	SSE (%)
0	142.11	10	0.011
1	4.41	20	0.0072
2	1.19	30	0.0064
4	0.10	40	0.0058
6	0.022	50	0.0053
8	0.013	100	0.0038

lower, a number of solutions with SSEs no larger than those in Fig. 6.3.1 have distinctly rounded tips to the phase distribution. The problem was sampled on 64 points; the use of 32 or 128 instead made little difference to the performance of the iteration. The inclusion in the sampled image of a zero-modulus region outside that considered above led to a much more difficult problem to solve, and the iteration was rarely successful.

Besides these variants on the problem, the use has also been explored of the acceleration technique outlined at the end of Section 5.3, in which the iteration is allowed to run for a few cycles at a time on a series of synthetic problems in which accentuation of the detail of the diffraction plane modulus distribution is progressively relaxed; its performance is illustrated in Table 6.3.2. A number of trials were made of the normal form and two accelerated forms; the difference between the two accelerated forms was the value of the parameter λ [see (5.3.10)] used to adjust those diffraction plane modulus values which were *greater* than the average value μ. The table gives the values used above and below as λ_+ and λ_-. Also given is the SSE value (*not* as a percentage) reached after about 100 cycles in each case. Broadly speaking, the phase solution was quite satisfactory for SSE values below 25×10^{-6} and poor for values much above 100×10^{-6}. Locked iterations were terminated prematurely ("T" on the table) for the normal form, but the accelerated forms were allowed to run their full length in the (disappointed) hope that the gradual adjustment of the problem data might succeed in freeing the locked condition. There appears to be little advantage in the use of the accelerated form for the case with a maximum phase shift of π, the case with the least detail in the back focal plane modulus to be accentuated; one solution obtained by the accelerated method is shown at the top right of the figure. For the 2π case, there *is* some value in the accelerated form with $\lambda_+ = \lambda_-$, and this is borne out by the few attempts

Table 6.3.2 Comparison of normal and accelerated iterations.

Mode	Maximum image phase shift	λ_+	λ_-	Final SSE values ($\times 10^{-6}$)			
Normal	π	–	–	2	30	0	4
				4	3	1	0
Accelerated	π	$1/\mu$	$1/\mu$	520	16	16	10
				8	8	4	5
Accelerated	π	0	$1/\mu$	0	5	534	4
Normal	2π	–	–	44	3906T	3920T	5944T
				104	53	4058T	31
Accelerated	2π	$1/\mu$	$1/\mu$	21	424	23	4020
				22	123	79	112
Accelerated	2π	0	$1/\mu$	24	3894	5852	3881
Normal	3π	–	–	4624T	8821T	2809T	18387T
Accelerated	3π	$1/\mu$	$1/\mu$	1384	1384	1340	22

to solve the case with a 3π maximum phase shift, in which the normal form failed completely while the accelerated form found one good solution, which is reproduced at the bottom right of Fig. 6.3.1. Subsequently, only the normal form of the iteration has been considered however.

Table 6.3.2 also illustrates another point, namely there are usually only a small number of different phase estimates that cause the iteration to lock; in the set of trials on the 2π maximum phase shift problem, for example, one particular such case is encountered several times (with an SSE of about 3900×10^{-6}).

As for the steepest descent method, the next problem investigated is the contrastless image with a double-peaked phase distribution as shown in Fig. 6.3.2. Several solutions are shown. The first (top center) is the result of 35 cycles of the iteration starting with a phase estimate ± 1 rad from the true phase and is given for direct comparison with Fig. 6.2.2A, the corresponding solution obtained by the steepest descent method. The comparison illustrates the fact that the iterative Fourier transform method produces *smoother* solutions, due simply to the low-pass filtering implied by the repeated modulus correction in the diffraction plane; the phase estimate after a single cycle of this method, though inaccurate, is always smooth for the same reason. The second (bottom center) required 40 cycles from a completely random starting point. Typical of rather higher SSE "solutions" are the other two curves shown; both were terminated after about 40 cycles from random starting points, but the SSE was dropping steadily at the time.

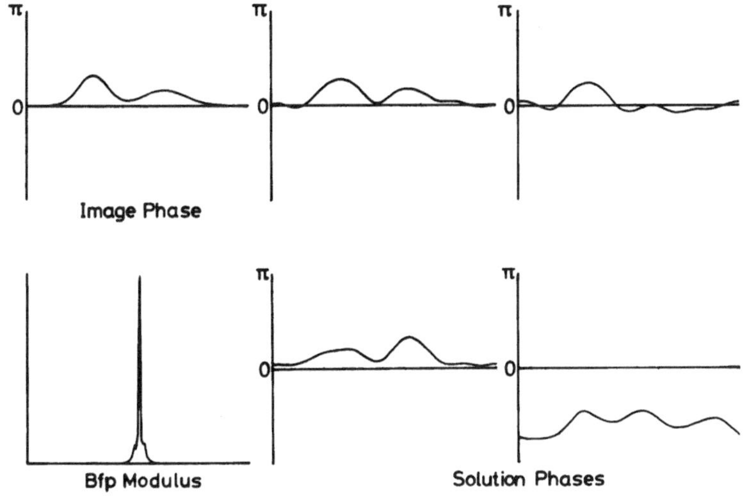

Figure 6.3.2 Four solutions to the problem of Fig. 6.2.2. SSEs: above center 0.002%, right 0.014%, below center 0.002%, right 0.011%.

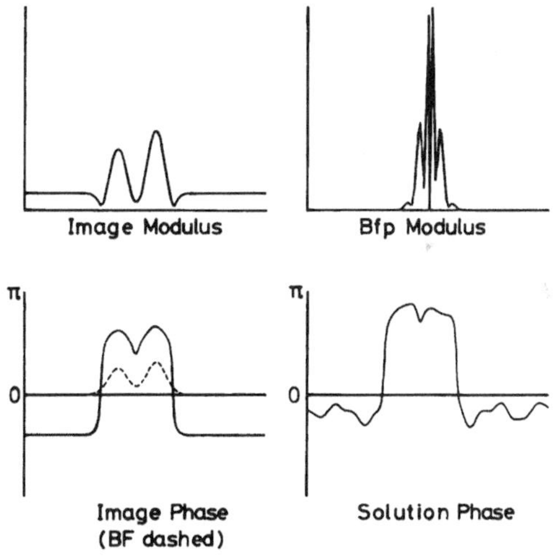

Figure 6.3.3 A dark-field problem. SSE: 0.001%.

The distribution of the initial phase values for that top right was restricted to $-\pi/2$ to $\pi/2$ rather than $-\pi$ to π; this seems to have had little effect.

Fig. 6.3.3 shows a similar problem in dark-field conditions; the phase solution shown was obtained in 25 cycles from a random starting point.

Given the basic characters now established for the steepest descent method and the iterative Fourier transform method, it is an obvious move

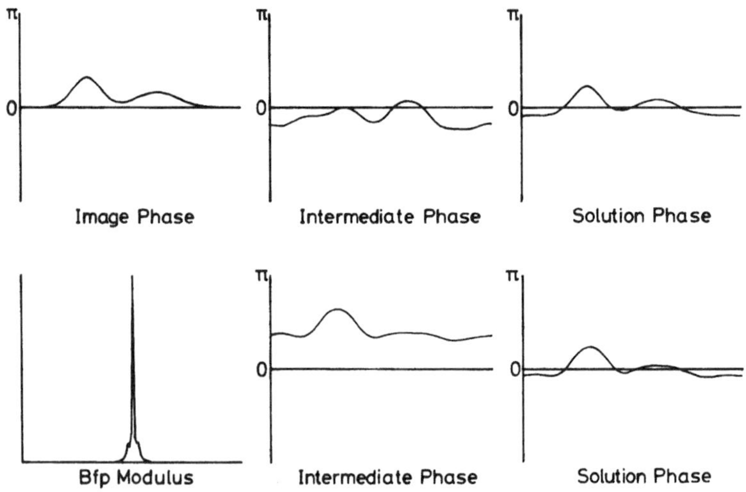

Figure 6.3.4 Combined iterative transform and steepest descent methods: two solutions to a bright-field problem. SSEs: above center 0.049%, right 0.001%, below center 0.043%, right 0.001%.

to use the stability of the latter to approach the neighborhood of a solution and then to switch to the former for its rapid convergence. The next two figures show that this can indeed be useful. Fig. 6.3.4 concerns the same problem as Fig. 6.3.2: a contrastless image with a double-peaked phase distribution. In the first case shown (top center and right) the Fourier transform method was run for 6 cycles, starting $+\pi/2$ from the true phase, after which its phase estimate was that designated the "intermediate phase" in the figure; the steepest descent method then reduced the SSE by a factor of 50 over 10 cycles leading to the final solution shown. The improvement is dramatic. The corresponding curves for a second case commenced with a random phase estimate are shown in the lower part of the figure; 20 cycles of the Fourier transform algorithm and eight cycles of the steepest descent method were used. The improvement by the steepest descent method is still huge in terms of the SSE, but its effect on the phase estimate is moderate, though certainly beneficial.

Fig. 6.3.5 shows the application of this combination method to a more complicated image phase model, the image again being contrastless. The initial phase estimate was completely random; only nine cycles of the Fourier transform method and six cycles of the steepest descent method were needed.

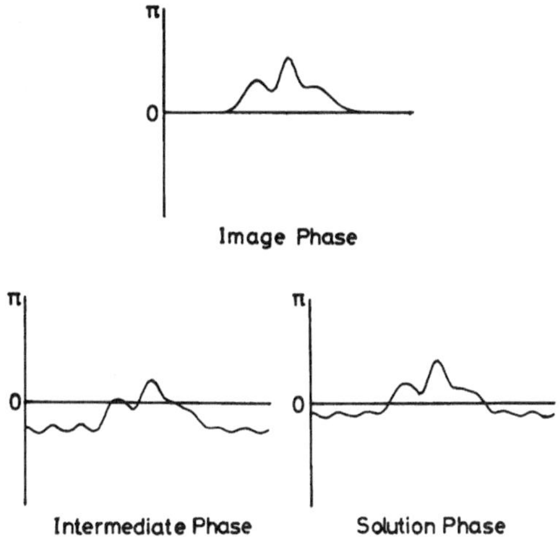

Figure 6.3.5 Combined iterative transform and steepest descent methods: a bright-field problem. SSEs: left 0.047%, right 0.001%.

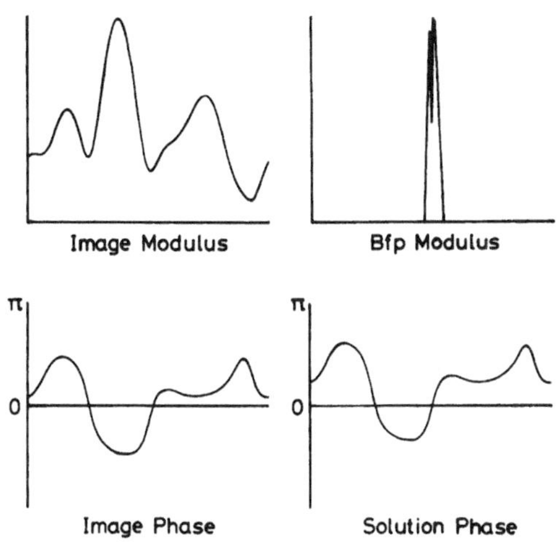

Figure 6.3.6 A problem with a narrow diffraction plane. SSE: 0.001%.

Returning now to the iterative Fourier transform method on its own, Fig. 6.3.6 shows a problem like that of Figs. 6.2.5 and 6.2.6, the width of the diffraction plane wave function being sharply restricted. The data were again generated by means of a random choice of modulus and phase for the 11 of the 128 diffraction plane sample values that were nonzero.

The solution shown was obtained in 15 cycles from a completely random starting point and is extremely accurate. The restriction on the

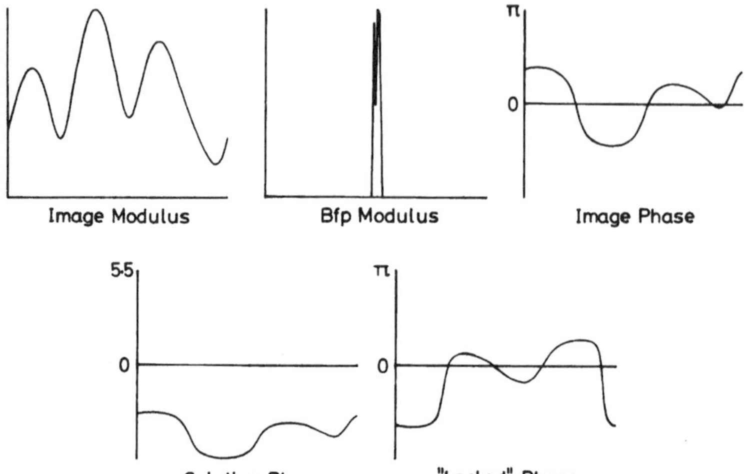

Figure 6.3.7 A solution and a "locked" phase for a problem with a narrow diffraction plane. SSEs: left 0.001%, right 7.34%.

diffraction plane width does not, unfortunately, prevent the iteration from "locking" on occasion however. Such a case is shown in Fig. 6.3.7 since the locked phase estimate is rather interesting; 7 diffraction plane sample values are nonzero, and a successful solution is shown left; the locked estimate is shown right, being very close to a shifted version of the true phase; the phase distribution has slipped across the modulus distribution from one maximum to another. This is not an isolated occurrence, but is difficult to interpret.

The remaining examples presented are two-dimensional; although the iteration was applied to a single cell of the image as previously (here recorded on a 32 by 32 array of sample points) the results are displayed as larger portions for realism and clarity. Fig. 6.3.8 shows six solutions for a single phase distribution, shown top left with two positive peaks at $\pi/2$ rad and a broader negative peak at -1 rad, the image once more being contrast free. The diffraction pattern for a single cell is shown top right in the form of its logarithm. Fifteen cycles of the iteration only were applied in each case, with random initial estimates; further iterations would doubtless have been beneficial as the SSE was still dropping steadily at termination, but the results are still encouraging. The solutions are arranged in order of increasing SSE: (A), (B), and (E) are useful solutions, but rotated by 180° and conjugated (cf. Section 4.2); (D) is in the original orientation; (C) and (F) would be unlikely to be of any real value.

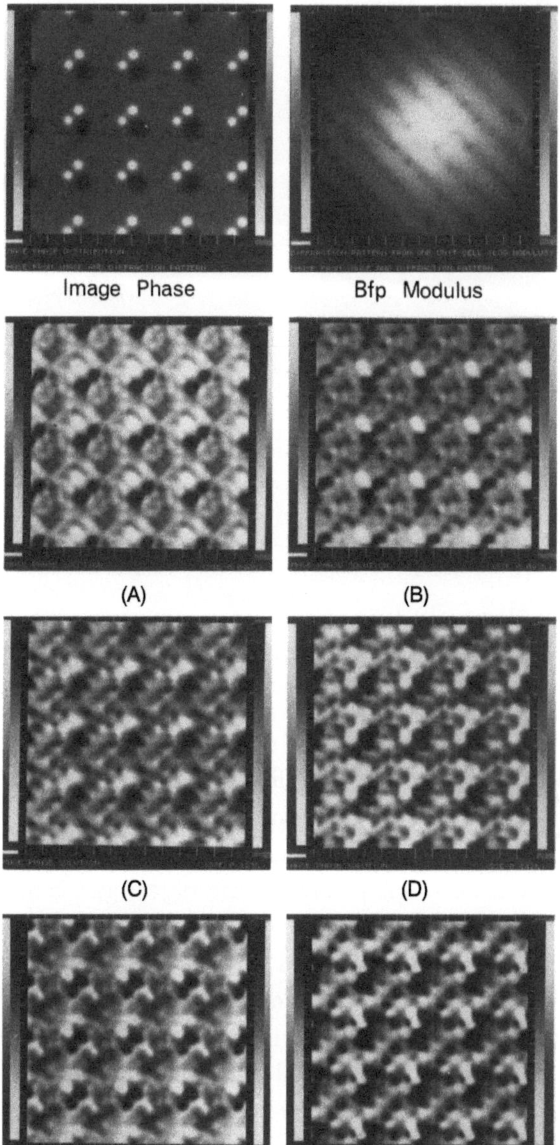

Figure 6.3.8 Six solutions to a two-dimensional bright-field problem. SSEs: (A) 0.090%, (B) 0.093%, (C) 0.123%, (D) 0.142%, (E) 0.143%, (F) 0.194%.

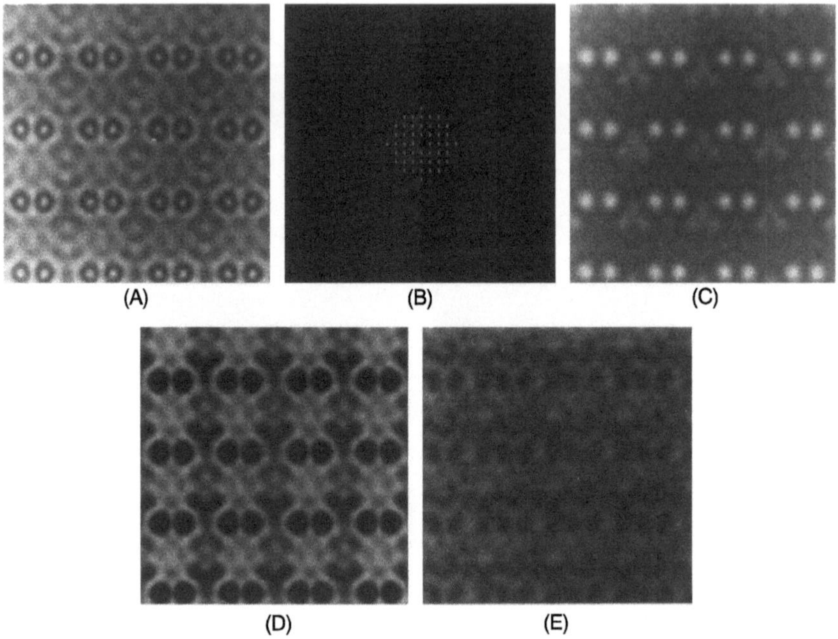

Figure 6.3.9 Two solutions to a bright-field problem. (A) Image modulus, (B) bfp modulus, (C) image phase, (D) phase solution (SSE: 0.001%), (E) phase "solution" (SSE: 0.02%).

Figs. 6.3.9 and 6.3.10 refer to a pure phase object with five distinct Gaussian peaks, the largest phase shift being about 1.5 rad. A diffraction limit has been imposed giving the image some modulus variation (scattering contrast). Here and subsequently, the diffraction pattern is displayed as if it too derived from a large area of the object and is not logarithmically scaled. The bright-field case is shown in Fig. 6.3.9 which includes one very acceptable solution obtained in 15 cycles from a *zero* initial phase estimate (now useful since neither modulus distribution is centrosymmetric); the other "solution" was one of many obtained from random starting points; but while the others had final SSEs no lower than 0.1%, this particular iteration achieved 0.02% in 50 cycles. Since the final estimate bears little resemblance to the actual phase distribution, this can only be regarded as disturbing. The situation is much more satisfactory in dark-field conditions (Fig. 6.3.10) where no starting point was found at all for which convergence to the true phase distribution was not achieved. The two solutions shown started from random phase estimates, but a zero starting point also proved successful. The first solution shows sharp transitions from

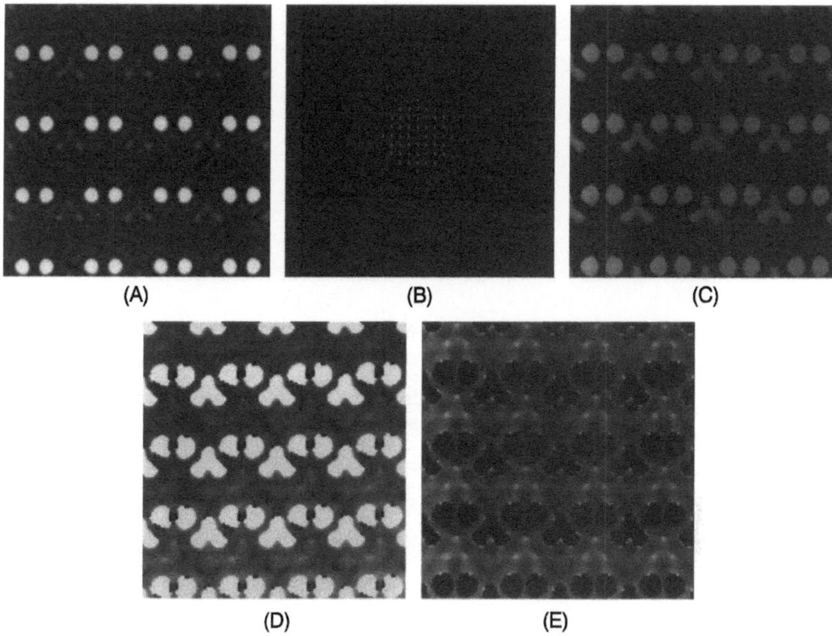

Figure 6.3.10 The problem of Fig. 6.3.9 in dark-field conditions. (A) Image modulus, (B) bfp modulus, (C) image phase, (D) phase solution (SSE: 0.020%), (E) phase solution (SSE: 0.032%).

white to black; these are merely due to slight circular overflow in the phase from $-\pi$ to π. The second solution shown appears more disturbing at first, since a conjugate solution is permitted only when the diffraction pattern is centrosymmetric, which is not the case here. In fact, however, the phase shift between the two approximately constant-phase regions of the image is π rad (cf. Fig. 6.3.3), so the apparent reversal can be achieved merely by adding π to all image phases.

The remaining figures refer to a more complicated pure phase object, with a maximum phase shift of 2.1 rad, made up from ten Gaussians, not all resolved by the diffraction plane aperture. In bright-field conditions (Fig. 6.3.11) it proved difficult to find the solution at all from random starting points, and the solution shown (the result of 25 iterations from a zero initial phase estimate) is the best that was achieved. Most iterations locked with an SSE of 1.5% or more, and so would at least be clearly identifiable as unsatisfactory in practice. No solution was achieved even from a zero initial estimate when 10% noise was applied to the data used. In dark-field conditions the situation was again rather better. Fig. 6.3.12 shows a

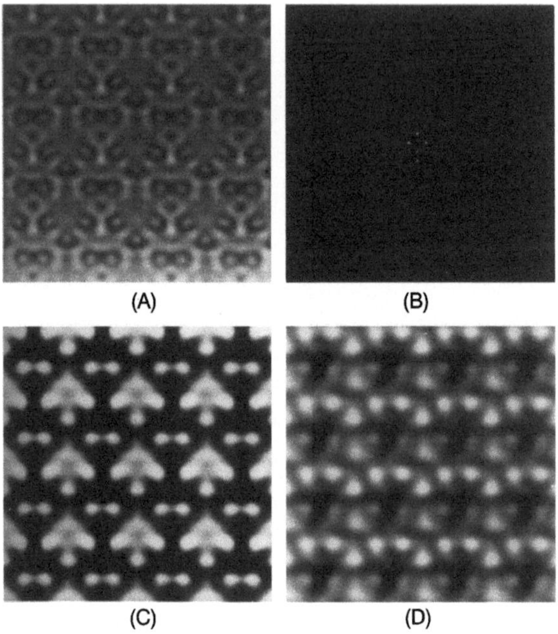

Figure 6.3.11 A bright-field problem. (A) Image modulus, (B) bfp modulus, (C) image phase, (D) phase "solution" (SSE: 0.22%).

solution obtained in 50 cycles from a random starting point, but a starting point *was* found this time for which the iteration locked with an SSE of more than 4%. Fig. 6.3.13, finally, repeats Fig. 6.3.12 from the same starting point but with 10% noise added to both data moduli (by means of a random multiplier uniformly distributed in 0.9 to 1.1); 50 cycles were again allowed, and the solution is extremely satisfactory.

This description of the iterative transform method with image and diffraction plane data is concluded with a reference to some other applications that have been made. Gerchberg (1972) applies the method to a fairly thick (60 nm) stained catalase crystal, including diffraction orders to about 8 Å (a higher resolution than is possible in the image itself) and finding little difference between solutions obtained assuming a pure phase object and those using experimental image data. His final SSE of 1.6% is not low for a bright-field determination, allowing an rms difference of 13% between observed and predicted moduli; it is thus understandable that the actual image modulus data were of little importance, since the rms variation in these was only 8.5%. The rms phase error of 0.13 rad, though not small,

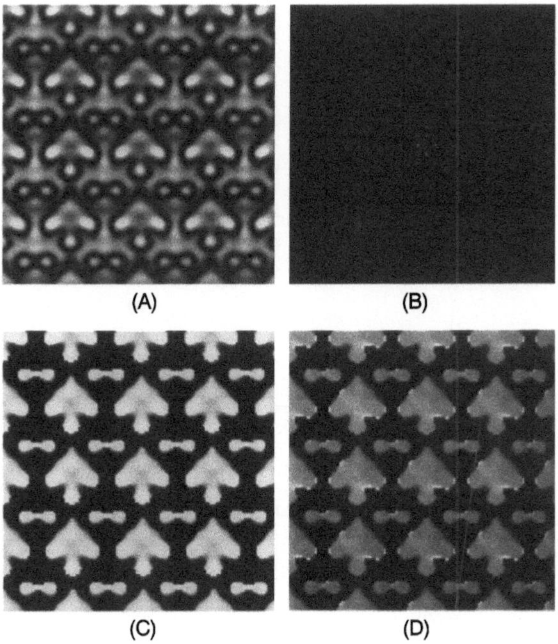

Figure 6.3.12 The problem of Fig. 6.3.11 in dark-field conditions. (A) Image modulus, (B) bfp modulus, (C) image phase, (D) phase solution (SSE: 0.083%).

is at least much lower than the largest phase shifts found in the solution (about a radian).

Gonsalves (1976) reports encouraging results for small one-dimensional problems using the iterative transform method, and also a further method in which the phase distribution is approximated by an expression with a relatively small number of parameters which are adjusted to minimize the SSE by a standard routine.

Toms et al. (1972) describe an extremely interesting application to Lorentz microscopy, in which the field distribution in thin film magnetic domains causes large phase shifts in the wave leaving the specimen and for which a severe interpretation problem exists; the phase shifts may extend to many radians, although the distributions are usually simple in shape. This work has been pursued extensively by Chapman (1975) who proposes a modification to the method in which the moduli are not fully corrected to the observed values, since these may be distorted by noise, but are simply brought to within a certain distance (determined by the noise level) of them. He presents a number of model calculations and analyzes exper-

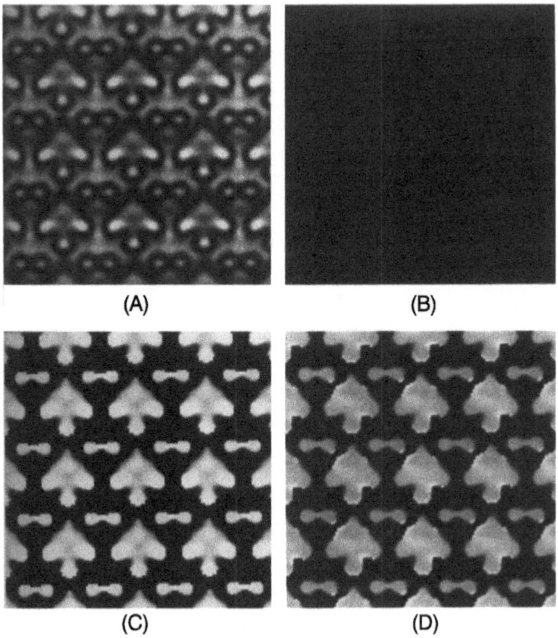

Figure 6.3.13 The problem of Fig. 6.3.12 with noisy data. (A) Image modulus, (B) bfp modulus, (C) image phase, (D) phase solution (SSE: 0.52%).

imental diffraction patterns (using the assumed contrast-free object plane modulus to complete the data) for stripe domains in permalloy and cobalt, finding phase variations of from 3 to 32 rad. It is unfortunately the *derivative* of the phase that reveals the magnetic field distribution, which puts stringent requirements on the smoothness of the phase solution which are difficult to meet. Nevertheless, in this context the method is providing useful information not otherwise available.

6.4. The matrix inversion method

Some results are now presented based on the conventional matrix inversion approach (5.2.7) applied to the residues of (5.2.4). The large quantity of information used about the variation of individual residues with individual phases leads to very rapid convergence near the solution, but the method is wildly unstable elsewhere. It has been used here to find *exact* solutions to the problem, confirming some of the conclusions in Chapter 4.

Table 6.4.1 Results for 8-point problems.

Initial phase spread	Original phase found	New solutions found	Failures
$\pi/1000$	230	0	0
$\pi/100$	104	3	0
$\pi/20$	151	22	6
$\pi/10$	56	11	6
$\pi/2$	0	8	85
Random	0	3	63

The method *can* be applied in two dimensions, though the necessary programming is rather tedious; its performance is extremely disappointing, and it has been possible to do no more than verify that it has been correctly programmed by achieving the expected rapid convergence in a 16 by 8 point problem starting within ± 0.2 rad of the root; attempts to tackle more realistic problems invariably diverged immediately and irrecoverably.

What is presented is accordingly one-dimensional. The data were generated by choosing the phase at random over one plane, taking the modulus to be unity, and computing the corresponding data for the other plane by discrete Fourier transformation. They have no visual significance and are not therefore presented graphically. It is with statistics that we are concerned, large numbers of attempts being made to locate phase solutions (if any) besides the original. Small sets of data points are of course cheap in terms of processor time, and since their uniqueness properties are just as significant theoretically as those of larger sets, many more trials have been made at this end of the scale. Given the instability of the method, the majority of trials fail to converge at all unless started quite near the solution; when they do converge, however, whether to the original phases or to a new solution, they stabilize in two or three cycles to about ± 0.001 rad at all image points, with an SSE many orders of magnitudes lower than that normally achieved by any of the other methods described in Chapter 5. These facts encourage the belief that the "solutions" obtained are in fact independent solutions to the data and not simply approximations to the original phase that would become indefinitely closer if the SSE were pushed progressively lower.

Tables 6.4.1–6.4.4 summarize the results of a series of trials of 8-, 16-, 32-, and 64-point problems. Each trial is based on a freshly generated set of data, and the iteration is commenced with a phase estimate distributed about the true phase with a width as given in the first column of the ta-

Table 6.4.2 Results for 16-point problems.

Initial phase spread	**Original phase found**	**New solutions found**	**Failures**
$\pi/1000$	67	0	0
$\pi/100$	62	1	0
$\pi/20$	27	3	3
$\pi/10$	11	4	10
$\pi/2$	0	0	63
Random	0	0	16

Table 6.4.3 Results for 32-point problems.

Initial phase spread	**Original phase found**	**New solutions found**	**Failures**
$\pi/1000$	30	0	0
$\pi/100$	29	0	0
$\pi/20$	5	0	2
$\pi/10$	0	2	6
$\pi/2$	0	0	3
Random	0	0	3

Table 6.4.4 Results for 64-point problems.

Initial phase spread	**Original phase found**	**New solutions found**	**Failures**
$\pi/1000$	6	0	0
$\pi/100$	2	0	1
$\pi/20$	1	0	1
$\pi/10$	1	1	12
$\pi/2$	0	0	4
Random	0	0	23

bles. First, it is unequivocally demonstrated that there do in general exist solutions to the data distinct from the true phases; second, it appears, rather curiously, that the density of solutions near the true phase is higher than elsewhere, since the trials with relatively low initial phase spreads find solutions distinct from the original phase much more often than trials with random starting points. The tables do not show how far from the original phases the new solutions are, but few of them are in fact much more than 0.1–0.2 rad away at any point. The number of trials of 32- and 64-point problems is too small for it to be possible to say whether there are more or fewer additional solutions to the larger problems than the smaller.

Table 6.4.5 Seven phase solutions to one problem.

1	2	3	4	5	6	7
−0.160	1.009	−1.779	−0.303	0.198	−2.712	0.153
−0.434	2.838	1.284	−0.907	−0.746	−0.846	1.430
2.175	2.235	1.919	1.282	1.724	−1.469	2.902
1.442	−1.859	2.980	−2.895	3.000	0.786	0.942
−1.307	0.246	2.707	−2.138	−1.810	0.015	2.685
1.363	1.004	−0.968	1.903	2.512	2.617	1.862
−1.857	−1.239	−1.700	−1.183	−2.082	0.541	−1.373
−1.221	1.958	1.834	−2.042	−2.850	1.067	−2.317

Besides the large series of trials of *different* problems, a small number of problems were selected for examination to see how many different solutions they possessed. Ten trials from different starting points with spreads ranging from $\pm\pi/10$ to $\pm\pi$ from the original phase were made on each of eleven 64-point problems without finding any solutions at all. At the other extreme, two 8-point problems were also examined; in 99 trials of one, starting from a phase spread of $\pm\pi/10$, the original phase was found 93 times and a distinct solution once; more significantly, in 57 trials of the other, with an initial phase spread of $\pm\pi/2$, 47 diverged, three found the original, and seven found a different solution. On examination, two of these were found to be the same solution shifted laterally, but the remainder were apparently independent. These are reproduced in Table 6.4.5 together with the original phase values. Recapitulating, if eight complex numbers are formed with unit moduli and phases taken from one column of the table, the moduli of the discrete transform values are the same no matter which column is selected.

Section 4.5 predicted a unique solution for periodic or discrete phase problems if at least half of the discrete transform values vanish. With this condition, the image values can no longer all have unit modulus, and the problems were generated instead by choosing random phases and moduli for the nonvanishing transform elements and obtaining the corresponding image values by discrete transformation. Tables 6.4.6 and 6.4.7 summarize the results for some 8-point and 16-point problems; sets of trials of a single problem are indicated by footnotes. The total number of trials is rather less than that represented in Tables 6.4.1 and 6.4.2, but it is nevertheless clear that the restriction on the number of nonzero values has led to overdetermination of the phase with consequent elimination of additional solutions to the data.

Table 6.4.6 Results for overdetermined 8-point problems.

Initial phase spread	Original phase found	New solutions found	Failures
$\pi/10$[a]	66	0	24
$\pi/10$[a]	22	0	6
$\pi/10$	52	0	20
$\pi/4$	11	0	18
$\pi/2$[a]	16	0	63
$\pi/2$	12	0	63
Random[a]	7	0	21

[a] Sets of trials of a single problem.

Table 6.4.7 Results for overdetermined 16-point problems.

Initial phase spread	Original phase found	New solutions found	Failures
$\pi/20$[a]	7	0	4
$\pi/10$[a]	20	0	11
$\pi/10$[a]	7	0	4
$\pi/10$	18	0	11
$\pi/4$[a]	3	0	6
$\pi/2$[a]	0	0	30
$\pi/2$	0	0	29

[a] Sets of trials of a single problem.

This concludes the computational results to be presented in connection with the image and diffraction plane problem; they demonstrate clearly that the data are sufficient, in the type of problem studied, to determine the phase distribution, and that it is feasible to establish the phase numerically, though a very low SSE is often necessary, corresponding to very accurate measurement of the data moduli; even for higher SSE values, however, the broad features of the actual phase distribution are reproduced. Only in one case (Fig. 6.3.9) was an alternative phase solution found with an SSE too low for the "solution" to be clearly recognizable as spurious.

The requirement that the image be oversampled unfortunately runs counter to that for accurate measurement of the image, and the two will clash when the specimen is radiation sensitive, as is the case for most biological materials. In such cases, there may be some value in the use of a small objective aperture passing a relatively small number of beams only, resolution being traded for knowledge of the diffraction plane phases; the process could of course be repeated (with fresh specimen preparations) until the entire diffraction plane was covered (Nathan, 1971).

CHAPTER SEVEN

Alternative data for the phase determination☆

W. Owen Saxton
Murray Edwards College, University of Cambridge, Cambridge, United Kingdom
e-mail address: wos1@cam.ac.uk

Contents

7.1. Defocus pairs

It was with image and diffraction plane pairs that the direct wave function reconstruction approach to image interpretation for strong objects was originally proposed, and these data have been accordingly studied at some length in the preceding chapters. However, they are by no means the only data that may be suitable, and this chapter reviews what is known of some of the alternatives. We begin with the most obvious of these, namely pairs of micrographs made at different defocus values.

Such pairs are well known to permit a full restoration of a *weak* complex object (see Section 1.5) under reasonable conditions, and so the data will not admit the kind of uniqueness failure which was so damaging for image/diffraction plane data in the weak or near weak object case. While it is still necessary to achieve exact alignment between the two micrographs used, this can be done in most cases by standard correlation methods (see Section 9.4) and does not present the problem it does with image/diffraction plane data. Accordingly, there is no restriction to periodic objects as was found to be the case for practical purposes previously, though of course periodic objects will remain a particularly attractive proposition because of the low noise levels attainable. The coherence requirements are moreover lower for defocus pairs than for image/diffraction plane pairs; for the latter,

☆ Reprinted from Advances in Electronics and Electron Physics, Supplement 10, 1978.

Advances in Imaging and Electron Physics, Volume 214
ISSN 1076-5670
https://doi.org/10.1016/bs.aiep.2020.04.007

the entire object field must be within the coherence area of the illumination, while for the former the coherence area need not extend over much more than the point response function for the image formation system. Finally, we need not expect to find, as we did with the previous data, increasing difficulty in obtaining a solution as the information content of the image grows; whereas previously local data in one plane was affected by the data throughout the other plane, the interconnection is now a purely local process, so that two separate problems can be solved side by side without any mutual interference.

A particularly important probable application for the methods of this section is likely to be the interpretation of high resolution crystal lattice images [e.g., Iijima (1971, 1973); Allpress & Sanders (1973)]. Crystals with moderately heavy atoms present cease to be weak objects when the thickness exceeds about 1 nm at 100 kV, rendering the detailed interpretation of the images extremely difficult. However, for a useful range of thickness beyond this, the electron wave below the specimen is still obtained simply by imposing on the incident beam phase shifts obtained by projecting the crystal potential (see Section 1.2), the object being thus a pure phase, though not a *weak*-phase object. Within this range, a knowledge of the electron wave function at the object thus provides direct information on the crystal potential in a way that is not possible when multibeam calculations are required to represent the object adequately, and the techniques explained below, which enable this wave function to be recovered, should accordingly be of great assistance in structural and defect studies of such crystals. Lynch et al. (1975) give a useful table of the ranges of resolution and thickness for which different approximate treatments of the scattering process are adequate, and state for example that the pure (strong) phase object approximation is adequate at 100 kV for thicknesses up to 5 nm and resolutions down to 0.35 nm. These figures allow its use in a large number of cases of interest, and the model has been exploited previously in approximate theories of the image formation such as the charge density approximation (see Section 1.4) or that given by Cowley and Iijima (1972). These however lose their simplicity, and therefore their usefulness, at high resolution because of the effect of spherical aberration, and it is anticipated that the methods below will be necessary to establish the phase shifts at the object with any real reliability.

The theoretical uniqueness of the phase solution for defocus pair data has been the subject of several recent papers. Schiske (1975) treats the case of (two-dimensional) periodic objects with a finite diffraction plane aperture, and shows by developing a solution similar to the first of the three approaches given below that uniqueness is the rule in this context. Drenth et al. (1975) consider the opposite extreme, namely, one-dimensional objects of finite extent, for which the diffraction plane wave function, besides being of finite extent, is also analytic (see Section 3.1); they find a unique solution provided that the diffraction plane intensity is known at the edge of the aperture (though they surmise that this may in fact be unnecessary), and that the defocus difference between the micrographs used is limited by condition (7.1.7). Huiser and Ferwerda (1976) lift the assumption of analyticity. Such incorrect phase solutions as arise in practice must, it seems, be near but not exact solutions to the data, or derive from observational noise in them. As an example, we might cite the case of a weak-phase object when one image plane is aberration- (and defocus-)free, which is admittedly impossible in high resolution microscopy: the presence of a finite objective aperture will prevent the wave function modulus from being exactly constant in this plane (see Sections 3.1 and 3.3), but in practice the data in this plane will provide no information, leaving us with the problem of restoring the object phase from a single micrograph, which is hampered by transfer function zeros as explained in Section 1.5.

Each of the three classes of numerical method proposed in Chapter 5 for the image and diffraction plane data has its analog for the present defocus pair data, and the analogs prove to be quite remarkably exact, except that one of the major drawbacks of the direct method, the time involved in solution tree generation, no longer applies. We outline the three methods in turn, beginning with this direct method, appropriate to band-limited periodic functions. We shall work in one dimension, the extension to two being made easily along the lines laid down in Section 5.1.

The wave function in one image plane is denoted by g_k, with a transform $\tilde{g}_j$ as usual; with

$$\widetilde{T}_j = \exp(i\delta j^2) \tag{7.1.1}$$

[cf. (1.3.2): δ is proportional to the known defocus difference between the two pictures], we can denote the transform of the wave function in the second image plane by $\tilde{g}_j \widetilde{T}_j$. The corresponding image plane intensities being, respectively, I_k and J_k, the data allow us at once to calculate the autocorrelations

$$\widetilde{I}_j = \sum_j \tilde{g}_l \tilde{g}^*_{l-j}, \qquad \widetilde{J}_j = \sum_j \tilde{g}_l \tilde{g}^*_{l-j} \widetilde{T}_l \widetilde{T}^*_{l-j}. \tag{7.1.2}$$

(We shall suspend the summation convention temporarily and write expressions in full.) If we make the same oversampling assumption as we did in Section 5.1, there is no overlap distortion in calculating these autocorrelations as discrete transforms of the image intensities. Again following that section, we shall suppose that the extreme nonzero values in the transform $\tilde{g}_j$ are $\tilde{g}_n$ and $\tilde{g}_{-n}$, so that the correlation value $\widetilde{I}_{2n}$ is simply

$$\widetilde{I}_{2n} = \tilde{g}_n \tilde{g}^*_{-n}, \tag{7.1.3}$$

while the equation for $\widetilde{J}_{2n}$ provides no further information. We fix a phase origin by assigning an arbitrary phase to $\tilde{g}_n$, and assign it also an arbitrary modulus, which we shall say is r times the true modulus; r carries through the solution and is determined at the end. A value for $\tilde{g}_{-n}$ follows from (7.1.3) and will be too small by a factor r. We proceed then to the next pair of correlation values,

$$\begin{aligned} \widetilde{I}_{2n-1} &= \tilde{g}_n \tilde{g}^*_{-(n-1)} + \tilde{g}_{n-1} \tilde{g}^*_{-n}, \\ \widetilde{J}_{2n-1} &= \tilde{g}_n \tilde{g}^*_{-(n-1)} \widetilde{T}_n \widetilde{T}^*_{-(n-1)} + \tilde{g}_{n-1} \tilde{g}^*_{-n} \widetilde{T}_{n-1} \widetilde{T}^*_{-n}. \end{aligned} \tag{7.1.4}$$

These form a pair of simultaneous linear equations for $\tilde{g}_{n-1}$ and $\tilde{g}^*_{-(n-1)}$, and so the next two $\tilde{g}_j$ values are easily determined unless the pair are singular: $\tilde{g}_{n-1}$ will be too large and $\tilde{g}_{-(n-1)}$ too small by a factor r. Discounting the possibility of singularity initially, we can pursue the sequential solution of the $\tilde{g}_j$ values inward from the periphery of the diffraction plane, until the equations

$$\begin{aligned} \widetilde{I}_n &= \tilde{g}_n \tilde{g}^*_0 + \tilde{g}_0 \tilde{g}^*_{-n} + \text{other known terms}, \\ \widetilde{J}_n &= \tilde{g}_n \tilde{g}^*_0 \widetilde{T}_n \widetilde{T}^*_0 + \tilde{g}_0 \tilde{g}^*_{-n} \widetilde{T}_0 \widetilde{T}^*_{-n} + \text{other known terms} \end{aligned} \tag{7.1.5}$$

are reached; the values obtained are too large for positive j and too small for negative j by the factor r. If we solve (7.1.5) also as a pair of simultaneous equations for $\tilde{g}_0$ and $\tilde{g}^*_0$, the ratio of the two values obtained for $\tilde{g}_0$ will immediately reveal r^2, allowing us to correct all the $\tilde{g}_j$ values accordingly to complete the solution.

The condition for (7.1.4) or any subsequent equation pair to be singular is easily stated: the equation introducing $\tilde{g}_j$ and $\tilde{g}^*_{-j}$ is singular if the coefficient determinant $(\widetilde{T}_n \widetilde{T}^*_{-j} - \widetilde{T}_j \widetilde{T}^*_{-n})$ vanishes, or, using (7.1.1), if

$$\sin(\delta(n^2 - j^2)) = 0. \tag{7.1.6}$$

This can easily be prevented by choosing δ, and hence the relative defocus of the two pictures, sufficiently small; the requirement that $\delta < \pi/n^2$ is adequate, and this can be reexpressed in terms of the actual defocus D, the electron wavelength λ, and the smallest object period passed by the objective aperture d, as

$$D < d^2/\lambda. \tag{7.1.7}$$

This relatively small defocus limit however (less than half a wave length over the significant part of the diffraction plane) implies that the two images will be relatively similar and the conditioning of the data may be correspondingly poor. With larger defocus differences, the condition (7.1.6) is unlikely to hold exactly for any value of j, and the advantage of better conditioning at most points may well offset the disadvantage of a small coefficient determinant at a few. The numerical examples presented below all use substantially larger defocus differences with considerable success.

This method is proposed, albeit in rather different terms, by Schiske (1975), who gives some attention to the exceptional cases of failure, proposing the use of a third defocus value to resolve these. A similar approach is given with numerical examples by Van Toorn and Ferwerda (1976). As Schiske acknowledges, the weakness of the approach is that, although it is fast, it is very sensitive to observational noise in the data, like the method in Section 5.1, particularly since it begins with what will usually be the least accurately known correlation values at the outside of the diffraction plane. There *are* additional data permitting crosschecking ($\widetilde{I}_j$ and $\widetilde{J}_j$ for $0 \leqslant j < n$) but it is not obvious how to use these to improve the estimate of $\tilde{g}_j$ already achieved. My own numerical trials show catastrophic divergence from the true solution with no more than 7 nonzero $\tilde{g}_j$ values if the outermost $|\tilde{g}_j|$ values are halved.

We proceed now to steepest descent methods, the analogs of those of Section 5.2. We shall, in the light of the extended discussion there, only give expressions appropriate to one of the many possible choices of residue. For want of a better notation, primes are used to indicate defocused quantities; thus the wave function g in one image plane becomes

$$g'_m = g_k T_{m-k} \tag{7.1.8}$$

in the other [with T_k as defined by (7.1.1) and our summation convention restored]. The inverse operation is

$$g_k = g'_m T^*_{m-k}. \tag{7.1.9}$$

Denoting the observed modulus of g'_m by j_m, we can define residues

$$f_m(\theta_k) = |g'_m| - j_m \tag{7.1.10}$$

as functions of the phase θ_k of g_k [cf. (5.2.6)]. Differentiating this with respect to θ_p gives

$$\frac{\partial f_m}{\partial \theta_p} = \mathrm{Im}\left\{ g_p^* \frac{g'_m}{|g'_m|} T^*_{m-p} \right\} \quad \left(\sum\right) \tag{7.1.11}$$

[cf. (5.2.10)]. Defining a sum square residue $E = f_m f_m$, we can establish its gradients

$$E_p = -2\,\mathrm{Im}\left\{ g_p \frac{g'_m}{|g'_m|} j_m T^*_{m-p} \right\} \quad \left(\sum_p\right) \tag{7.1.12}$$

[cf. (5.2.14)]. This expression allows us to implement the steepest descents iteration of (5.2.12) efficiently, four Fourier transformations being necessary to compute E_p; two are used to obtain g'_m from g_k [the convolution (7.1.8) being of course evaluated by Fourier methods] after which a further two are used to evaluate the correlation sum in (7.1.12), which corresponds to restoring the quantity $g'_m j_m / |g'_m|$ to the first defocus plane.

Protection against catastrophic divergence, should E have a minimum but no actual root, is provided as in Section 5.2 by performing the steepest descents process not on E but on $B = E_p E_p$. The gradients of this are

$$B_p = 4\,\mathrm{Re}\left\{ g_p^* \left(E_p a_p - \frac{1}{2}\left(b'_m + b^*_m \frac{g'^2_m}{|g'_m|^2} \right) \frac{j_m}{|g'_m|} T^*_{m-p} \right) \right\} \quad \left(\sum_p\right), \tag{7.1.13}$$

in which $a'_m = g'_m j_m / |g'_m|$ and $b_k = E_k g_k$ [cf. (5.2.18)]. Calculation of this requires a further four Fourier transformations beyond those necessary for the calculation of E_p. Two-dimensional extensions of the preceding expressions are obvious and straightforward.

The steepest descents method outlined above remains to be evaluated in numerical trials, though it is reasonable to expect its performance to be comparable to that of its analog in Section 5.2 and perhaps to improve on it in view of the better conditioned data used here. In a different extension of the Newton–Raphson method (5.2.7), Van Toorn and Ferwerda (1976) derive an integral equation for the approximate difference between the true phase and a close estimate of it, iterative solution of which provides a method successful in noise-free one-dimensional trials, though apparently very slow.

Finally, the analog of the third basic method of Chapter 5, the iterative transform method, is presented. This is not so much an analog as a twin, since the only change necessary is the replacement of Fourier transformation as the operation carrying the wave function from one data plane to another by the convolution operation (7.1.8) representing defocusing [Misell (1973a); Gerchberg & Saxton (1973b)]. It was pointed out in Section 5.3 that the convergence properties of the algorithm depend only on the linearity and unitarity of the transformation between the two data planes; now these properties apply to the transformation between the wave functions in any two planes of the microscope except where apertures intervene; the linearity follows from the linearity of the Schrödinger equation (1.1.7) while unitarity is equivalent to the requirement of electron flux conservation. It follows that the algorithm may be directly applied to defocus image pairs, in either bright- or dark-field illumination conditions and may be expected to converge in the same manner as before.

In this application then, the iterative transform method involves carrying a wave function estimate to and fro between the two data planes [using Fourier methods to evaluate the convolutions (7.1.8) and (7.1.9)], correcting the modulus to the observed value in each plane, but preserving the phase. A suitable error criterion for a wave function estimate in one plane is again furnished by the sum square difference (SSE) between observed and predicted moduli in the other plane (the function E defined above), but a more useful normalization factor, in view of the much greater similarity between the two data modulus distributions than is present between an image and diffraction pattern pair, is the sum square difference between the data moduli rather than the summed intensity in either plane. While a random initial phase estimate is still quite suitable, the arguments given in Section 5.3 against taking all phases equal initially lose their force in the present application, and a real wave function is a simple and viable starting point.

Misell (1973b) gives some examples of the use of this approach in solving model problems in dark-field conditions with phase distributions made up from heavily overlapping Gaussians, besides an example of a bright-field *weak*-phase image where the maximum phase shift is 0.05 rad, which is soluble though understandably noise sensitive. We give here a few examples of the solution of model problems with strong phase objects in bright- and dark-field conditions, aberrations again being discounted.

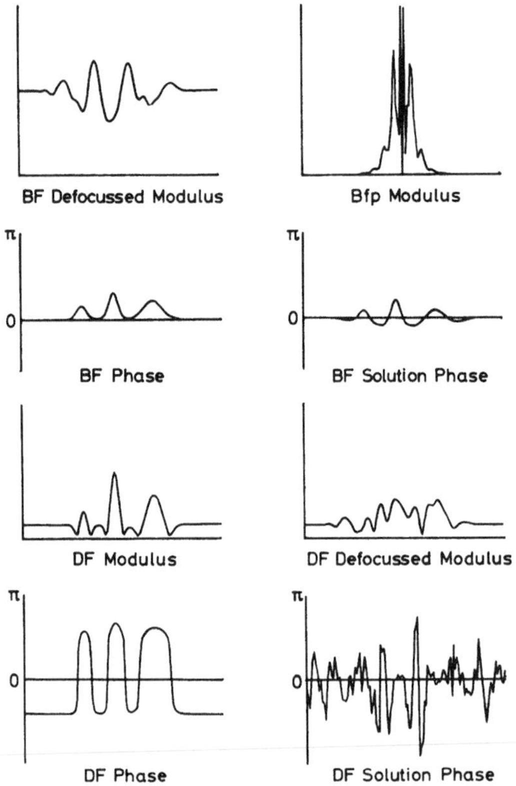

Figure 7.1.1 Solutions for a problem in bright-field (BF) and dark-field (DF) conditions. SSEs: BF 0.5%, DF 0.6%.

Fig. 7.1.1 involves an object with a peak phase shift of a radian. In the top half, a solution is given for the bright-field phase obtained from the defocused image modulus shown, supplemented by an assumed contrast-free in-focus image; the defocus used was equivalent to two full cycles of phase shift on the outermost significant transform values. The high frequencies are accurately reproduced, but the low frequency variation is less well determined, presumably because defocus affects the low frequency components of the wave function relatively little. The initial phase estimate was zero everywhere, and the solution shown was reached in 10 cycles, the SSE decaying as in Table 7.1.1. Unfortunately, the same problem translated into dark-field conditions, as in the lower half of the figure, was not solved by iterations started both at zero and at several random phase estimates. The phase estimate after 30 cycles of one such iteration is shown: its SSE (0.6%)

Table 7.1.1 SSE decay for the iterative transform method with defocus pairs.

Iteration number	SSE (%)	Iteration number	SSE (%)
0	100.0	6	0.9
2	5.0	8	0.7
4	1.6	10	0.5

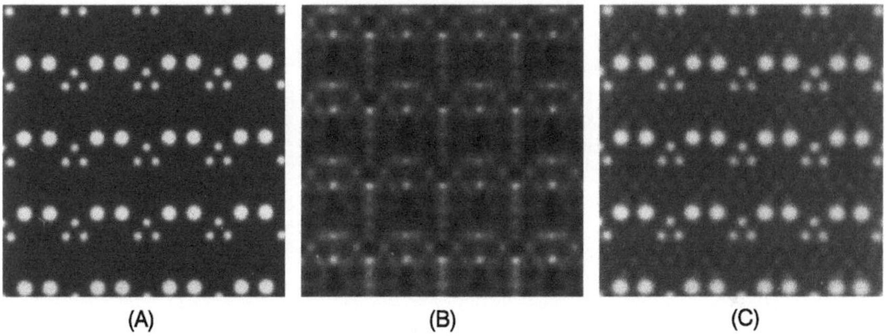

Figure 7.1.2 A bright-field problem with the model of Fig. 6.3.9. (A) In-focus image phase, (B) defocused image modulus, (C) phase solution (SSE: 0.09%).

is as low as that normally obtained with successful solutions, and so it seems likely that, notwithstanding the theoretical results reviewed above, there may sometimes be ambiguities in the solution for practical purposes.

In Figs. 7.1.2 and 7.1.3 the object is the same as that used in Figs. 6.3.9 and 6.3.10, except that no objective aperture has been included this time; the in-focus image has a peak phase shift of 1.5 rad. Fig. 7.1.2 refers to bright-field conditions, showing an excellent solution with an SSE of 0.09% obtained in 15 cycles from a zero initial phase estimate, from data consisting of the defocused image shown and an ideal contrast-free in-focus image. Fig. 7.1.3 refers to dark-field conditions; both modulus distributions and the actual in-focus image phase are shown with a solution having an SSE of 0.4% obtained from a zero initial phase estimate in 15 cycles. This looks very unsatisfactory but only because the phase values vary wildly where the modulus is very small. Images at a third defocus value calculated with the actual image phase and that obtained in the iteration are also shown in the figure and agree very well, confirming that the phase solution is in fact perfectly acceptable. At the same time, an iteration commenced with these data and a random initial phase estimate failed to reach an SSE below 19% in 40 cycles.

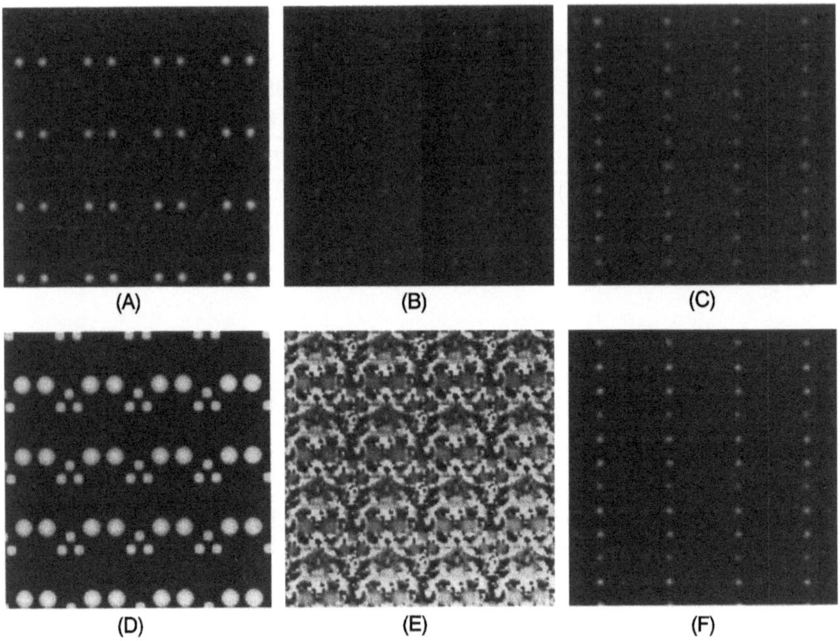

Figure 7.1.3 The problem of Fig. 7.1.2 in dark-field conditions. (A) In-focus image modulus, (B) defocused image modulus, (C) further defocused image modulus, not used as data, (D) in-focus image phase, (E) phase solution (SSE: 0.39%), (F) "cross-check" image calculated from (E) for comparison with (C).

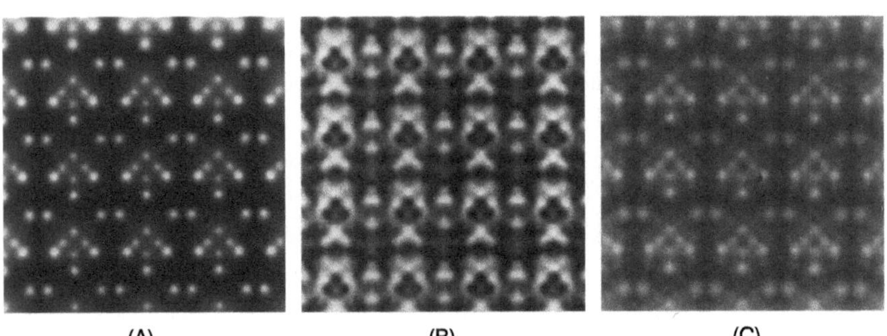

Figure 7.1.4 A bright-field problem with the model of Fig. 6.3.11. (A) In-focus image phase, (B) defocused image modulus, (C) phase solution (SSE: 0.45%).

The remaining three examples derive from the object model used in Figs. 6.3.11–6.3.13, again without an objective aperture, the peak phase shift being 2.1 rad. The bright-field case is given in Fig. 7.1.4, using a

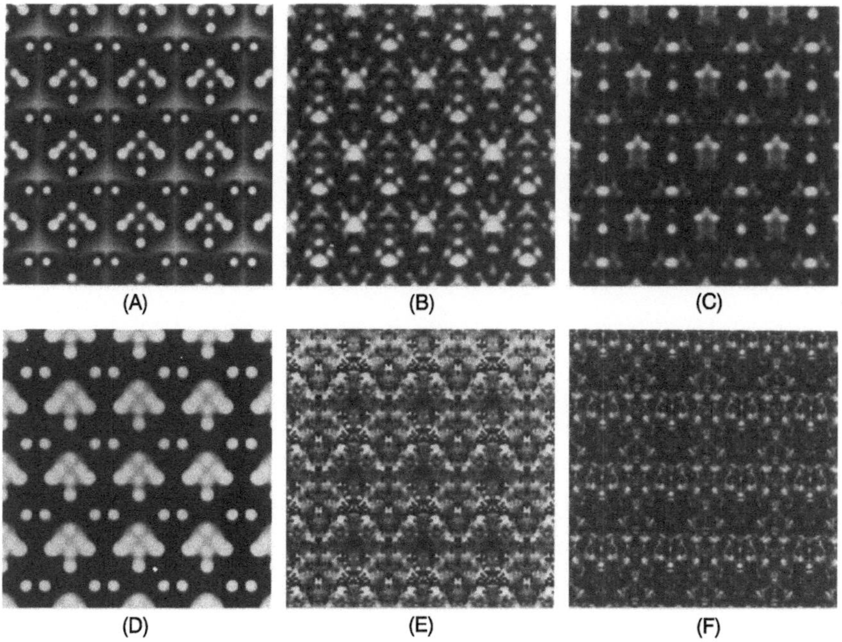

Figure 7.1.5 The problem of Fig. 7.1.4 in dark-field conditions. (A) In-focus image modulus, (B) defocused image modulus, (C) further defocused image modulus, not used as data, (D) in-focus image phase, (E) phase solution (SSE: 1.61%), (F) "cross-check" image calculated from (E) for comparison with (C).

contrast-free in-focus image as the second set of data. The solution has an SSE of 0.5% which required 15 cycles from a zero initial phase estimate and is very satisfactory. The dark-field case is however again troublesome (Fig. 7.1.5), the SSE dropping only to 1.6% in 31 cycles and the supplementary images calculated at a third defocus value for the actual and best estimated phases diverging substantially.

For all the difficulties encountered in dark-field conditions (in which the present data differ from the image and diffraction plane data, which appeared *more* reliable in dark-field conditions) the bright-field case seems very satisfactory. The last example, perhaps the most exciting of all the examples given of strong object restoration, involves the repetition of the bright-field solution with noisy data, obtained by multiplying the exact data values by a random factor distributed uniformly between 0.8 and 1.2 (Fig. 7.1.6). The solution is clearly influenced by the noise but nevertheless compares very well with the actual phase and is an extremely impressive deduction from the very confusing data images at this realistic noise level.

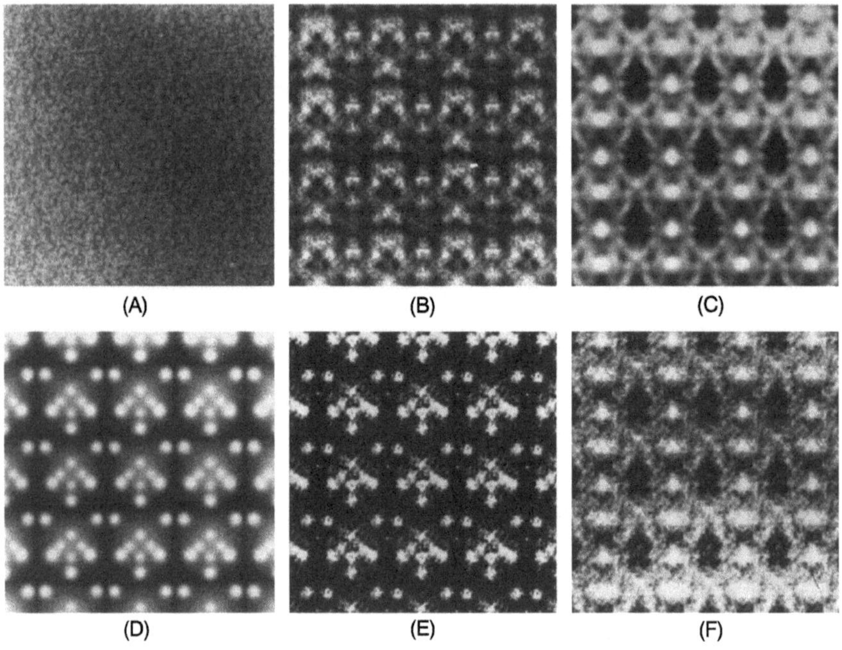

Figure 7.1.6 The problem of Fig. 7.1.4 with noisy data. (A) In-focus image modulus, (B) defocused modulus, (C) further defocused image modulus, not used as data, (D) in-focus image phase, (E) phase solution (SSE: 3.21%), (F) "cross-check" image calculated from (E) for comparison with (C).

Supplementary images at a third defocus value confirm the correctness of the solution.

In spite of the fact that a great deal of high frequency variation can develop in the dark-field convergence failure cases from a zero initial estimate (cf. Fig. 7.1.1), it is tempting to feel that even a very modest knowledge of the diffraction plane modulus sufficient to justify the use of low-pass filtering to smooth the phase estimate could be extremely beneficial. This has not been explored however in practice. Misell (1973b) pursues in numerical trials the complications of inaccurate relative alignment of the data moduli, an inaccurate value for the relative defocus between the planes, and background intensity due to inelastically scattered electrons with encouraging results; Lannes (1976a) compares the method with a version of the one-sided pair method (Section 3.5) in finding the phase distribution for a model one-dimensional problem simulating the image of a dislocation dipole.

7.2. Bright-field/dark-field diffraction pattern sets

In the interests of eliminating practical, if not theoretical, ambiguities in the phase solution, and providing reconstructions less influenced by the observational noise in the two data micrographs so far assumed, one particular choice of *three* data planes to be exploited simultaneously has been examined in numerical trials also. The combination of a bright- and (central stop) dark-field image pair has already been mentioned in Section 1.5 where it was noted that a twofold choice is possible for the image phase at every point in general, but that continuity arguments, based on a restriction of the diffraction plane width, might be used to reduce this considerably. The proposal explored in this section is the use of the diffraction plane modulus data to resolve the ambiguities; in comparison with the image and diffraction pattern problem, we have an additional datum modulus distribution now and may accordingly expect no real uniqueness problems. This proves to be the case in practice, but there remain problems sometimes in actually finding the solutions.

The only numerical method that has been applied to these new data is the iterative transform method, which is easily adapted because of the simple conditions required for its convergence. The modification necessary is merely to arrange that individual cycles of the algorithm alternate between the bright-field and the dark-field image, with the diffraction plane being of course the intermediary. Modulus corrections are made in turn in the bright-field image wave function, the diffraction plane wave function, the dark-field image wave function, and the diffraction plane wave function again. Switching between diffraction plane wave functions corresponding to bright- and dark-field conditions is effected by zeroing or forcing the central point. Two error criteria (SSEs) are defined, one for the bright- and one for the dark-field stages, and it is no longer impossible for these ever to increase, for while the bright-field SSE cannot increase during the operation of the bright-field stages of the algorithm, it may do so during the dark-field stages, and vice versa. In practice, this does not appear to occur more than occasionally. "Locked" conditions do still arise, and an interesting example of this is included among those presented in what follows, but when the solution is approached closely convergence is usually much more rapid than it is for the data previously considered.

Four examples of wave function determination from these data complete the numerical trials to be presented. Fig. 7.2.1 considers an aberration-free image of a pure phase object with a peak phase shift of 1.0 rad. The

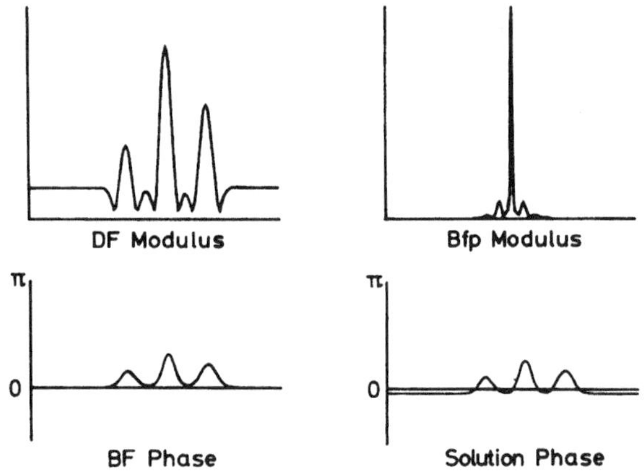

Figure 7.2.1 A three-data-plane problem. SSE: 0.000%.

Table 7.2.1 SSE decay with three data planes.

Iteration number	SSE (%)		Iteration number	SSE (%)	
	Dark-field	Bright-field		Dark-field	Bright-field
0	54.494	3.206	3	0.248	0.006
1	38.362	0.490	4	0.016	0.000
2	2.141	0.094	5	0.002	0.000

contrast-free image modulus distribution is taken together with the dark-field and diffraction plane moduli shown, and these lead rapidly to the (apparently exact) phase solution shown from a random initial estimate. Table 7.2.1 gives the SSE decay for the two image planes as a percentage of the sum square modulus in each plane.

Fig. 7.2.2 involves the model used in Figs. 6.3.9–6.3.10 again. Of six iterations commenced with random initial estimates, two converged, the others locking with SSEs between 0.3% and 1.8%. The solution shown reached an SSE of 0.06% in 10 cycles. Fig. 7.2.3 tackles the model of Figs. 6.3.11–6.3.13. The same success rate of one in three was found for random initial estimates. A solution obtained in 10 cycles with an SSE of 0.8% is shown.

Finally, Fig. 7.2.4 repeats the problem of Fig. 7.2.3 but with multiplicative "noise" applied to the data moduli first. Those random initial estimates that were successful previously continue to be successful, and a solution obtained in 15 cycles for a 10% noise level with a final SSE of 0.11% is shown. Of particular interest however is the estimate (*f*) obtained after 35 cycles

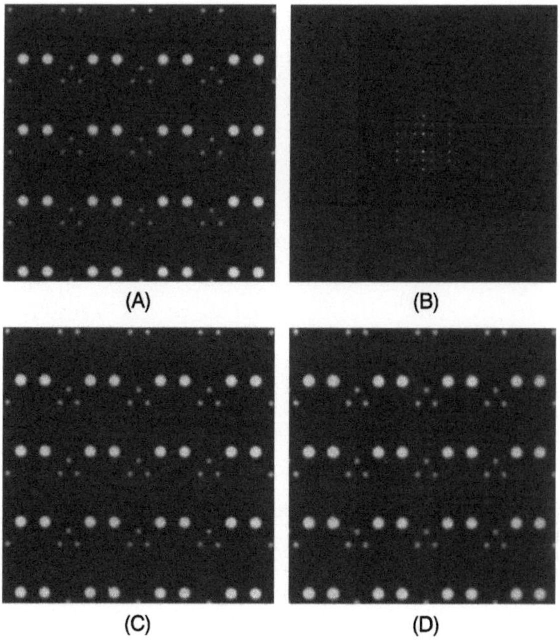

Figure 7.2.2 A three-data-plane problem with the model of Fig. 6.3.9. (A) Dark-field image modulus, (B) bfp modulus, (C) bright-field image phase, (D) phase solution (SSE: 0.06%).

from a zero initial estimate with a 20% noise level, in which the SSE settled at 2.1%. The chief reason for the convergence failure is clearly the fact that one of the "atoms" has the wrong sign. The sign change allows the estimate to remain consistent with both image plane modulus distributions, though not with the diffraction plane modulus distribution; it is easy to appreciate how this leads to a local minimum in the SSE, since a gradual reversal of this "atom," although leading to a lower ultimate SSE, will necessitate an initial rise, and it is just such a rise that our present algorithm avoids. Inspection of other convergence failures confirms the pattern of a signal reversal over a portion of the picture, and it might seem as if we are little better off than we would be with the two image moduli alone since these define regions (bounded by $2n\pi$ phase contours) within which a twofold ambiguity exists. However, our convergence rates are clearly much higher than could be expected from a random choice for each of the ambiguities, and we are certainly able to confirm that the three modulus distributions taken together do define the phase solution uniquely. Moreover, it should be possible to make use of image plane data for which the bright- and

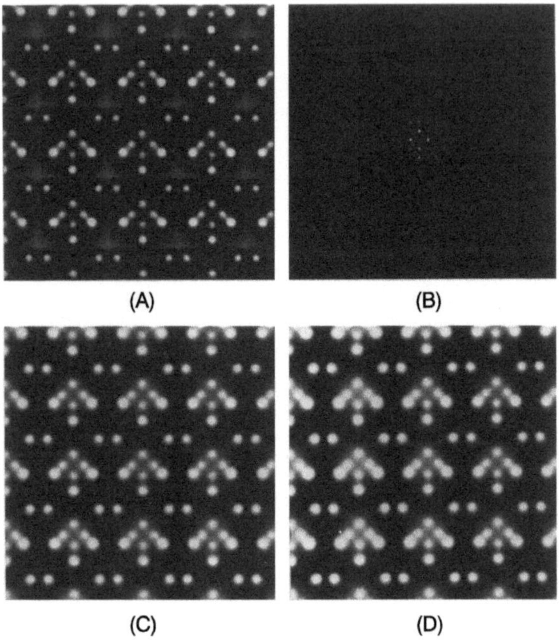

Figure 7.2.3 A three-data-plane problem with the model of Fig. 6.3.11. (A) Dark-field image modulus, (B) bfp modulus, (C) bright-field image phase, (D) phase solution (SSE: 0.07%).

dark-field pictures are at different defocus values, which is not possible in the bright-field/dark-field method as originally proposed.

7.3. Further possible data

In view of the simple conditions of linearity and unitarity required for convergence of the iterative transform algorithm of Sections 5.3 and 7.1, it is of interest to speculate as to any further data, beyond those of the last section, to which it might usefully be applied. This section mentions a number of candidates, each of which could be easily explored in numerical trials without any further analysis in the confident expectation of a converging iteration. The mere fact of the convergence of the iteration, of course, in no way implies the uniqueness of the phase solution for any particular data, this problem requiring independent investigation for each case, nor is any information as to the conditioning of the data immediately provided, though both of these questions could be explored by a sufficient number of

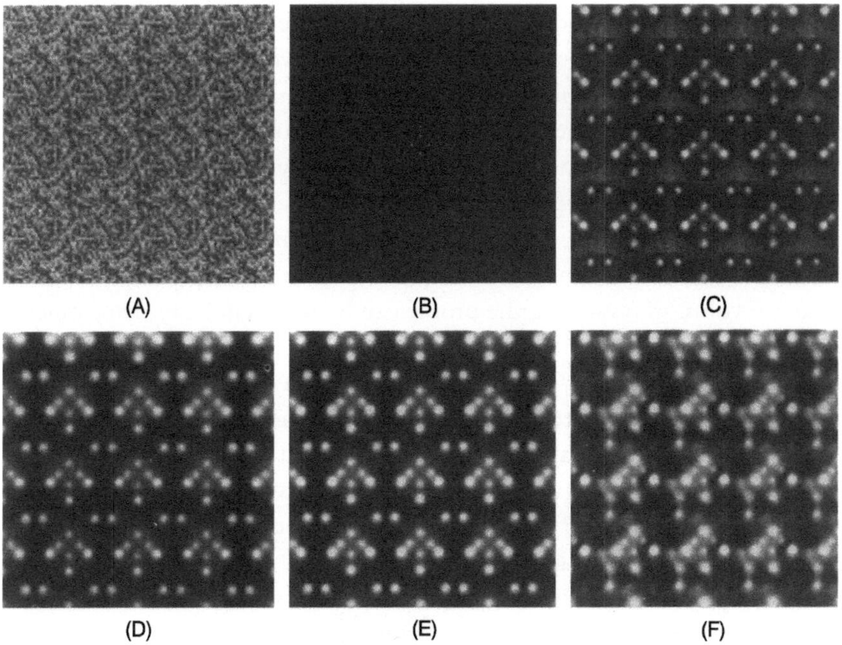

Figure 7.2.4 The problem of Fig. 7.2.4 with noisy data. (A) Bright-field image modulus, (B) bfp modulus, (C) dark-field image modulus, (D) bright-field image phase, (E) phase solution (SSE: 0.11%), (F) phase "solution" (SSE: 2.1%).

numerical trials. In some cases, the other methods presented in Chapter 5 can also be adapted for the new data with little difficulty; however, we shall not mention further in detail *any* specific numerical methods.

The general transformation between the wave functions in different planes of the microscope is the integral (1.1.9) which involves nothing worse in principle that multiplication by quadratic phase factors and Fourier transformation. In practice the transformations cannot be carried out with reasonable economy when the quadratic phase factors vary too rapidly, and this limits the planes of serious interest to those in the neighborhood of the object plane, the diffraction plane, and the Gaussian image plane.

The wave function cannot of course be observed near the object, but the object can still be a data plane of interest since in some cases the wave function is known a priori to be unimodular on leaving the specimen: this is true in Lorentz microscopy and lattice imaging over a useful range of thickness, for example. The use of this plane has been mentioned in several places in the previous chapter, explicitly and implicitly, so this is not wholly a new suggestion. However, replacing image plane by object plane data has three distinct advantages worth noting. First, it avoids the object exposure involved in obtaining a micrograph; second, the phase solution for an object plane is automatically free of the effect of instrumental aberrations; and third, the spatial coherence requirements of the whole process are greatly reduced since the object plane information is obviously independent of the illumination conditions, and a diffraction pattern used as the other data plane can be corrected for the finite profile of the effective electron source by deconvolution (assuming the profile can be independently determined). Admittedly, the use of this a priori information deprives us of the benefits of most of the analysis so far made of the question of the uniqueness of the phase solution, since virtually all of this has assumed a finite extent for the wave function in one plane, which is inconsistent with the wave function in the other plane being unimodular (see Sections 3.1 and 3.3). In spite of this, the value of the a priori information has been amply demonstrated in the examples previously given.

Moving on to the neighborhood of the diffraction plane, a defocus pair of diffraction patterns presents an obvious possibility. The analysis would probably be simplest for periodic objects, the diffraction patterns of which consist of arrays of isolated spots which would spread and overlap on defocusing, the resulting interference revealing their phase, but would still be possible for more general objects. There is an obvious relationship here to Hoppe's ptychography (see Section 1.5), but the mechanism by which the different parts of the diffraction pattern are made to interfere is different and somewhat simpler.

A possibility similar to the use of defocus image pairs is the use of micrograph pairs made with different angles of tilt in the illumination system, provided the objective aperture and tilt angles are such that the whole of the object spectrum is passed in both cases. The transformation between the two tilted image wave functions consists of replacing in the diffraction plane wave function the lens aberration phase shifts at one tilt angle by those appropriate to the other, thus involving two Fourier transformations, like defocusing.

Allowing the use of more than two micrographs, we might propose the use of defocus series in bright- or dark-field conditions, with or without a diffraction pattern. Again, in discussing the possibility of phase determination from one-sided dark-field micrographs, which by themselves present exactly the same problem as two-sided dark-field micrographs (see Section 3.8), Misell and Greenaway (1974) suggest the use of a two-sided dark-field image *together* with the one-sided pair, but do not propose a technique for extracting the phase from these data; this problem too might be solved by a modification of the iterative transform method.

Before concluding the discussion of numerical methods for phase determination, it may illustrate the flexibility of the iterative transform approach to note that modified forms of it have been applied successfully to quite different problems from those for which it was originally devised. Gerchberg (1974) tackles the superresolution problem (extrapolating the known portion of a diffraction plane wave function rendered analytic by a finite object assumption); see also Högbom (1974) and De Santis and Gori (1975). Saxton (1975, Section 6.2) shows how possible source brightness distributions can be conveniently deduced from modulus measurements only on the far-field mutual coherence (the data being well-known commonly to permit considerable ambiguity in the brightness distribution), and how to restore images formed by a linear system with a finite object when only part of the image has been recorded (Section 6.4: the problem is stated there in formal terms with no particular physical application in mind).

7.4. An assessment of phase determination

The survey of interpretation methods for use outside the linear theory of the weak object approximation being now complete, it is appropriate to attempt a brief comparative summary of the capabilities and requirements of the three chief two-micrograph methods available, namely the image/diffraction pattern (ID) method (Chapters 4–6), the defocus pair (D) method (Section 7.1), and the one-sided pair (O-S) method (Sections 3.5–3.8).

Unlike the other two methods, the ID method requires periodic objects. The O-S method requires that the object is not *too* strong, and the ID method that it is not too weak, but the D method appears to be of general application. The ID and D methods both require a substantial degree of oversampling in the image plane, but the O-S method requires this only when the object is particularly strong. The O-S method requires bright-field micrographs, but the other methods will also accept dark-field data. All three methods can make use of a single micrograph, the ID method for a strong pure phase object, the D method for a pure phase object, and the O-S method for a pure weak-phase or a centrosymmetric object. The ID and D methods involve iterative searches for a solution, the former becoming progressively less reliable as the image information content grows, while the O-S method offers a direct (closed form) solution. The ID and D methods are reasonably noise tolerant; the O-S method has a high noise tolerance except for objects near its strength limit. There are no serious difficulties in obtaining the data required by the ID and D methods, but several practical problems hamper the O-S method. Finally, the ID method is the only one to have so far produced useful results from real micrographs.

Experience with weak object reconstruction (Frank, 1972) suggests that with practical noise levels, two micrographs may well be inadequate in any case for reliable restoration of complex objects. In that event, we must look for the further development of series reconstruction methods achieving a least-squares fit to several micrographs. Nevertheless, a sufficient basis of techniques now exists for it to be worth while attempting practical restoration of strong objects, and the field of crystal lattice images, especially with faulty crystals, seems a promising area for future work.

CHAPTER EIGHT

The hardware of digital image handling☆

W. Owen Saxton
Murray Edwards College, University of Cambridge, Cambridge, United Kingdom
e-mail address: wos1@cam.ac.uk

Contents

8.1. Optical or digital manipulation?

A rotating drum microdensitometer and filmwriter, with a small computer with magnetic tape and disc storage, may easily cost 30 times as much as a good quality optical bench with a laser source and photographic equipment. In view of this huge difference, it is worth considering the capabilities of the two methods of processing: both have their peculiar compelling advantages.

The comparison in cost being wholly to the advantage of the optical methods, consideration is given first to the other great advantage of these, namely their immense information capacity. A reasonably fine grained photographic emulsion may comfortably hold 1000 picture elements in both directions in a 2.5 cm square, and an optical bench with a camera length of 1.25 m can easily produce the Fourier transform of such a picture with the same number of elements spread over the same area. In contrast, such a picture in floating point representation (see Section 8.2) would require about 2×10^6 (16-bit) words of disc store or a fifth of a standard length (2400 ft) magnetic tape (at 800 bits per inch), the former being a larger disc than is possessed by many small computers, and the latter implying extremely slow access times. It is rarely possible for even the largest computers to hold larger than 250 element square pictures in core storage, and small computers must rely entirely on backing store with consequent overheads due to access times (see Section 8.2).

☆ Reprinted from Advances in Electronics and Electron Physics, Supplement 10, 1978.

Advances in Imaging and Electron Physics, Volume 214
ISSN 1076-5670
https://doi.org/10.1016/bs.aiep.2020.04.008

The advantages of optical methods in this respect have however been considerably overstated in the past [e.g., Stroke (1972)] and it is anticipated that this chapter will redress the balance to some extent. Even the present consideration (capacity) does not wholly favor optical methods. While transformation is undeniably slow to execute digitally, other operations such as image subtraction are extremely rapid.

Cost and capacity once mentioned, other comparisons favor digital methods. These are free of noise problems, being capable of performing lengthy sequences of operations without introducing any additional noise in the data; there are no linearity problems and no alignment problems such as arise in photographic methods, though subtle forms of distortion not present in optical systems may arise in calculating transforms (see Section 9.3). There are no dynamic range problems (compare, for example, the output of optical and digital processes for producing a power spectrum) and complete quantitative information is available at every stage. Once established, digital methods are relatively trouble free and easily applied by "amateurs," in contrast to optical methods in which considerable care is necessary throughout. Above all, however, digital methods are flexible and of wide scope.

The flexibility of digital methods is illustrated by the ease with which parameters such as defocus and noise level can be varied simply by entering different numbers with a teletype; such parameters vary almost continuously between different micrographs, and optical processing necessitates a completely new set of masks for every possible combination of them—a range of masks covering a very conservative range of defocus, astigmatism, spatial and chromatic source size, and image noise level, for processing bright-field phase contrast images of weak objects would extend to at least 25,000 masks. Their scope depends on the direct numerical representation used which makes possible any mathematical operation whatever, involving as many pictures as desired, applied to regions of any size down to individual points. Thus, for example, the logarithm of a picture, or the maximum value at corresponding points throughout a set of pictures, extremely difficult to obtain optically, become immediately available as tools. Processing entirely ceases to be directed and limited by the ingenuity of the optical bench design and becomes free to explore a variety of techniques that would be quite impossible to implement optically. The ingenuity persistently displayed in the field of optical processing (particularly the use of "carrier waves" to realize masks with complex transmittance—"holographic filters"—and the use of electronic/CRT systems in nonlinear situations) deserves considerable admiration. Even so, no one would seriously propose to explore optically an iterative procedure of the type described in Chapter 5.

Finally, it is worth nothing that there are many incidental uses for both optical and digital systems, and it is therefore preferable that both systems should be available. Optical benches, for example, permit extremely rapid spectral analysis of micrographs, useful both for aligning microscopes and detecting periodic structures, while digital systems provide a means both of calculating images of model objects under various conditions, thus assisting visual interpretation of actual images and providing information on the relative importance of different effects involved in image formation, and of manipulating data in other than two-dimensional form—performing three-dimensional structure determination, for example.

A useful review of the technology of optical processing for electron microscopy is given by Mulvey (1973), while Van der Lugt (1974) reviews some sophisticated devices developed recently for interconverting optical and electrical signals.

8.2. Digitization and regeneration

Postponing until the next section discussion of the hardware for manipulating the digital images, we review briefly in this section how micrographs can be conveniently converted to and from digital form.

Conventional transmission electron microscopy results in an image on photographic emulsion, either on a glass plate or on film. A simple application of Poisson statistics predicts that for low exposures the specular optical density of the developed plate is linearly dependent on the electron exposure; this is confirmed to within a few per cent experimentally up to optical densities of about 1.5 [see, e.g., Valentine (1966)]. Transmission microdensitometry naturally measures specular rather than diffuse optical density and is the obvious tool. The optical arrangement necessary is simple, the light flux transmitted forward being measured by a suitable detector when a small (usually circular) region of the emulsion (the "sampling aperture") is illuminated. The area illuminated is scanned across the emulsion, and the detector reading is recorded at an interval which is normally equal to the diameter of the sampling aperture. It is relatively easy to achieve a satisfactory optical system; the problem lies in recording the data at reasonable speed.

Manual densitometers with meter displays are of use in measuring diffraction pattern intensities where a relatively small number of intensities need to be recorded but are obviously of little value for scanning images. Adaptations are available of originally one-dimensional scanning microdensitometers with digital output at rates of around 20 points/sec. Data rates of many thousands of points per second are possible, however, with instruments specifically designed for fast digitization on a two-dimensional grid, interfaced with computers or magnetic tape decks.

Four main types of these are currently in use. In "flying spot" scanners [e.g., Billingsley (1971)], a cathode ray tube spot forms a moving light source demagnified onto the emulsion while a photomultiplier tube or similar detector records the transmitted flux, the resulting signal being passed through a logarithmic amplifier (to make it proportional to optical density) to an A/D converter interfaced with a computer. The scan may be directed by the computer allowing search procedures. Data rates of up to 50,000 readings/sec can be achieved, the limiting factor being the time required for phosphor decay and spot movement. The chief problem of these instruments is the geometric aberrations in the CRT optics which distort the scanning grid and cause the spot size to vary between the center and edge of the scanned region; the former effect may be compensated by appropriate adjustment of the coordinates at the software level (Arndt et al., 1968).

Flying spot scanners are not yet available commercially, but several manufacturers will supply the other three types of scanner. Rotating drum scanners preserve a very accurate scanning grid over large areas, using an incremental encoder on the shaft of a drum carrying the emulsion to define a grid in one direction, and a stepping motor driving a lead screw carrying the optical assembly for the other. High data rates are again achieved during the active part of each revolution. The drawbacks are the impossibility of digitizing micrographs recorded on glass plates instead of film and the necessarily serial nature of the scan.

Flat-bed scanners are rather more expensive but offer the advantages of both the above systems at somewhat slower data rates (about 5000 readings/sec), the limiting factor now being the mechanical settling time of the emulsion-carrying stage, which is moved below a stationary optical assembly by stepping motors driving lead screws and controlled by linear optical encoders mounted on the stage.

These systems typically offer 8-bit digitization with a resolution of a few microns over a wide field. Camera tube systems [reviewed by Chien & Snyder (1975)] do not match this performance normally, being restricted to 480 or 625 lines, with a 4–6 bit digitization capability matching the human eye's performance, and being prone to a nonuniform spatial response ("shading"). The ready availability of inexpensive TV equipment makes them attractive, however, where their limitations are acceptable; and several schemes exist for effecting the reduction in data rate necessary for digitization and computer storage, including the use of video disc as intermediate storage and the extraction of a *column* from the TV signal during each scanning frame.

In each of these systems it is possible to identify a "sampling aperture" with which the transparency of the micrographs is convolved before sampling. (It will be a physical aperture in the optical system for flat-bed scanners, for instance, and the CRT spot profile in flying spot scanners.) Provided the transparency varies reasonably slowly (not necessarily true in high resolution microdensitometry of dark-field micrographs), the convolution of the transparency with the aperture is easily shown to be equivalent to a convolution of the optical density with the aperture; the result is therefore an attenuation of high spatial frequency detail in the digitized signal, the transform of which is multiplied by the transform of the aperture. Although this effect can be substantially compensated by subsequent high frequency enhancement, it can be compounded by an unfortunate phase relationship between sampling and sampled lattices. A 20 μm aperture and sampling interval, for instance, will not cause very serious attenuation of a 40 μm sinusoidal density variation if the sample points fall on the maxima of the sine wave but will fail to detect it altogether if they fall on the zeros. The use of a sampling aperture smaller than the sampling interval is unlikely to be helpful for electron microscopical purposes since the film granularity will lead to high noise levels in the digital values. It is normal therefore to use a sampling aperture whose transform is substantially wider than that of the micrograph to be digitized and a sampling interval of the same size. Square apertures are probably slightly preferable to the usual circular ones, since fewer electrons striking the emulsion will thereby be passed over and "wasted."

A certain amount of noise is inevitably added to the micrograph intensity record by the digitization process. An obvious source is the forcing of the optical density values onto discrete levels. This corresponds to adding numbers rectangularly distributed with a maximum modulus half the digitization step size, thus giving an rms noise level about a third of the step size; the values are not independent from point to point, however, and the damage can be serious in very low contrast signals. Typical commercial scanners spread 256 levels between density limits of zero and two, so the "discretization" noise for a bright-field micrograph with a mean density of one is about 0.25%, though it can rise to more than 1% for less dense images. The latter figure implies a significant extra noise contribution to the image of low contrast specimens.

Other noise sources include amplifier noise, thermal noise, and photon shot noise. The last is proportional to the square root of the number of photons transmitted, i.e., to the square root of the transmission factor t. The optical density D being $-\log_{10} t$, we have for small changes

$$\delta D = -0.4\delta t/t = -c\sqrt{t} = c10^{D/2},$$

so that the noise on the density values (and by implication the electron intensity values) rises exponentially with density. Good commercial scanners are accordingly limited by the high density shot noise level to about ±3 in 256, with better figures applying to less dense images.

There is increasing interest currently in the use with conventional microscopes of image intensifier/camera tube systems for recording electron images as an alternative to photographic emulsions. These would yield the serial analog signal so convenient in the STEM. The advantages include the possibility of operation at beam current levels too low to be visible to the naked eye, prolonged study of an image replayed by a video recorder without further damage to the specimen, electronic signal processing (e.g., for background removal), immediate sight of the permanent record (i.e., no delays for film development), immediate digitization for further processing with the possibility of interaction between the microscope and the computer, and the ready availability of multiple images of a given area free of relative alignment problems. The price paid for these is a restriction in the field of view to a few hundred resolvable elements in each direction, about 50 μm in size, and a certain amount of "shading."

Various designs are in use, and several commercial systems are available by now. Typically, the electron image is converted to photons at a phosphor screen or clear plastic scintillator (Kuo & Glaeser, 1975), amplified by an image intensifier and projected on to the faceplate of a high sensitivity TV camera tube, fiber optic coupling being used throughout. [Imura et al. (1971) have explored the alternative use of a silicon diode array target within the microscope column.] This arrangement is sensitive enough to detect individual electrons and will provide images at beam currents a thousand times lower than a dark-adapted eye can see; large numbers of photons being produced by each incident electron, the output signal can remain limited for low signal levels by the shot noise of the original image (Herrmann et al., 1972). The camera faceplate will accumulate the incident signal for short periods so that lower shot noise levels can be achieved by slower than TV rate scanning [e.g., Siegel & Hertel (1968)]. The integration capacity is however limited, and the large number of particles contributing to bright-field images presents problems: reducing the camera sensitivity means that its noise level (around 2% of the maximum signal) becomes a significant addition to the image noise. The ideal system may be of the type described by Herrmann et al. (1976) in which the camera is operated at normal scanning rates and full sensitivity, the signals due to individual electrons being detected by suitable discriminator electronics and accumulated *digitally* in fast (semiconductor) storage.

At the other end of a digital process is the problem of presenting the processed data in a visually assimilable form, i.e., regenerating images from digital signals. There are many possibilities here, all available commercially. The least satisfactory, but probably the most widely used nevertheless, is the conventional lineprinter, several passes of overprinting being used with carefully chosen characters to achieve a suitable variation in the density of the ink. The resolution is low and anisotropic (about 120 by 60 elements to a page typically), and only a moderate number (8–16) of reliably distinguishable gray levels can be obtained, but lineprinters are at least almost universally available (see Fig. 8.2.1). An enormous improvement can be achieved by the use of electrostatic matrix printers (Fig. 8.2.2), as fast as, and cheaper than, conventional line printers, but not yet common; in these, powdered ink fixes a dotted charge pattern applied to the moving paper by a row of individually addressable electrodes (4–8 mm^{-1}). Taking a block of 4 by 4 dots for each picture element, for example, and varying the number blackened, allows a 16 level display of a picture with 256 elements in each direction in a 5 in. square of paper.

Conventional pen plotters are well-suited to contour map displays (see Fig. 8.2.1 again), which can offer precise quantitative information without altogether sacrificing visual effects, but not to gray level simulations; though they have been used for the latter purpose by varying the length of a spiral drawn in each picture element, the time needed can easily extend to more than an hour.

A computer-interfaced oscilloscope display of any kind is extremely valuable without necessarily being excessively expensive. The cheapest units are of the "refreshed" type; since it is not possible, without special hardware provision, to refresh the display sufficiently rapidly to make the

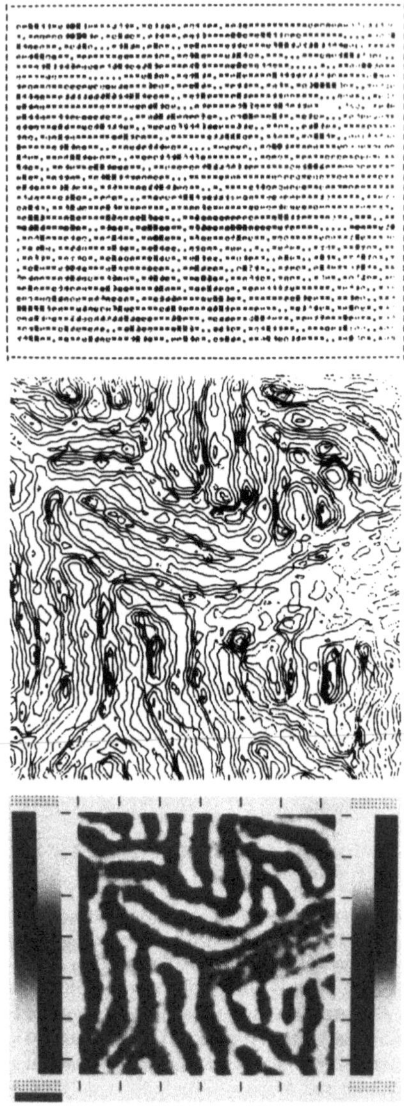

Figure 8.2.1 Three forms of pictorial output from computers. Top, line printer; center, pen plotter (contour map); bottom, filmwriter (all 64 elements square).

whole picture visible at once, a photograph has to be made of the screen while the spot traces out the full picture. Storage oscilloscopes, at slightly greater cost, will retain a display once traced for long periods. Again, the cheaper units have binary displays only, and a range of gray levels has to

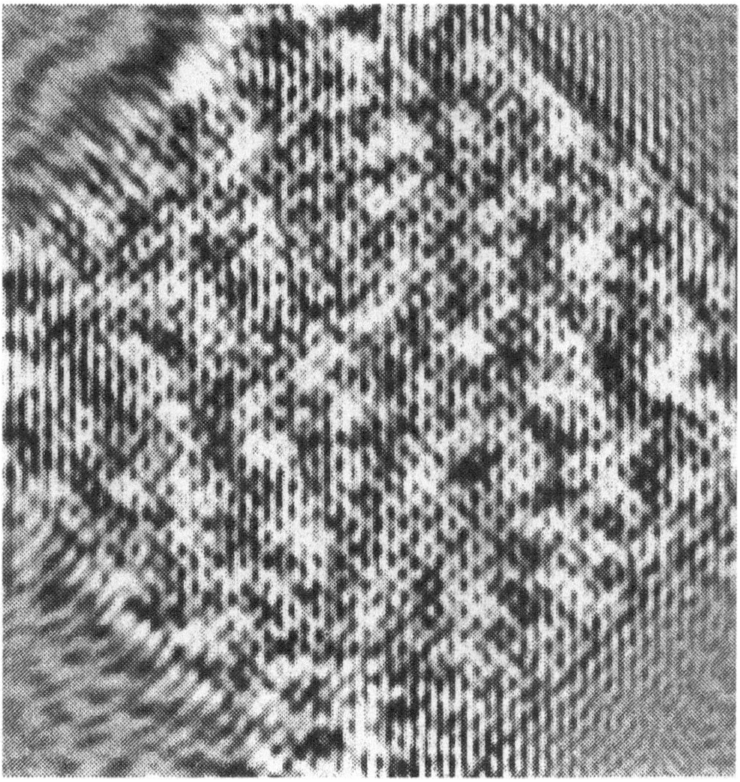

Figure 8.2.2 Pictorial output with a matrix printer (a Versatec 1200A). Calculated bright-field surface image of gold film; 512 elements square. [By courtesy of W. Krakow and T. Holladay (Xerox Corps.) and R. Peterson (Versatec).]

be achieved through varying the number of points displayed within a small area associated with each picture element (Fig. 8.2.3) which commonly reduces the number of picture elements that can be displayed to 128 or 256 square, or (for refreshed displays only) varying the time for which the spot rests on each element. More expensive storage oscilloscopes offer multilevel storage with superposed movable cross wires. At the upper end of the video display range are raster-scanned refreshed screens taking data directly from the computer main store and sustaining a 512 element square display in 256 gray levels with their own scaling and rotation facilities. The better displays make effective interaction with the processing computer possible and are also of considerable value for the display of other kinds of information too, such as histograms, or program text.

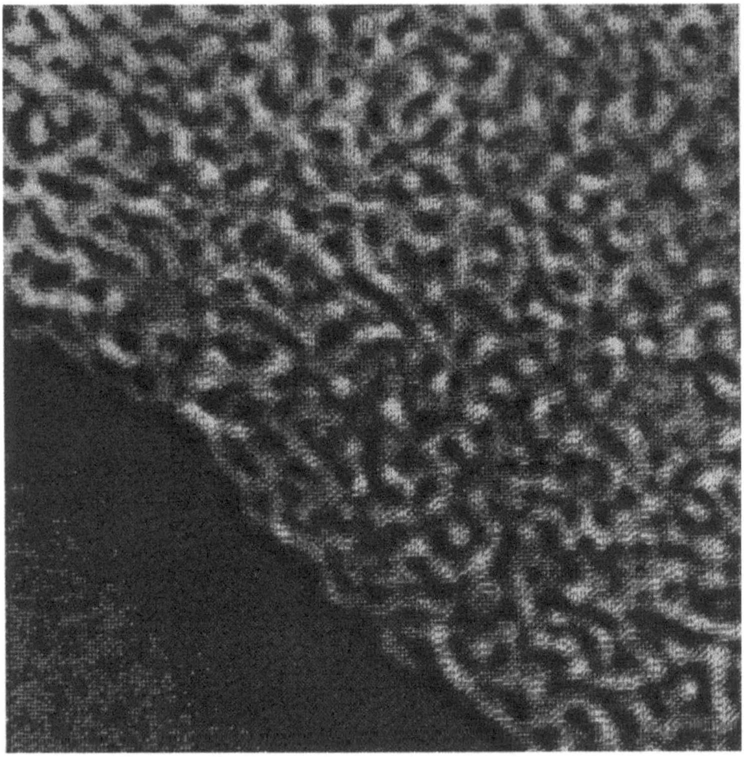

Figure 8.2.3 Pictorial output with a binary storage oscilloscope (a Tektronix 611). Bright-field image of germanium film; 256 elements square. [Micrograph by courtesy of A. Mistry.]

The best quality permanent output currently available is achieved by filmwriters analogous to the flying-spot and rotating drum microdensitometers described above, the micrograph to be digitized being replaced by an unexposed photographic emulsion, and the brightness of the light source being varied to control the exposure given to the emulsion. The filmwriters exactly complement the microdensitometers in their gray level capability, resolution, and data rate (scanners and filmwriters frequently sharing the same CRT or drive) and their chief drawback is simply the time required to develop the emulsions. Unlike the corresponding scanners, flying-spot filmwriters are readily available commercially as standard computer peripherals in the form of microfilm plotters.

Numerous examples of output from a rotating drum filmwriter appear throughout the present work (see also Fig. 8.2.1 again). A special hardware

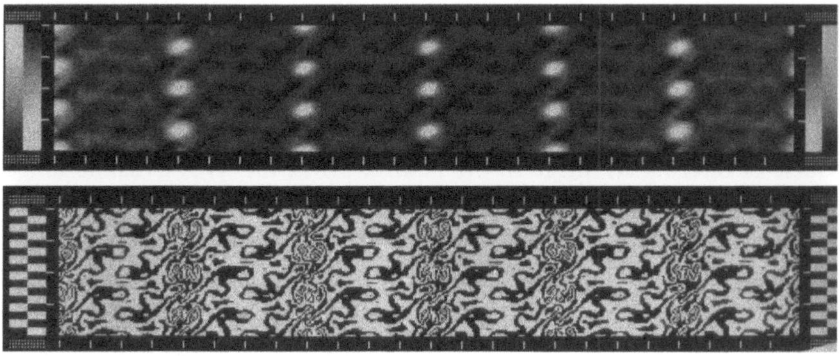

Figure 8.2.4 Hardware "contour mapping" with a filmwriter. Above, normal display; below, the 2^4 bit only displayed.

feature of this particular instrument (an Optronics "Photowrite") allows a bit mapping to be applied to the data before they are passed to a D/A converter. Fig. 8.2.4 illustrates its use to produce a contour map with alternate bands shaded black and white by connecting all the output bits to one input bit.

8.3. Processors, representation, and storage

The very rapid spread of dedicated minicomputer systems means that processing that might formerly have been attempted only on large computers shared with many other users is now frequently carried out on small machines. Since in the former case, the hardware configuration must usually be accepted as it is and the problems are likely to be simply the all-too-finite turnaround time typical of multiuser operating systems and the difficulties of arranging an interface with a scanner and filmwriter, which may well necessitate using magnetic tapes as intermediaries, we are subsequently concerned chiefly with the latter case, where there are a number of points deserving discussion.

The first of these concerns the internal representation of the optical density supplied in the first instance typically as an 8-bit integer. This is interrelated with the selection of a processor. The most convenient way to handle these might at first appear to be as integers, with a number of bits fixed by the processor word length (usually 12, 16, or 18 bits). This is economical on space and permits rapid manipulation. Such a representation is quite satisfactory for geometric stretching, contrast enhancement, picture subtraction and similar processes, where the order of magnitude of the data remains unchanged; and if sufficient care is taken in normalization, it can be used extensively. However, it implies a very limited dynamic range in the data, and consequently severe rounding errors in calculations. In the course of filtering an image in the Fourier plane, the large dynamic range in this plane might well cause rounding errors that outweighed the effect of the filtering. For general purposes, therefore, a better representation is required. A possible compromise is the use of double length integers, particularly if the computer hardware supports these, which greatly reduces the dynamic range problem, though by no means removing it. Frequent careful normalization of data is still necessary, and data storage requirements are doubled (probably not serious), but basic manipulation may remain rapid. Undoubtedly the most convenient representation however (and the one almost certainly used on a large computer) is as "floating point" numbers, a mantissa and exponent being stored separately for each numerical value; this removes the dynamic range problem completely for most purposes and ensures the same proportional accuracy at all points. Storage requirements are again increased slightly, but the real problem may be the increase in computation time—by any factor from 1 to 200 depending on the hardware (if any) available in the computer for manipulating floating point numbers. The latter figure is obviously prohibitive. An additional consideration in connection with the representation chosen is that while nearly all small computers have at least software to support floating point numbers as well as single length integers, relatively few support double length integers, so that programming a system in this form can be substantially more tedious than in other forms.

There are, of course, many other ways of representing pictures digitally, of a rather more sophisticated nature. The digital increments might define a nonuniform intensity scale, for example, there being a higher density of digital steps for low intensity than for high, thus providing the same proportional accuracy for all points. Computation based on such representations is obviously cumbersome however. Savings in storage space can be made by holding picture differentials (DPCM), since the differentials usually display a smaller dynamic range and can therefore be coded in fewer bits; again, there is a loss of computational speed. It is possible to pursue the process further unless the picture has a vanishingly narrow autocorrelation peak. "Linear predictor" methods anticipate the picture value at any given point by linear extrapolation from the previous point and code only the deviation from the expected value (almost a second differential of the picture) typically using 3 bits per point. With variable length codes, visually acceptable pictures can be defined with as few as 1.5 bits per point on average. A good review of these techniques is given by Connor et al. (1972). Although they permit considerable economies in space, they involve storing pictures in a form not readily susceptible to computation; processing time being usually a much more stringent requirement than storage space, we are obliged to adopt the most computationally convenient representations for the data. On the other hand, the techniques above could be useful for archive purposes, or for transmitting pictures along low data-rate lines (e.g., between a large computer and a local video display terminal).

Next to the internal data representation permitted, the most important processor characteristic is simply how fast it performs the basic arithmetic operations, multiplication and division being as important as addition and subtraction. Also important is a reasonably flexible and efficient addressing scheme, since a large number of array operations is likely to be performed. Special supplementary "hard-wired" processor units are available which will perform operations such as matrix product calculation and Fourier transformation on integer data held in the main processor's core without interfering with the main processor's activities, thus increasing dramatically the speed of some procedures. These are fairly expensive, however, and cannot of course reduce the time spent in input/output, which is frequently a large proportion of the processing time (see Sections 9.2 and 9.3).

After the processor, we may consider storage. No small computer is able to hold full pictures in core storage so core requirements amount only to whatever is necessary to hold a processing program and at least two lines of a picture at once; this is quite modest. Backing store is a greater problem, and we must provide for processing programs as well as for pictures: a processing system is not viable without a relatively rapid means of loading processing programs in turn.

It will probably be necessary to have at least one magnetic tape deck; this provides the only realistic form of long term picture storage, a disc being unlikely to hold more than a handful of scans simultaneously; it provides ample storage for programs, which may share the tape with picture files; otherwise, it may form the link between the computer and the scanner and filmwriter, where these use independent decks. Many problems arise however if the tape is the *only* storage unit. It is impossible to move a file from one tape to another. Unless a "preblocked" format is used, which is not always convenient or possible, it is very difficult to remove "dead" space on a tape filled by old versions of programs or no longer required data. Still

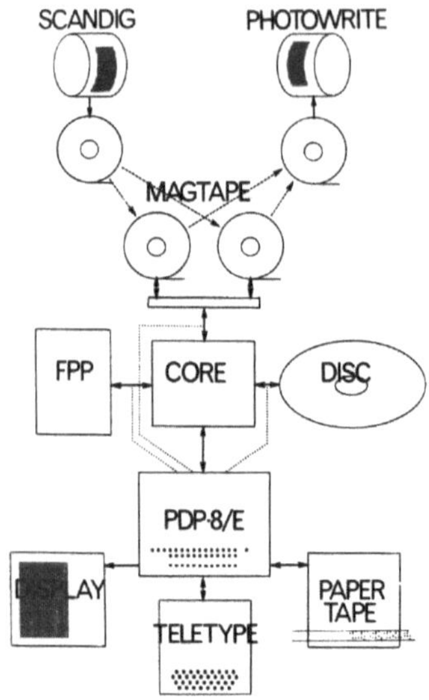

Figure 8.3.1 The configuration of a PDP 8/E based processing system (see text for details).

more seriously, there is nowhere to direct the processed lines of a picture being read from one part of the tape (although a preblocked format would at least allow a picture to be *overwritten* by the processed version).

A disc unit is a compelling choice for the second unit, offering equally rapid access to any line of the picture, and being almost essential to two-dimensional Fourier transformation (see Section 9.3). Its access time should obviously be as small as possible since many operations involve a great deal of input/output, but the fastest discs (fixed head discs) are sometimes too small in minicomputer ranges to be useful. A 256 element square picture in floating point representation (two 16-bit words per number) requires 128 K words of storage, and the disc must provide program storage on top of this, or the tape administration problems of the one-unit system will persist. A disc of reasonable size will also accommodate a third logical unit, which is necessary for such operations as picture subtraction or cross correlation, where two source pictures are involved at once. Although the discussion of relative merits has been so far limited purely to image processing criteria, the incidental advantages of a disc-based system for software development may in fact affect dramatically what is actually achieved with a system.

The system in use in the Electron Microscopy section of the Cavendish Laboratory (Fig. 8.3.1) illustrates a viable arrangement, though several different choices would probably be made now. A PDP 8/E processor shares 20 K of 12-bit core with a FPP 12/AB floating point processor (multiply time $\simeq 25$ μs), the former handling input/output and the latter the bulk of the computation. A 1.6 M word moving head disc drive (random access time $\simeq 50$ ms) provides program and picture storage backed up by two $\frac{1}{2}$ in. 800 bpi magnetic tape units which form an off-line link with an independent rotating drum microdensitometer and filmwriter. A Tektronix 611 storage oscilloscope (binary, with somewhat more than 512 resolvable elements in each direction) provides an interactive display, from which Fig. 8.2.3 was taken, and is heavily used for text display by the operating system.

The dependence of processing time on the hardware available may be roughly illustrated by the time required to transform a 256 element square complex picture under various circumstances. On the system just described, 9 min are needed; the figure would be about 50 min if the data were tape-rather than disc-based. If no floating point hardware were available, very similar figures could be achieved using 24-bit integer data (for which there is partial hardware provision), but an additional 30 min would be needed if floating point arithmetic were carried out by subroutines. On a large computer (an IBM 370/165), the transform requires 27 sec for disc-based data, or about 7 sec with the entire array held in core storage, which is quite feasible on such a machine.

CHAPTER NINE

Basic software for digital image handling☆

W. Owen Saxton
Murray Edwards College, University of Cambridge, Cambridge, United Kingdom
e-mail address: wos1@cam.ac.uk

Contents

9.1. A processing system

The sections following the present one discuss a range of different techniques relevant to different stages of processing a micrograph, each of which can be neatly encapsulated as an independent program or routine. This section outlines the considerations involved in the linking together of such routines to form a coherent whole with a convenient control system. Although this is in itself a somewhat tedious business, it more than repays the effort in terms of the suite's ultimate ease of use.

The central part of such a system is a convenient means for directing the application of the different routines and supplying them with information as to the data upon which they are to operate, the optional parameters of the process and so on. One possible system, described in detail in Chapter 10, involves the machine generation from the user's program of a Fortran main program containing appropriate calls to a precompiled subroutine library. The main program is compiled, the subroutine links are effected, and the resulting module is run in the usual way. While this allows a very flexible control language, the technique is unsuitable for interactive or small computer use. The alternative approach is for all the routines to return control

☆ Reprinted from Advances in Electronics and Electron Physics, Supplement 10, 1978.

Advances in Imaging and Electron Physics, Volume 214
ISSN 1076-5670
https://doi.org/10.1016/bs.aiep.2020.04.009

on terminating to a program which interprets the next line of the user's program and calls an appropriate routine to give it effect. In the VICAR system at the Jet Propulsion Laboratory of the California Institute for Technology (Efron, 1968), for example, the interpretation is done by a special machine supervisor program, the routines being stored on disc as independent core images, ready to run (in spite of the duplication of many minor routines involved). In other systems, such as those written by P. R. Smith at the Biozentrum in Basle, P. H. Buszler at the Max Planck Institute for Biochemistry in Munich (Hawkes, 1974b), or M. Horner at the Cavendish Laboratory in Cambridge (Horner, 1976), the interpretation is done instead by a Fortran main program that uses a massive computed GOTO instruction to select an appropriate subroutine call. This requires the linking together of all the routines available, whether or not they are required in any one particular job, which necessitates the use of overlay methods, routines being swapped in and out of core at run time by the Fortran system. An intermediate arrangement is also possible if overlay facilities are not available in the form of "chaining." The routines are stored as independent core images, but their loading is performed by a Fortran interpreter which is reloaded by each routine on termination.

At almost the same level as the passing of control between routines is the management of data. The possible options were described in the previous chapter, and it is only necessary here to note that a convention must be established between the routines covering data types, coordinate systems, and access methods. When data are stored on peripherals rather than in core, it may be convenient to route all access requests through a single subroutine, thereby making processing routines independent of the medium on which the data is stored. In multiaccess systems, temporary storage (for the duration of the job only) may greatly exceed permanent storage space, and it may be worth providing special facilities to exploit this, while some form of internal directory system for permanent data may be useful in any system.

Conventions will also need to be established between the routines and supervisor/interpreter as to such matters as the use of core working space, the initialization and/or rewinding of input/output units, the possible maintenance of a program "log" recording the operations performed as the job proceeds, the signaling of error states (through an error "flag" in core or by using multiple subroutine return facilities), the passing of parameters between the routines and the user via the interpreter (through subroutine arguments, or core-resident tables, or both), and the provision of default values (by the routines or by the interpreter). It is not essential for the routines all to be written in the same language, although the subroutine calling sequences must be compatible among different languages used. There are some operations (particularly input/output) worth coding directly at assembler level, but for most purposes the convenience of high level languages will outweigh the inefficiency of the generated code, particularly if an optimizing compiler is available for once-only use when a routine is error free.

If it offers a sufficiently comprehensive set of facilities, the form of the instructions accepted by the interpreter may reasonably be said to define a control language for the processing system. If it is to be convenient to use, and this is particularly important for interactive use, it will be important that the syntax is consistent throughout the language, that numbers may be entered in relatively free formats, and that an attempt is made to provide a more intelligible form of parameter specification than a list of numbers in predefined fields or separated by commas—perhaps by the use of keyword/value pairs in the command syntax, with a comprehensive system of defaults. Different pictures should be identified by names or numbers of the user's choice only, associated information as to type, size, storage medium, etc. being stored by the system. The language should be sufficiently flexible to allow operations not provided for by specific routines to be effected by suitable ingenuity at control level. In particular, local variables should be available with arithmetic facilities, testing and jumping instructions, and some form of access at control level directly to individual picture elements. In spite of the formidable appearance of this list of attributes, it is possible to achieve every feature mentioned within an interpreter written only in Fortran.

As mentioned earlier, the following chapter treats one particular processing system in some detail; the rest of this chapter is concerned with individual processing techniques and routines.

9.2. Input, output, and data selection

This section deals with some programming considerations relevant to the selection of image data for processing and its subsequent visual display.

While some input devices (programmable scan flat-bed microdensitometers, for example) will provide data points one at a time to the computer, most will provide it at a row at a time, in the form of packed integers—as four 8-bit integers in each 32-bit word of core, for example. The unpacking of these is most efficiently performed by an assembler code routine, but if necessary the integers can be accessed in Fortran by successively dividing the 32-bit integers by 256 in the example quoted. Conversion to floating point representation is then straightforward. Similar considerations apply to the repacking for output where the peripheral concerned requires it (as is the case with the commercial filmwriters, for example), the integers now being packed by successive multiplication by 256.

If the input medium is magnetic tape, the Fortran input facilities are usually quite inadequate to allow efficient or even convenient use of the tape. Data selection is likely to involve input from more than one file, not necessarily in serial order, and may well involve repeated access to some regions. The necessary freedom to move between source files, backspacing or rewinding as necessary is not available in Fortran, and here too an assembler code routine is useful permitting tape access at the lower level. The input routine must of course maintain a note of the current position of the tape. Similar considerations apply to output to magnetic tape in many cases.

A further advantage in the use of an assembler code routine for tape access applies also to disc store access. Fortran systems do not allow the transfer of data directly between the peripheral and the specific core area, but rather transfer it in the first instance to an independent "buffer" core area and only thence to its destination; assembler code routines will eliminate this source of inefficiency completely, thereby speeding up input/output significantly.

A problem common to all visual picture output routines is the selection of a suitable data scaling procedure. The simplest automatic approach is of course to search the image data for maximum and minimum values and scale these to the extreme output levels. The case of displaying power spectra illustrates the limitations of this method, however, since a small number of points near the origin of a power spectrum may be very much larger than those throughout the bulk of the data array, and scaling the display to cope with such extremes reduces contrast over the rest of the picture to a low or even zero level. A useful alternative procedure is therefore to prepare a histogram of the data to be displayed and then to use this to find maximum and minimum values such that a small fraction (say, 1%) of the data fall outside these extremes; in this way a small number of points have no real effect on the scaling chosen. Other expedients may also be helpful with power spectra, including removing the central point and radial weighting of the data.

When the output is being broken into a relatively small number of levels, for a contour map or for a line printer gray level simulation, there are some advantages in classifying the data into the different levels by a binary testing process (four tests being required to select one of 16 levels, for example) rather than by direct scaling. With such an arrangement, the level boundaries can be varied at will at no extra cost, so that various effective transformations of the source data can be made at output time without actually altering the data at all. Spacing the level boundaries geometrically instead of arithmetically between the minimum and maximum chosen will for example generate a display of the data logarithm without the expense of computing it, and this offers a further approach to the problem of data with a large dynamic range.

A common irritant once the image data has been selected is the presence of long range variation in the background level of the image, due perhaps to nonuniform specimen illumination in the microscope. This can make it very difficult to see the signal and causes a number of other problems—in cross correlation for example (see Section 9.4). To some extent, the problem can be cured by removing the lower order Fourier components of the image at the cost of two Fourier transforms, but in many cases the data have to be modified directly in real space to achieve a satisfactory result. A steady gradient across a picture, for example, implies a sharp discontinuity in passing off one edge of a picture to the aliased edge, and this will produce *high* order Fourier components. A very rapid procedure which has proved satisfactory for most practical purposes is to subtract from the data a least-squares fitted plane, which removes at least the linear part of the background variation at the cost of only two direct passes through the data. A variation in background level however implies a corresponding signal level variation if it is due to nonuniform specimen illumination. This will mean that after background level adjustment the micrograph will have different transfer characteristics at different places, and while this will not matter for such purposes as lattice averaging (see Section 9.6), it makes least-squares object reconstruction (see Section 9.7) impossible. While more elaborate ad hoc procedures can be devised for adjusting the signal level appropriately at different parts of the picture, nonuniform noise statistics will result, and it is almost certainly simpler to seek better micrographs in the first place if the background variation is substantial.

Preliminary inspection of the data often reveals also a number of obvious dust particles that have settled on the micrograph during microdensitometry if this is the input method used, and these can be prevented from causing further inconvenience by a pass through the data during which values outside a suitable threshold (say, three standard deviations for bright-field data) are reset to the picture mean. This process and the least-squares plane fitting, if applied, are to some extent interdependent and may need to be applied more than once as a result.

Next we examine how a particular image area, not necessarily aligned with the original sampling grid, is conveniently extracted from its less interesting surroundings. The input routine will itself offer a measure of area selection in that the user should be able to specify the size and position of the required area within the already defined sampling grid, and it will allow undersampling of the data, if the grid proves to have been unnecessarily fine, by integral factors, averaging the source data values in appropriate square groups. More general adjustments of the sampling grid to effect rescaling by nonintegral factors or rotation of the image require interpolation techniques since the required sampling sites will not coincide with the available grid points.

The most satisfactory method of interpolation from a mathematical point of view is probably the use of (2.2.9) which was based on the assumption that the image had a transform of finite extent and thus provides an optimally smooth interpolation. The drawback of this "sinc" interpolation is that it involves a sum over a large number of the source data points for every output point obtained. (Note however that the sine occurring in each term of such a sum is the same, so that it is at least unnecessary to calculate a trigonometrical function with each term.) Consequently, the method is justified only occasionally, for example, in estimating peak values in the transforms of periodic images (see Section 9.6).

Bilinear interpolation is less satisfactory mathematically, but rapid computationally, and tolerable for most practical purposes when the image is sampled somewhat more rapidly than the width of its transform suggests is strictly necessary (as is usually the case, allowance being made for the distortions of digital transform processes; see Chapter 2 and the next section). Here the interpolated value depends only on the source data at the four corners of the sampling grid square in which the required point lies. Fig. 9.2.1 illustrates the procedure. Linear interpolation is applied in one dimension between source data f_{00} and f_{10} to provide a value at A, a value at B being similarly obtained from f_{01} and f_{11}, after which the value g is obtained by linear interpolation between A and B. The same value is obtained if the vertical interpolation *precedes* the horizontal; the result is neatly expressible as a sum of the four source data each weighted by the product or horizontal and vertical distances to the *opposite* corner of the square, the distances being expressed as fractions of the sampling interval, i.e.,

$$g = (1-x)(1-y)f_{00} + x(1-y)f_{10} + (1-x)yf_{01} + xyf_{11}. \qquad (9.2.1)$$

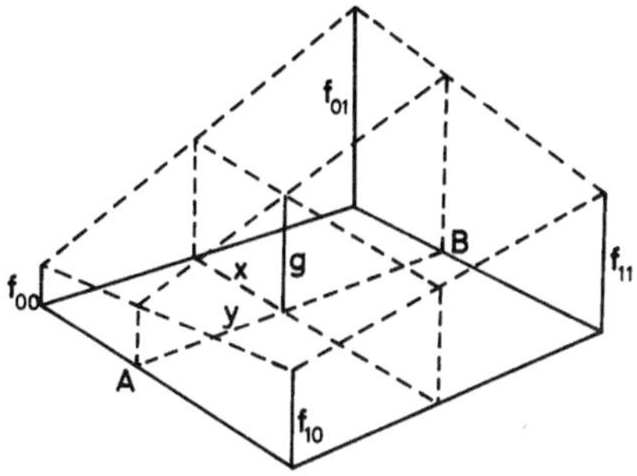

Figure 9.2.1 Bilinear interpolation of a value g between four known values f_{00}, f_{10}, f_{01}, and f_{11}.

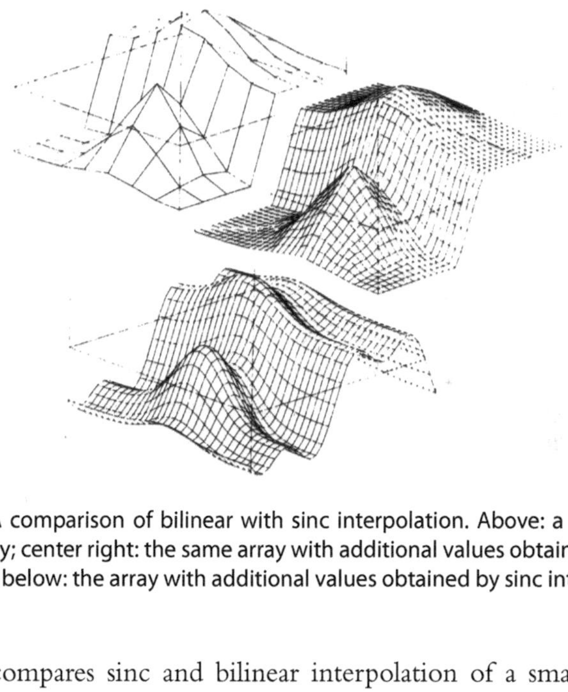

Figure 9.2.2 A comparison of bilinear with sinc interpolation. Above: a coarsely sampled data array; center right: the same array with additional values obtained by bilinear interpolation; below: the array with additional values obtained by sinc interpolation.

Fig. 9.2.2 compares sinc and bilinear interpolation of a small array and makes clear the different characteristics of the two approaches.

Rotation of image regions is particularly expensive with small computer systems, and efforts should be made to achieve a suitable orientation at scanning time as far as possible. In the worst case when only two rows of a picture will fit in core at once, every output row obtained involves reading several source rows in succession, and rotating a picture by a fifth of a radian in these circumstances can easily lake longer than a Fourier transformation. The process also becomes more difficult to program in these circumstances since one no longer has the four source data points available at once. It is the source rows rather than the output points that must be taken in turn, and each of these may contribute to several output points.

In two special cases, there is some advantage in effecting interpolation by manipulating the Fourier transform of the image rather than the image itself. One of these is the case of shifting an image by a fraction of a sampling interval in both directions, in which case an effective sinc interpolation can be achieved by applying a suitable linear phase factor to the transform [see (2.1.13)]. The other is the case of rescaling a picture to use a larger number of sample points but preserving the field of view exactly; here an effective sinc interpolation can be achieved by surrounding the transform with a bed of zeros, thus increasing the aliasing period with a consequent decrease in the image sampling interval (see Section 2.3).

A final stage in the data selection process may be the suppression of background detail outside a frame enclosing the region of interest, or simply beyond a certain distance from the origin. This process, straightforward in principle, is discussed in the relevant contexts in subsequent sections.

9.3. Transformation

The basic mathematics of Fourier transformation with discrete data have already been presented in Chapter 2. This section outlines the "fast" algorithm that has made digital transformation and filtering a practical proposition, the calculation of power spectra and autocorrelation functions, Hilbert transformation, and the "trade off" between direct and Fourier methods of convolution.

We begin by considering the one-dimensional discrete transform of a sequence of N (in general complex) values,

$$\tilde{x}_j = x_k W_{jk}. \tag{9.3.1}$$

Direct evaluation of this involves computing N sums, each comprising N terms which in turn necessitates a complex multiplication and a complex addition for their accumulation. $8N^2$ real operations are thus necessary, treating addition and multiplication as equivalent. Moreover, $\tilde{x}_j$ must be stored separately from x_k since the latter cannot be overwritten until the last sum has been evaluated.

The "fast" algorithm (Cooley & Tukey, 1965) exploits the periodic property of W_{jk}, and achieves a substantial reduction in the number of operations to be performed, besides permitting the transform to be carried out in situ, i.e., entirely within the storage occupied by x_k. We shall require that N be a power of 2 in the first instance, and will express subscripts as a series of binary digits, so that

$$k = 2^m k_m + 2^{m-1} k_{m-1} + \ldots 2k_1 + k_0$$

and

$$x_k = x_{k_m k_{m-1} \cdots k_1 k_0}.$$

With this notation, the transform becomes

$$\tilde{x}_{j_m \cdots j_0} = \sum_{k_m} \cdots \sum_{k_0} x_{k_m \cdots k_0} W_{jk}, \tag{9.3.2}$$

each sum extending over the two values 0 and 1. We reorder the sequence x_k into "bit-reversed" order, defined by

$$x^b_{k_0 \cdots k_m} = x_{k_m \ k_0}, \tag{9.3.3}$$

and note that W_{jk} is periodic in jk with period N, so that we may replace jk in (9.3.2) by

$$j_0 k_m 2^m + (2j_1 + j_0) k_{m-1} 2^{m-1} + \ldots j k_0.$$

With this decomposition, (9.3.2) may be factorized into the form

$$\tilde{x}_{j_m \cdots j_0} = \sum_{k_0} W_{jk_0} \left\{ \sum_{k_1} \cdots \left\{ \sum_{k_m} W_{j_0 k_m 2^m} x^b_{k_0 \cdots k_m} \right\} \cdots \right\}. \tag{9.3.4}$$

We can break the evaluation of this up into $\log_2 N$ stages as follows:

$$x^{(0)}_{k_0 \cdots k_{m-1} j_0} = \sum_{k_m} W_{j_0 k_m 2^m} x^b_{k_0 \cdots k_m}$$

$$x^{(1)}_{k_0 \cdots k_{m-2} j_1 j_0} = \sum_{k_{m-1}} W_{(j_0 + 2j_1) k_{m-1} 2^{m-1}} x^{(0)}_{k_0 \cdots j_0}$$

$$\tilde{x}_{j_m \cdots j_0} = \sum_{k_0} W_{jk_0} x^{(m-1)}_{k_0 j_{m-1} \cdots j_0}.$$

At each stage, dependence on some k_i is replaced by dependence on some j_i; the sums for the two values of j_i are stored in the locations addressed by the two k_i being replaced so that the calculation proceeds entirely

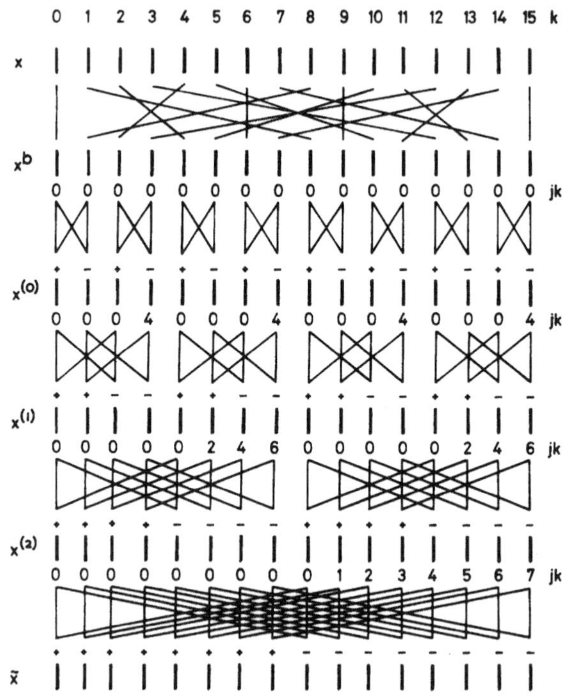

Figure 9.3.1 The fast Fourier transform algorithm with 16 points.

within the storage occupied by x_k. The scheme is conveniently illustrated in Fig. 9.3.1 which shows the sequence of operations in a 16-point transform from top to bottom and incorporates an optimization of the evaluations of the sums at each stage. The short vertical lines represent the data at each stage; after the initial bit reversal, each set of data is followed by *jk* values representing W_{jk} factors multiplied onto half the data points, the resulting data being added and subtracted in pairs to yield the data for the next stage.

In order to count the total number of operations required, we note that each of the $\log_2 N$ stages involves $N/2$ pair combinations, each of which in turn necessitates a (complex) multiplication, addition, and subtraction, i.e., 10 real operations (treating multiplication and addition/subtraction as equivalent) per pair. The total is thus $5N\log_2 N$ operations, which obviously compares very favorably with the $8N^2$ operations required for direct evaluation. Computing time is more nearly proportional to the number of data points than to its square.

These estimates exclude the calculation of W_{jk} itself. It is possible to tabulate values of this initially, subsequent reference to the table being made at negligible cost, regardless of whether "direct" or "fast" evaluation of the transform is being employed. If tabulation is not chosen, however (for reasons of space, or because it puts an ultimate limit on N), W_{jk} will be evaluated recursively for each sum in the direct method or at each stage in the fast method. Because there are only $\log_2 N$ stages in the latter, compared with N sums in the former, and only half the data require W_{jk} factors in any case, the fast method clearly involves many fewer operations than direct evaluation in this respect too.

Numerous variants on the basic algorithm now established are possible. Bases other than 2 may be used for the factorization. With base x, the number of operations required per data point is proportional to $x \log_x N$, which has a minimum at $x = e$ independently of N, is almost identical for $x = 2$ and $x = 4$, and is slightly lower for $x = 3$. The restriction that N is a power of the base is always irksome but is more so for base 3 than for base 2. Base 4 offers some advantages in array access, particularly for two-dimensional transforms (as discussed later) and in trigonometrical recursion, and since an extra base 2 iteration can convert an odd power of 2 transform into an even, allowing N to remain a power of 2, base 4 is a common practical choice. Ingenious combinations of stages with different bases can permit the restriction on N to be relaxed still further but at increasing computational cost.

An important adaptation of the basic algorithm allows the calculation of the transform of a $2N$ point *real* sequence in one iteration following a transform on N complex points, the total time required being roughly halved. If

$$y_k = x_{2k} + ix_{2k+1} \quad (0 \leqslant k < N),$$

taking x_k real, we have

$$\tilde{y}_j = (x_{2k} + ix_{2k+1})W_{jk}, \qquad \tilde{y}_{-j} = (x_{2k} + ix_{2k+1})W^*_{jk},$$

from which the $2N$ point transform may be obtained as

$$\tilde{x}_j = x_{k\,2N}W_{jk} = (\tilde{y}_j + \tilde{y}^*_{-j})/2 + (\tilde{y}_j - \tilde{y}^*_{-j})_{2N}W_{j1}/2i.$$

The process thus involves an extraction of conjugate symmetric and antisymmetric parts before a final base 2 iteration. Normally, $\tilde{x}_j$ will be evaluated only for $0 \leqslant j \leqslant N/2$, since the remaining values will follow from the conjugate symmetry (2.1.9). The number of storage locations required rises by one complex value above that occupied by the original real sequence x_k, though the number of independent values remains at $2N$, since x_0 and $x_{N/2}$ are both real. Inverse transformation of $\tilde{x}_j$ may be effected by a similar base 2 iteration preceding a transform on N complex values. Details of this, and other related variants on the basic transform algorithm, will be found in a collection of papers edited by Rabiner and Rader (1972).

We turn next to the calculation of two-dimensional transforms. These can be broken up into a product of independent one-dimensional transforms (2.3.1). We must transform the rows and the columns of the picture in turn in either order. If sufficient storage is available to permit the entire picture to be held in core at once, there are no special problems involved. If on the other hand storage is strictly limited, then a convenient decomposition of the data is to store picture rows independently on a mass storage device, which in the first instance we shall assume to be of the direct access type (a disc) allowing picture rows to be read or written in any order. The minimum core requirement is sufficient space to hold two rows at once. Fig. 9.3.1 may now be taken to refer to rows instead of individual elements, and the column transforms are performed by reading in rows in pairs, applying the relevant operations to corresponding elements throughout the entire row (thus virtually eliminating W_{jk} recursion time), and then writing out the rows again to the mass storage device. The row transforms are performed in core the first time each row is read in or just before its final writing. For N rows, the total number of read/write operations involved is $2N \log_2 N$, which commonly represents a time larger than that of the computation itself; the number can be roughly halved if the transform is factorized with the base 4 and sufficient core is available for 4 picture rows to be held simultaneously; still larger amounts of core storage can also be exploited usefully.

If a disc is not available, the transformation can be carried out on data held on magnetic tape, but the time required is usually considerably greater. If the tape is in a "preblocked" format permitting random access, the same scheme as the preceding can be adopted, except that every access will take a much longer time. If serial writing only is possible, a different technique must be employed, the rows being transformed before and after a transposition of the entire picture, and the resulting transform being transposed again to restore the normal orientation of the axes. Unless one is prepared to space the tape to and fro between input and output areas every time a row is transformed, a second tape deck is essential, but even this does not obviate the real difficulty presented by the transposition operation. If only two rows will fit in core simultaneously, then a complete pass through the input data is necessary to construct a single output row so the operation is extremely slow, though it will speed up in proportion to any additional core available. It is probably sensible to omit the second transposition accordingly, accepting the penalty of interchanged axes in the Fourier plane.

The inconvenience of having the origin at the $(0, 0)$ element rather than at the center of the picture (see Section 2.2) is felt in two dimensions still more strongly than in one. The problem can be cured in several ways. A circular shift in both directions by half the corresponding dimension, an operation sometimes called "scrambling," can be applied before and after the transform, for example. The prescrambling can be replaced by a sign reversal of the transform at alternate points in both directions, which might be called "postgridding" [a special case of (2.1.13)], so that if desired both operations can be effected after the transform. Bit reversal may also be postponed until the end of the transform iteration at some cost in W_{jk} recursion and may then be omitted completely in some circumstances, since point-by-point operations on the transform can be carried out regardless of its ordering, particularly the multiplication involved in calculating correlation functions. Scrambling and gridding can also be carried through bit reversal, the former becoming an exchange of diagonal nearest neighbors and the latter a sign reversal in alternate picture quadrants.

Chapter 2 dealt in some detail with the mathematical implications of the use of discrete data in terms of aliasing and consequent overlap in both planes. In the transform plane, the effect is greatest along the two axes; this rather diffuse artifact is unavoidable (see Fig. 2.3.1) but not usually troublesome. The aliasing also leads to anisotropic resolution in both planes, the resolution being poorer along the axes than diagonally. More serious distortion (not artificial) may arise when the data to be transformed is surrounded by a field of zeros to make up an array whose dimensions are convenient for fast transformation. If the mean level of the data within the original boundary is large compared with its variation, the computer transform will be completely dominated by the transform of the boundary shape itself, which is not usually what is of interest. It is necessary in these circumstances to remove the step at the boundary by setting the data outside it to the mean value of the data on it or within it, or to subtract this value from the data within it. The process has been called "floating" the data (De Rosier & Moore, 1970). The problem disappears when the boundary fills the array completely; changing the mean level then affects only the central point of the transform (2.3.4); the boundary shape transform is a sinc sampled at its zeros. A related distortion in the transform sometimes arises as a result of the use of finite square data arrays; the selection of a small square image area for transformation amounts to multiplication of the full image by a square mask and corresponds to convolution of the full image transform with a crossed sinc pattern (the transform of the mask) which can be much more effective than the transform plane aliasing in drawing the transform out along the two axes. The computed transforms even of images with completely isotropic spatial statistics, such as micrographs of thin amorphous carbon films, can thus show pronounced lines along the two axes. When the effect is troublesome, it is necessary to float the original picture outside a circular region before transformation, the process being subsequently referred to as "circular masking."

Closely related to the Fourier transform are the power spectrum (the squared modulus of the transform) and its image plane counterpart, the autocorrelation function (the Patterson function of crystallography). The former provides a map of the predominant spatial frequencies in the image, revealing periodicities that may not be obvious, and sometimes permitting identification of the microscope transfer function in the form of an overall modulation (see Section 9.7). The latter may also reveal periodicities since peaks will occur whenever the displacement brings a periodic structure into register with itself and might also be exploited in assessing the imaging conditions (Frank, 1975a). Both functions are useful for image alignment where there is an initial difference in orientation and/or magnification (see Section 9.5).

Calculation of a power spectrum (PS) obviously requires no more precautions than a possible floating of the data before transformation. The autocorrelation function (ACF) is most efficiently computed from this by inverse transformation (2.1.16), but the shifting in the data is performed *circularly*. This will not matter much for aperiodic images in which data at opposite edges of the picture are uncorrelated but may be troublesome in other cases. It can be largely eliminated by floating the image data beyond a third of the picture dimension from the center, as circular overlap will then contribute nothing to the ACF until a relative displacement by the same distance. A minor economy is possible in calculating an ACF through the omission of the bit reversal in the transforms, as mentioned previously. Normalization is conveniently effected by zeroing the central point of the PS and dividing the ACF by its central value (cf. the next section).

The effect of image noise on correlation functions is investigated extensively in the next section, but we should note here that for the important case of additive uncorrelated noise, the image ACF consists of a superposition of signal and noise ACFs. The latter can be largely eliminated from the ACF by calculating instead the cross-correlation function between two images recorded consecutively and thus having identical signal but uncorrelated noise components.

Hilbert transformation has been discussed extensively in Chapter 3. The one-dimensional convolution implied is easily implemented as a Fourier plane filter, the data to either side of the "common line" being multiplied by i and $-i$, though the appropriate treatment for the data on the common line itself depends on the particular application. If the common line is oriented along the columns, the transform may be effected as a series of one-dimensional transforms only: the filter being then independent of row number, the transforms on the columns that are part of the two-dimensional transform are clearly irrelevant and can be omitted. For a picture with N_2 rows of N_1 points, the number of operations is then proportional to $N_1 N_2 \log_2 N_1$ instead of $N_1 N_2 \log_2(N_1 N_2)$, a substantial saving in any case, and one made considerably larger on small computers by the elimination of the disc access requirements of the column transforms. Logarithmic Hilbert transforms are effected of course simply by replacing the initial data by their logarithms.

A further note is relevant at this point in connection with the neglect of alternate points in the convolution when Hilbert transforms are evaluated in this way by a simple sign filter [see Section 3.5, especially (3.5.15)]. The use of a modified filter

$$\begin{array}{ll} i((2j/N)-1) & (1 \leqslant j \leqslant N-1) \\ 0 & (j=0) \end{array}$$

which has an inverse Fourier transform $\cot(\pi k/N)/N$ (except only for $k=0$, when it is zero) is equivalent to a convolution which uses all source points equally (M. Wheeler and M. A. Fiddy, personal communication), and would thus appear to give a closer description of the Hilbert transform integral than the simple sign filter. It is as easily implemented as the latter and may have advantages in some applications. However it cannot be used to generate a sequence with a one-sided discrete transform; nor, if the sequence is taken to comprise sample values from a band-limited periodic function, is the corresponding Fourier series one sided; nor is the discrete

Hilbert transform described by this filter symmetric in the sense that a double application restores the original data (apart from a sign change).

As the cases of Hilbert transforms and autocorrelation functions illustrate, convolution is much more rapidly effected by multiplication of the discrete transforms than by direct evaluation. The former involves a number of operations roughly proportional to the total number of data points and the latter a number proportional to its square. If the convolvend is sufficiently narrow, however, direct evaluation may become advantageous, at least in one dimension. We illustrate this by considering a convolution with real data, assuming that the convolvend is available in either plane equally readily. The two Fourier transforms involved will amount to $5N \log_2 N$ operations, neglecting the multiplication of the transforms, while direct evaluation will require $2Nn$ operations, if n is the number of nonzero elements in the convolvend, both expressions being the same whether the process is one- or two-dimensional. Direct evaluation will be advantageous if $n < 2.5 \log_2 N$ therefore. Some typical limits on n are accordingly:

N	n	N	n
256	$20(> 4^2)$	256^2	$40(> 6^2)$
64^2	$30(> 5^2)$	1024^2	$50(> 7^2)$.

The limiting size of the convolvend may thus be too small to be useful in two dimensions but is quite reasonable in one. Optimal filtering with a limited number of nonzero convolvend values is discussed by several authors (e.g., Frieden, 1974). We have excluded mention of the number of disc operations necessary in the two methods however. This will be relevant in two dimensions only, and the number is of course core-dependent for either method. If only two rows can be held in core at once, the Fourier method requires about $4\sqrt{N} \log_2 \sqrt{N}$ accesses for square pictures, and the direct method $\sqrt{nN}$; the latter is advantageous if $\sqrt{n} < 4 \log_2 \sqrt{N}$, i.e., if the convolvend is no larger than 24^2, 32^2, and 40^2 points for pictures with 64^2, 256^2 and 1024^2 points, respectively, so the limit on the convolvend size for the direct method to be advantageous is clearly considerably higher than suggested earlier. It becomes higher still as more core is made available. If $\sqrt{n}$ rows can be held simultaneously, direct convolution is possible in a single pass, and while the Fourier method will also improve, it will not normally be able to match this. Further core still, however, eliminates the need to consider disc access, and the limiting value of n drops to its earlier level.

Similar considerations apply to the calculation of a small number n of values of the convolution of two large pictures each with N points, the previous expressions in fact all being exactly the same. Billingsley (1971) describes a system making extensive use of direct convolution, while Frank (1975a) discusses the use of the central region of a directly calculated ACF for focusing purposes.

It should be noted briefly that in estimating real transformation times allowance should be made for the inevitable overheads of the calculation—the time spent by the processor, roughly speaking, in deciding what to do next at each stage and where to find the data. This may be usefully illustrated by the case of a transform program written by M. Horner for a PDP 8/E computer (cf. Fig. 8.3.1) with hardware floating point operations requiring about 25 μsec, and random disc access about 50 msec. The expressions above predict a computation time of 170 sec (including W_{jk} recursion along the rows, but neglecting it for the columns), and a disc access time of 102 sec (the transform being factorized in fours) for a complex data 256^2 point transform, making a total of about 4.5 min, whereas the actual running time is about 8 min.

The Fourier decomposition is a natural one for electron microscopy because of the role played by the Fourier transform in the image formation mechanism (see Section 1.4). However, other inherently discrete decompositions have been proposed in this context (Hawkes 1973a, 1974a) that are already used in other fields (notably electrical signal processing), that provide real transforms of real data, and that are often far more rapid computationally than Fourier transforms. These might be used to advantage in some applications, such as processing localized features within a picture, since some decompositions involve localized functions, unlike Fourier components which cover the entire picture area, and can provide localized convolution through transform products. They can always be interconverted with Fourier transforms through an appropriate (slow) matrix multiplication, though more terms are commonly required for the representation of a continuous image with a given rms accuracy than are needed with Fourier series (Blachman, 1974). These decompositions have yet to be exploited in electron microscopy to any real extent, and are not considered further here.

9.4. Correlation and lateral alignment

This section treats the use of correlation functions in achieving lateral alignment of different images of corresponding object areas with a view to implementing any of the numerous techniques involving more than one image of a given area. It covers weighted correlation, the accuracy of alignment attainable, and the circumstances in which different images will correlate successfully in terms of noise levels and instrumental parameters. The following section pursues the more general alignment problem arising when the relative orientation and magnification of the images is also unknown initially.

There are two main reasons for the widespread use of correlation functions for alignment. The first is the rapidity with which the cross-correlation coefficient between two pictures may be calculated as a function of relative lateral displacement by means of Fourier transforms, as described later, and the second is their remarkably high tolerance to additive noise—we shall see that it is possible to achieve good alignment even when the noise is many times as high as the signal (depending of course on the array sizes that can be handled). The high noise levels common in electron microscopy, coupled with the change in appearance of fine image detail with differing imaging conditions, effectively rule out the use of "markers" common to the two pictures for alignment purposes in this field. Correlation functions are discussed extensively by Frank (1970, 1972, 1973a).

The cross-correlation function (XCF) between two pictures f_{jk} and g_{jk} is defined by

$$\rho_{pq} = \frac{\operatorname{cov}(f_{j-p,k-q}, g_{jk})}{(\operatorname{var}(f_{jk})\operatorname{var}(g_{jk}))^{1/2}}, \tag{9.4.1}$$

the averages being taken spatially over the array, rather than over an ensemble of pictures f_{jk} and g_{jk}, i.e.,

$$\rho_{pq} = \frac{\sum_j \sum_k f_{j-p,k-q} g_{jk} - \bar{f}\bar{g}N}{(F-\bar{f}^2 N)^{1/2}(G-\bar{g}^2 N)^{1/2}}, \tag{9.4.2}$$

where

$$\bar{f} = (1/N)\sum_j \sum_k f_{jk}, \quad \text{etc.},$$

$$F = \sum_j \sum_k f_{jk}^2, \quad \text{etc.},$$

and N is the total number of points in the picture. ρ_{pq} is easily seen to be invariant under linear transformations of either picture of the form

$$f'_{jk} = a f_{jk} + b,$$

i.e., it does not depend on the relative contrast or background value of the two pictures. It lies between -1 and 1, as is easily verified by applying the $b_2 < 4ac$ inequality to the positive definite quadratic in λ,

$$\mathrm{var}(f_{j-p,k-q} - \lambda g_{jk}). \tag{9.4.3}$$

When $\rho_{pq} = \pm 1$, the variance vanishes, implying that

$$f_{j-p,k-q} - \lambda g_{jk} = \text{const},$$

λ having the same sign as ρ_{pq}, so that the pictures differ *only* in contrast and background value. Summed over p and q, (9.4.2) will be seen to vanish, so that ρ_{pq} is evenly disposed about zero. Interchanging f_{jk} and g_{jk} simply reflects ρ_{pq} in the origin.

If $g_{jk} = f_{jk}$, (9.4.2) defines the normalized autocorrelation function (ACF) a_{pq} of f_{jk}, the familiar measure of its spatial statistics. The half-width of its central peak may be taken as a practical measure of the resolution of the picture (Frank, 1975b). If g_{jk} is a laterally shifted replica of f_{jk}, then ρ_{pq} clearly provides the ACF shifted laterally by the same amount; i.e., if

$$g_{jk} = f_{j-m,k-n}$$

then

$$\rho_{pq} = a_{p-m,q-n}.$$

The location of the correlation peak (m, n) thus identifies the relative displacement between the pictures, and it will do so with an accuracy at least as good as the resolution of the pictures. (If noise levels are not excessive, the symmetric character of the autocorrelation peak allows alignment with a considerably higher accuracy still.)

This then is the basis of lateral alignment through the calculation of XCFs (Fig. 9.4.1). Loosely speaking, the appropriate alignment is identified by the high correlation between correctly registered pictures, and if we understand the distance at which different image elements become independent (i.e., cease to be correlated) as defining the useful resolution of the image, the accuracy is clearly at least as good as the image resolution.

In the form (9.4.2), the correlation function depends equally strongly on all the source picture values. A more general form can be derived, however, in which the relative weights of different picture areas can be varied, and this will be used in the next section when the alignment problem is generalized to include the orientation and magnification problems. We suppose

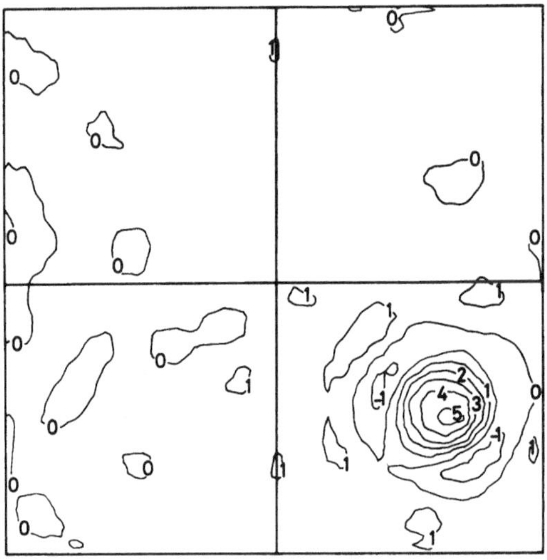

Figure 9.4.1 The central region (4.1 nm square) of the XCF between two micrographs of an amorphous germanium film at generalized defocus values 1.00 and 1.58. The displacement of the peak from the origin identifies the relative misalignment of the micrographs to within less than 0.1 nm, while the departure from circular symmetry reflects the residual astigmatism of 0.10.

a weighting function w_{jk} to be given, real and positive, defining the relative weight to be given to different picture points in the cross-correlation function. From the inequality

$$\sum_j \sum_k w_{jk}(f_{j-p,k-q} - \lambda g_{jk})^2 \geqslant 0$$

[cf. (9.4.3)], we find that the weighted XCF,

$$\rho_{pq} = \frac{\sum_j \sum_k w_{jk} f_{j-p,k-q} g_{jk} - \bar{f}_{pq} \bar{g} W}{(F_{pq} - \bar{f}_{pq}^2 W)^{1/2} (G - \bar{g}^2 W)^{1/2}}, \tag{9.4.4}$$

with

$$\bar{f}_{pq} = (1/W) \sum_j \sum_k w_{jk} f_{j-p,k-q}, \quad \text{etc.,}$$

$$F_{pq} = \sum_j \sum_k w_{jk} f_{j-p,k-q}^2, \quad \text{etc.,}$$

and

$$W = \sum_j \sum_k w_{jk},$$

has most of the properties of the simpler form (9.4.2). It is invariant under linear transformations of either picture; it lies between −1 and 1, assuming those values only when the two pictures are linearly dependent. However, it does not have a zero sum in general, and so does not lie evenly disposed about zero; moreover the fact that the weighting function w_{jk} is associated with the coordinates of one picture rather than the other (here g_{jk}) means that the cross correlation is no longer exactly symmetrical in the two pictures f_{jk} and g_{jk}. However, neither of these effects interferes seriously with the use of (9.4.4) for image alignment, as will be demonstrated in the next section.

The lengthy step in evaluating (9.4.2) is the cross sum,

$$\sum_j \sum_k f_{j-p,k-q} g_{jk}. \tag{9.4.5}$$

This will be recognized as the inverse transform of the product $\tilde{f}^*_{lm}\tilde{g}_{lm}$ (2.1.15) since f_{jk} and g_{jk} are both real, so that it can be evaluated by means of three discrete Fourier transforms. Since, in general, the function is required for a large range of (p, q) values, this will be much faster than direct evaluation (see the discussion at the end of the previous section). It does however mean that the correlation function will be presented as a function of *circular* displacement of the pictures, and this has two consequences worth noting. For aperiodic pictures, data at opposite edges are uncorrelated, and their overlap in the circular cross correlation will contribute very little; in other cases, however, it may be desirable to "float" the data beyond $\frac{1}{3}$ of the picture dimension from the center, as this will eliminate the circular overlap contribution to (9.4.5) until (p, q) is also more than this distance from the center (cf. the calculation of ACFs in the previous, section). Second, as the relative displacement between f_{jk} and g_{jk} grows, the cross-correlation peak moves outwards, until, as the displacement exceeds half the picture dimension, the peak jumps to the opposite edge as it overflows itself circularly. Consequently, the position of the cross-correlation peak may be aliased and misleading unless the pictures are known a priori to be within half the picture dimensions of each other.

The sums $\bar{f}$, $\bar{g}$, F, and G of (9.4.2) are straightforward to evaluate in principle. In practice, the high background value and low signal level present in many micrographs can lead to serious rounding errors in the denominator of (9.4.2) and in the calculation of the cross sum; these are most conveniently cured by subtracting $\bar{f}$ and $\bar{g}$ from the source data before calculating ρ_{pq} leaving data with a zero mean; and the most convenient way of effecting the subtraction is to zero the central points of $\tilde{f}_{lm}$ and $\tilde{g}_{lm}$ when calculating the cross sum (9.4.5); F and G can then be evaluated by a sum over the Fourier plane according to Parseval's theorem (2.1.16). A further useful economy can be the omission of bit reversal in the calculation of $\tilde{f}_{lm}$ and $\tilde{g}_{lm}$ (cf. the preceding section).

The form (9.4.4) is in general substantially slower to evaluate since the normalization requires sums $\bar{f}_{pq}$ and F_{pq} which are themselves correlation sums and necessitate four further Fourier transformations in addition to the three already required. We can simplify the expression at the price of the background level invariance. The form

$$\rho_{pq} = \frac{\sum_j \sum_k w_{jk} f_{j-p,k-q} g_{jk}}{\sqrt{F_{pq}}\sqrt{G}} \tag{9.4.6}$$

is independent of a constant factor multiplying f_{jk} or g_{jk}, but not of a constant added to them. This degree of invariance is however sufficient for the applications presented subsequently and a saving of one Fourier transform is made.

Next, we investigate the conditions under which the cross-correlation peak will be clearly distinguishable in the function. There are many relevant considerations. Long range background variations in the source pictures can overshadow it; its shape is affected by actual differences in the signal between the two pictures; its height is affected by the source picture noise levels and by the area of overlap of the two picture frames when the images are aligned, while its detectability depends on the level of the fluctuation produced in the XCF by the source picture noise. Postponing to the end of the section discussion of signal differences between the two pictures, we shall quantify these for the time being by the cross-correlation coefficient between the aligned noise-free signals, and discuss the other effects first.

Background variation in the source pictures is a common problem due, for example, to nonuniform object illumination. The variation frequently has a magnitude greater than that of the signal, and in such cases the XCF is dominated by the background ACF, the signal peak appearing as a small local summit on a hillside. Fortunately, this problem is easily cured by the methods discussed in Section 9.2 for removing background variation.

A useful and general expression is easily obtainable for the reduction in cross-correlation peak height due to additive noise in the source pictures if the noise is signal independent. This is a good approximation in bright-field microscopy of thin specimens, even at low exposure levels, where the low signal does not seriously affect the mean number of electrons per resolution element at the image (see Section 1.6 for the statistics of the imaging process) but is not true of dark-field imaging modes. We suppose the pictures f' and g' to be obtained by adding (uncorrelated) zero-mean noise distributions n and m to the signals f and g, so that

$$f' = f + n, \qquad g' = g + m.$$

We assume, for simplicity only, that any background level has already been removed from the pictures. The peak value of the cross-correlation peak will then be

$$\rho = \frac{\langle f'g' \rangle}{\langle f'^2 \rangle^{1/2} \langle g'^2 \rangle^{1/2}}, \tag{9.4.7}$$

the angle brackets denoting spatial averages. Because of the independence of signal and noise, we may expand the averages to obtain

$$\langle f'g' \rangle = \langle fg \rangle + \langle fm \rangle + \langle gn \rangle + \langle nm \rangle = \langle fg \rangle$$

and

$$\langle f'^2 \rangle = \langle f^2 \rangle + 2\langle fn \rangle + \langle n^2 \rangle = \langle f^2 \rangle + \langle n^2 \rangle,$$

with a similar expression for $\langle g'^2 \rangle$. We thus obtain

$$\rho = \frac{\langle fg \rangle}{(\langle f^2 \rangle + \langle n^2 \rangle)^{1/2} (\langle g^2 \rangle + \langle m^2 \rangle)^{1/2}}. \tag{9.4.8}$$

The noise-free signal cross-correlation peak would be

$$\rho_{\text{ideal}} = \frac{\langle fg \rangle}{\langle f^2 \rangle^{1/2} \langle g^2 \rangle^{1/2}} \tag{9.4.9}$$

and (9.4.8) is clearly lower than this because of the larger denominator. If we make the further assumptions that the two pictures have the same signal and noise levels so that

$$\langle f^2 \rangle = \langle g^2 \rangle = \langle s^2 \rangle, \qquad \langle n^2 \rangle = \langle m^2 \rangle = \langle \sigma^2 \rangle,$$

we can simplify (9.4.8) still further to read

$$\rho = \langle fg \rangle / (s^2 + \sigma^2) \tag{9.4.10}$$

so that

$$\rho = \rho_{\text{ideal}} \alpha^2 / (\alpha^2 + 1), \tag{9.4.11}$$

where α is the signal-to-noise ratio s/σ for the source pictures. Finally we note that if $f = g$, we obtain

$$\rho = \alpha^2 / (\alpha^2 + 1) \tag{9.4.12}$$

since the noise-free value would then be unity. The height of the cross-correlation peak between two pictures of a given object area made under identical conditions can thus provide a convenient measure of the signal-to-noise ratio of the data (Frank & Al-Ali, 1975).

A substantial reduction in height also accompanies movement of the peak away from the origin of the computed XCF, due to the reduction in overlapping image areas when alignment occurs. If the peak occurs at (p, q), the numerator of (9.4.8) is reduced by a factor

$$(N - |p|)(N - |q|)/N^2,$$

N being the linear dimension of the picture. The factor represents the ratio of the overlapping to the total area. For small displacements, the fractional reduction in height is thus approximately

$$(|p| + |q|)/N \tag{9.4.13}$$

and a simple correction is thus easily applied to the estimated cross-correlation value if the peak is noncentral.

We spend some time considering the effects on an XCF of additive signal-independent noise in more detail, taking into account the finite array sizes and obtaining expressions for the mean and variance of an arbitrary point in a computed cross-correlation function. The results enable us to predict the array sizes that must be used to effect alignment for a given signal-to-noise ratio, the accuracy with which signal-to-noise ratios, ACFs or XCFs, can be estimated, and the noise levels at which they become unmeasurable.

The mean and variance can be defined in two different ways, namely, as spatial averages over the computed cross-correlation array or as ensemble averages over large numbers of different source picture noise distributions. It is the latter that we shall explicitly calculate, the corresponding results for the former being stated, largely without proof, at the appropriate points. If the cross-correlation sum is left unnormalized, the averages are in fact identical. We shall use the following well-known statistical relations:

$$E(x+y) = E(x) + E(y), \tag{9.4.14}$$

$$E(xy) = E(x)E(y) + \text{cov}(x, y), \tag{9.4.15}$$

$$\text{var}(x+y) = \text{var}(x) + \text{var}(y) + 2\,\text{cov}(x, y), \tag{9.4.16}$$

$$\text{var}(xy) = \text{var}(x)E^2(y) + E^2(x)\,\text{var}(y) + \text{var}(x)\,\text{var}(y) \tag{9.4.17}$$

(the last holding only if x and y are independent).

We assume the source picture noise to be stationary, with zero mean and variance σ^2, to be independent of the signal and uncorrelated from point to point. We seek the mean $E(\rho_{pq})$ and variance $\text{var}(\rho_{pq})$ of the cross-correlation array over an ensemble of picture pairs with signals f_{jk} and g_{jk} and statistically varying noise values n_{jk} and m_{jk}. Again we assume, for simplicity only, that the signals also have zero mean, and treat in the first instance the unnormalized cross-correlation sum,

$$c_{pq} = \sum_j \sum_k (f_{j-p,k-q} + n_{j-p,k-q})(g_{jk} + m_{jk}) \tag{9.4.18}$$

[the numerator of (9.4.2)]. The individual terms of this sum have means

$$f_{j-p,k-q}g_{jk}$$

and variances

$$\sigma^2(f^2_{j-p,k-q} + g^2_{jk} + \sigma^2).$$

Since the terms are clearly independent, we simply add these to obtain the mean and variance of c_{pq}, namely,

$$E(c_{pq}) = \sum_j \sum_k f_{j-p,k-q}g_{jk} = \rho N s^2 \tag{9.4.19}$$

(unsurprisingly), and

$$\text{var}(c_{pq}) = \sigma^2 \sum_j \sum_k (f^2_{jk} + g^2_{jk}) + N\sigma^4 = N\sigma^2(\sigma^2 + 2s^2) = N\sigma^4(1 + 2\alpha^2) \tag{9.4.20}$$

if the total number of points is N and we assume the same signal level s in both pictures. It can be shown by related methods that the same expressions apply to spatial averages in the sense that the difference between c_{pq} and its ensemble mean [(9.4.19), allowing ρ to vary with p and q] is also statistically stationary, with zero mean and a variance as given in (9.4.20), so that (9.4.20), besides measuring the likely error in any given coefficient, also quantifies the point-to-point fluctuations in the cross-correlation array.

Before considering the normalization of c_{pq}, we will use the expressions just obtained to estimate the maximum noise level that can be tolerated if alignment is to be achieved by cross correlation. A central peak will be clearly distinguishable in the cross-correlation array if

$$E^2(c_{00}) \gg \mathrm{var}(c_{00})$$

or, say, if

$$\rho^2 N^2 s^4 \geqslant 16N\sigma^4(1+2\alpha^2) \tag{9.4.21}$$

(ρ being the *signal* cross-correlation coefficient as before). In the small signal limit, this becomes

$$\alpha \gtrsim 2/(\rho^2 N)^{1/4}, \tag{9.4.22}$$

giving an idea of the tremendous capacity of cross correlation to detect common signals buried in noise; (9.4.22) implies that the cross correlation of 256-point square arrays could successfully identify a signal buried in noise with eight times the level. Saxton and Frank (1977) extend this result to the general case of signal-dependent noise.

The resolution dependence of the image noise level (Section 1.5) is reflected in a greater detectability of the cross-correlation peak if the image data is blurred by a local convolution (J. Frank, personal communication) or simply retabulated with a larger sampling interval. Grouping the original data points together in sets of p points square, to give an array of N/p^2 points at each of which the data is an average over p^2 original data points, gives an unchanged signal level but a noise level lower by a factor of p; α therefore rises by a factor of p, but its "target value" rises only by a factor of $\sqrt{p}$ (9.4.22), giving a much higher detectability than before the grouping (at a cost of alignment *accuracy*, of course).

It is an easy step from these results to obtain the *spatial* mean and variance of the normalized cross-correlation coefficient ρ_{pq} since the normalizing factor is the same for all points. The results for an ensemble average are much less easy to obtain since the denominator then varies randomly too, and the square root and quotient at first sight present substantial difficulties. However the denominator proves at least to be narrowly spread about its mean, allowing a series expansion approach. Except for points far from the peak where the signal cross-correlation coefficient vanishes, the variance proves to be sensitive also to the fourth moment of the noise distribution; the dependence is weak however except for highly correlated signals and low noise levels, and the skew coefficient of the noise continues to be irrelevant.

We write the normalizing denominator [cf. (9.4.2)] in the form

$$\sqrt{d} = \sqrt{FG}, \quad \text{with } F = \sum_j \sum_k (f_{jk} + n_{jk})^2 \quad \text{etc.}$$

The mean of F (and G) is easily seen to be

$$N\sigma^2(1 + \alpha^2),$$

but its variance depends on the noise distribution being

$$N\sigma^4(\mu - 1 + 4\alpha^2)$$

for a noise distribution with mean fourth power $\mu\sigma^4$. Since F and G are independent, the mean of d,

$$E(d) = N^2\sigma^4(1 + \alpha^2)^2, \tag{9.4.23}$$

and its variance,

$$\text{var}(d) = 2N^3\sigma^8(1 + \alpha^2)^2(\mu - 1 + 4\alpha^2), \tag{9.4.24}$$

follow easily from (9.4.15) and (9.4.17), a small term having been dropped from (9.4.24). The covariance of c_{pq} and d can also be evaluated, with some effort, as

$$\text{cov}(c_{pq}, d) = 4\rho N^2\sigma^6\alpha^2(1 + \alpha^2) + N\lambda^2\sigma^6, \tag{9.4.25}$$

the noise distribution being taken to have a mean cube $\lambda\sigma^3$ (this term will disappear subsequently) and a further small term having been omitted.

It is clear from (9.4.23) and (9.4.24) that the spread of d is small compared with its mean. We shall expand c_{pq} and d about their means $\bar{c}_{pq}$ and $\bar{d}$,

Table 9.4.1 Cross-correlation standard deviations.

α	ρ			
	0	**0.5**	**0.8**	**1.0**
0.1	1.000	1.000	1.000	1.000
0.5	0.980	0.968	0.950	0.933
1	0.866	0.818	0.742	0.661
2	0.600	0.538	0.424	0.280
5	0.275	0.239	0.170	0.054
10	0.140	0.122	0.085	0.014

writing

$$\rho_{pq} = (\bar{c}_{pq} + \gamma)/(\bar{d} + \delta)^{1/2} \simeq \bar{c}_{pq}/\bar{d} + (\gamma - \bar{c}_{pq}\delta/2\bar{d})/\bar{d}.$$

The mean of this is easily obtained from (9.4.19) and (9.4.23) as

$$E(\rho_{pq}) = \rho\alpha^2/(\alpha^2 + 1), \tag{9.4.26}$$

which confirms that the use of a finite area does give an unbiased estimate of (9.4.11). To evaluate the variance, we use (9.4.16), recalling that

$$\mathrm{var}(\gamma) = \mathrm{var}(c)$$

etc., which, together with (9.4.19), (9.4.20), and (9.4.23)–(9.4.25), leads finally to

$$\mathrm{var}(\rho_{pq}) = \{(1 + 2\alpha^2)(1 + \alpha^2)^2 - \rho^2\alpha^4(2\alpha^2 + (9 - \mu)/2)\}/N(1 + \alpha^2)^4, \tag{9.4.27}$$

the term in (9.4.25) involving λ having become insignificant. Some limiting forms of this expression are

$$\begin{aligned} \mathrm{var}(\rho_{pq}) &= 2(1 - \rho^2)/N\alpha^2 && (\alpha \gg 1) \\ &= (\mu + 1)/2N\alpha^4 && (\alpha \gg 1,\ \rho = 1) \\ &= 1/N && (\alpha \ll 1). \end{aligned}$$

A value of 3 is reasonable for μ, being the exact value for Gaussian noise, and holding for Poisson noise also until the signal averages less than one electron per picture element. Table 9.4.1 gives some values of the standard deviation of ρ_{pq} as predicted by (9.4.27) for $\mu = 3$ and $N = 1$, so that the values quoted should be divided by the linear picture dimension for a particular application.

The corresponding expressions for the spatial averages are very easily obtained. The normalization factor will differ negligibly from $\sqrt{d}$, so we may use (9.4.20) and (9.4.23) to give the spatial variance of the difference between the noisy XCF and the now scaled "noise-free" value (9.4.26) as

$$(1+2\alpha^2)/N(1+\alpha^2)^2, \tag{9.4.28}$$

which is the value of (9.4.27) for $\rho = 0$, incidentally, and is therefore also available in Table 9.4.1.

All of these expressions have been verified as far as possible by trials with computer generated noise, rectangularly distributed ($\mu = 0.6$), without any clear inconsistency having been found.

We can immediately deduce the accuracy with which a signal-to-noise ratio can be estimated from the cross-correlation coefficient between two pictures with identical signals, from the relation

$$\rho = \alpha^2/(\alpha^2+1),$$

for this leads to

$$\Delta\alpha = (1+\alpha^2)^2\Delta\rho/2\alpha$$

for small changes in α and ρ. Replacing $\Delta\rho$ by the limiting forms of the standard deviation given following (9.4.27) we have

$$\begin{aligned} \Delta\alpha &= \alpha/2\sqrt{N} && (\alpha \gg 1) \\ &= 1/2\alpha\sqrt{N} && (\alpha \ll 1). \end{aligned}$$

The second of these limits allows us to determine the lowest value of the signal-to-noise ratio that can be measured in this way with a given size of picture, and the result is (not surprisingly) much the same as (9.4.22) (with $\rho = 1$). The high α limiting form indicates a constant fractional error in α, but in fact for $\alpha > 3$, when the cross-correlation coefficient exceeds 0.9, the factor actually determining $\Delta\rho$ is often not the finite source picture areas at all but the interpolation errors caused by the cross-correlation peak's failing to fall exactly at a sample point (because of a slight lateral shift between the two picture sampling grids). A possible alternative approach in this range may be to evaluate several coefficients of the cross-correlation array at some distance from the peak where the two signals have zero cross-correlation coefficient, and to compare the standard deviation of these coefficients with the predictions of the first column of Table 9.4.1, which will be noted to be very sensitive to α in this range. The use of finite areas will however mean that the signal contribution to the cross-correlation coefficients will fluctuate slightly about zero from point to point, and it is not yet clear whether the fluctuations are comparable with those, due to the noise, which we are hoping to measure.

The section concludes with a brief discussion of *signal* differences between the two pictures. Except for the purpose of reducing noise levels by superposition (see Section 9.6) it will be micrographs made in different ways that will require alignment, and so it is important to assess how far the different instrumental conditions will destroy the cross correlation between the two pictures.

Within the realms of the weak-phase object linear imaging theory, it is possible to be quite precise about the effect. In these circumstances, the transforms of two images i_1 and i_2 of an object o are

$$\tilde{i}_1 = \tilde{o}\widetilde{T}_1, \qquad \tilde{i}_2 = \tilde{o}\widetilde{T}_2,$$

where $\widetilde{T}_1$ and $\widetilde{T}_2$ are the two transfer functions, and tildes denote Fourier transforms as usual. The XCF between the two pictures is the inverse transform of the product

$$\tilde{i}_1\tilde{i}_2 = |\tilde{o}|^2\widetilde{T}_1\widetilde{T}_2^*,$$

which is of course equivalent to a real space convolution of the object ACF with the cross-correlation function of the two point responses [cf. (2.1.15)]. For ideal imaging, when T_1 and T_2 are both delta functions, the result is simply the object ACF, which is thus the best that can be expected of the cross-correlation peak. With less than perfect imaging, but $T_1 = T_2$, the object ACF peak is blurred by convolution with that of the point response, thus becoming wider and furnishing a somewhat less accurate alignment mechanism. In the general case with different T_1 and T_2, the blurring is by convolution with a function that is quite sensitive to the absolute defocus values of both pictures, and the effect, although simple to calculate for any given case is difficult to describe in useful general terms. A pattern of increasing peak width with relative defocus is to be expected, and negative peaks will indeed be expected at relative defocus values giving rise to contrast reversal between the two images for the dominant range of spatial frequencies. The problem has been investigated experimentally by Al-Ali (1976a, 1976b).

For objects with isotropic statistics, an important point that *can* be deduced generally relates to the symmetry character of the XCF peak. If T_1 and T_2 are centrosymmetric, as is the case for axial illumination modes, the XCF peak will also be centrosymmetric. This can considerably improve the reliability with which its center can be determined in the presence of noise. A further improvement occurs if the instrumental astigmatism is well corrected, in which case T_1, T_2, and therefore the XCF peak, are actually circularly symmetric.

For the simple cases previously described, the deleterious effects of the transfer functions can of course be largely eliminated by preprocessing each image so as to correct for the transfer functions (see Section 9.7). The effective transfer functions will then suffer only from small annular gaps. For many interesting cases, the object will in fact be complex, the two components being transferred by independent filters into the image intensity (see Section 1.3). In these circumstances, the XCF peak will contain superposed peaks from the two object components, together with cross terms between the two, and the behavior is more complicated still, although one component may well dominate the other (the phase component at high resolutions, for example), in which case the preceding arguments will still apply. In this way, alignment is possible for micrograph pairs made at (not too) different defocus values, or with (not too) different illumination tilt angles, and even for complementary half-plane apertured micrographs (in this last connection, see Downing & Siegel (1973) for a different approach).

No convenient formulations are available for the cases of strongly scattering objects or for dark-field images, where the imaging is nonlinear, nor has any systematic study been made experimentally. It seems likely that a substantial degree of correlation will continue to exist between pictures made in these circumstances, however, and it may well be that alignment accuracy is not seriously degraded since the squared terms in the image intensity will be less likely than the first order (interference) terms to cancel each other at different defocus values.

A final unrelated example is furnished by the case of establishing a common origin throughout a tilt series of projections with a view to three-dimensional reconstruction. Hunsmann et al. (1972) have shown experimentally that such projections show significant cross correlation for small angular separations, and the effect has been treated for thin objects consisting of isolated point atoms by Hoppe (1974). Recent work (Frank and Saxton, in preparation) indicates that for extended, statistically stationary objects, the XCF peak between two projections a given angle apart consists of the projected ACF peak of the object, blurred by convolution with a function dependent on the outline shape of the object and the angular separation of the projections, and that the accuracy attainable should be comparable with the resolution possible in a reconstruction from views with the given angular separation (Crowther et al., 1970).

9.5. Alignment in orientation and magnification

The previous section having discussed extensively the alignment of pictures initially displaced with respect to each other, this section treats the more general problem that arises when in addition to a lateral displacement, the pictures differ in orientation and magnification. Orientational differences may arise even between micrographs made in rapid succession unless extreme care is taken in mounting them on the microdensitometer, and both differences will be common between those made at different times and places. A digital procedure for establishing the parameters is therefore useful, provided that it shares the high noise tolerance of correlation methods for lateral alignment. Initially, the problem posed by orientational difference alone is considered, as this is computationally much more rapid than the combined problem.

Orientational alignment could be effected easily by computing the cross-correlation coefficient between the two pictures as a function of angle, were the lateral alignment already established; the two problems are, however, interdependent. The elegant (in theory) method of performing *three*-dimensional cross correlation between three-dimensional pictures in which each layer is obtained by rotating its predecessor through a small angle is ruled out by the necessary space and time for the computation. Three viable approaches, based on correlation functions, are now presented by means of which the orientational problem can be solved efficiently and quickly. The last, which is now in regular use, is capable of extension to the magnification problem.

The first method is the most direct—and the most laborious. An initial estimate is required of the relative orientation. With one of the pictures rotated suitably, the two are cross-correlated, and the correlation peak found; the procedure is repeated at two different angles not too far removed from the first, the height of the correlation peak, wherever it is, being noted at each stage; treating this height as a function of the orientation, parabolic

Table 9.5.1 Angular alignment by the first method.

Estimate of angle	Correlation peak	Displacement of peak
−0.4	−0.286	−4, 3
−0.2	−0.255	−14, −14
−0.6	0.522	−7, 8
−1.0	−0.201	1, 7
−0.634	0.478	−7, 8
−0.579	0.543	−7, 7
−0.537	0.563	−7, 7
−0.535	0.553	−7, 7
−0.545	0.564	−7, 7

Angle found: -0.545 ± 0.03 rad

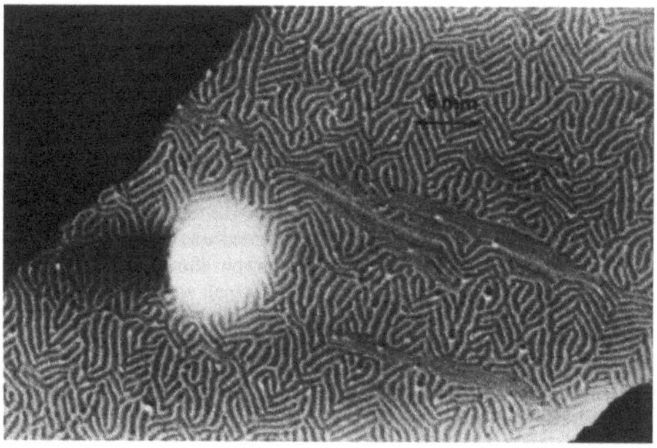

Figure 9.5.1 Magnetic stripe domains in permalloy (by courtesy of J. N. Chapman).

interpolation is used to choose an orientation at which the correlation peak should be higher—and so on, until the correlation peak height is maximized with respect to orientation. The final position of the correlation peak fixes the lateral shift between the pictures. The process requires two Fourier transformations at each stage (to compute the cross-correlation function; one picture is held throughout as its Fourier transform).

The method is illustrated in Table 9.5.1, which lists the successive estimates of the angle and position of one micrograph with respect to another as the alignment is found. The pictures being aligned are both taken from the permalloy magnetic stripe domain micrograph shown in Fig. 9.5.1 (for which I am indebted to Dr. J. N. Chapman), which was digitized twice on a 100 μm grid in different orientations. The only difference between the pictures is therefore due to the different scanning raster, the microden-

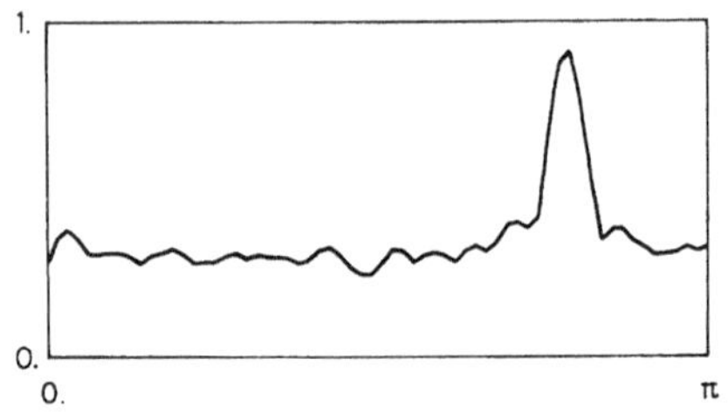

Figure 9.5.2 The cross-correlation coefficient between two micrographs as a function of relative orientation.

sitometer noise, any dust particles settling on the plate between the scans, and, of course, the misalignment. The pictures were both 64-points square in size; the need to rotate one, coupled with the restriction of the Fourier transform routine to pictures whose dimensions are powers of two, meant that the correlated regions were half this size (linearly), requiring initial alignment of the pictures laterally to within a quarter of the picture dimension; the process required 1.01 sec of computer time on an IBM 370/165. The accuracy that might be required in the initial estimate of the orientation is suggested by Fig. 9.5.2, which shows the correlation peak height as a function of angle for two similar, but laterally aligned, micrographs. It is not encouraging but represents an accuracy that might be achieved by eye without difficulty in some cases. The accuracy achieved by the routine is estimated in the table by calculating the displacement from the correlation peak necessary to cause a 5% drop in the height of a quadratic fitted to the peak. A similar figure is quoted for other cases in this section.

The second procedure is based on the suggestion made by Langer et al. (1970) that the problem of the lateral alignment might be eliminated initially by examining the autocorrelation functions (ACFs) of the pictures instead of the pictures themselves. The idea may be viewed statistically, anisotropy of the picture statistics (as described by the ACF) being exploited to define a direction independent of lateral translation of a picture. However the twin assumptions of stationarity and anisotropy implied here are unnecessary, and the technique is considerably more powerful than might seem likely, being capable of aligning portions of pictures with nonstationary, isotropic statistics too. In view of the finite frame size, the computed circular ACF even of a picture with isotropic statistics will show some angular dependence, and provided a second picture overlaps the first to a reasonable extent, its computed ACF should show sufficient similarity to that of the first picture to permit cross correlation for determining the rel-

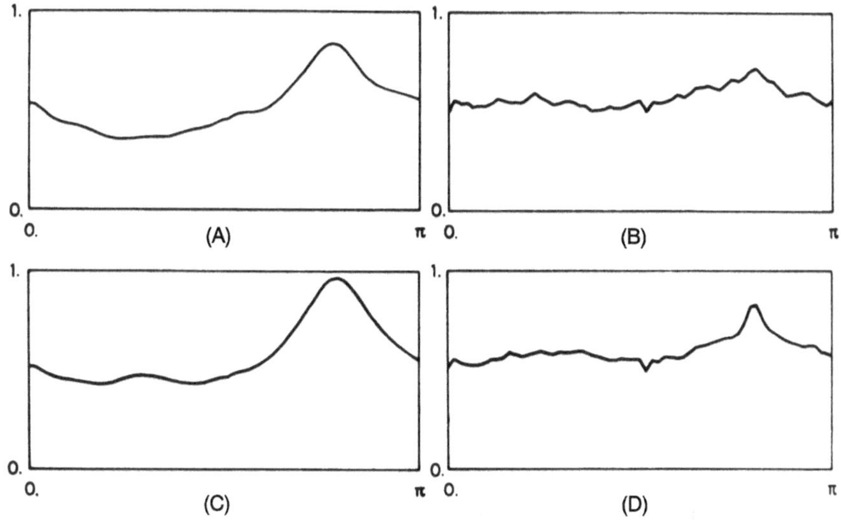

Figure 9.5.3 Angular correlation between ACFs and PSs. (A) ACF unmasked, (B) PS unmasked, (C) ACF masked, (d) PS masked.

ative angle. The various properties of an ACF discussed here (in particular its invariance under lateral translation of the original picture) are shared by the power spectrum (PS), which has the advantage of being more rapidly calculated by means of a single Fourier transformation instead of two; the method can therefore be alternatively applied to the PSs of two pictures instead of to their ACFs. The factors governing the choice are discussed subsequently. In both cases, the centrosymmetry of the functional introduces an ambiguity of π in the solution for the orientation.

The similarity of computed ACFs and PSs for misaligned pictures is demonstrated in Fig. 9.5.3: (A) shows the cross-correlation coefficient between 32-point square regions taken from the centers of the ACFs of the two micrographs used to generate Fig. 9.5.2 as a function of the angle at which one of them was extracted; (B) shows the same function for the PSs (in fact their logarithms, in view of the huge dynamic range involved); (C) and (D) show the corresponding functions when the original pictures were masked with circles prior to transformation (see Section 9.3) and show slightly better peaks in both cases.

The next question is how best to determine the relative rotation between the two ACFs or PSs. One possibility is again to use a maximization routine to find the maximum of the curves of Fig. 9.5.3. This requires at each stage only an interpolation on to a rotated grid in one picture, and

Table 9.5.2 Angular alignment by the second method.

	Estimate of angle	Correlation peak
ACF:	−0.4	0.755
	−0.2	0.515
	−0.6	0.912
	−0.576	0.926
	−0.558	0.929
	−0.556	0.929
	Angle found: -0.556 ± 0.08 rad	
PS:	−0.4	0.638
	−0.2	0.568
	−0.6	0.741
	−1.0	0.553
	−0.641	0.703
	−0.547	0.738
	−0.575	0.745
	Angle found: -0.575 ± 0.06 rad	

the few summations necessary to find the cross-correlation coefficient. An initial estimate is again required. In Table 9.5.2, the successive estimates of the required rotation provided by such a routine, operating on the same problem as that of Table 9.5.1, are listed for both ACF and PS approaches. The former required 0.94 and the latter 0.85 sec of computer time. The difference is mainly the additional time required to calculate the ACFs compared with the logarithms of the PSs. Detailed examination shows that the maximization search takes a time quite comparable with that involved in computing the ACFs and PSs. In comparing this method with the earlier method, it should also be remembered that the time involved in computing ACFs or PSs alone would allow several cycles of the earlier method, and the latter in fact is sometimes faster than this second method in consequence.

Search procedures compare unfavorably with survey procedures in terms of reliability, however, depending heavily on the initial estimate, and providing a rather limited indication of the margins of error in the results. Moreover, the repeated interpolation onto a rotated sampling lattice that is required by both of the methods described, though rapid enough for pictures held in core, is very time consuming if the data is held in small slices on disc, as it would be in a small computer system (see Section 9.2). There are consequently substantial advantages in the third method of orientational

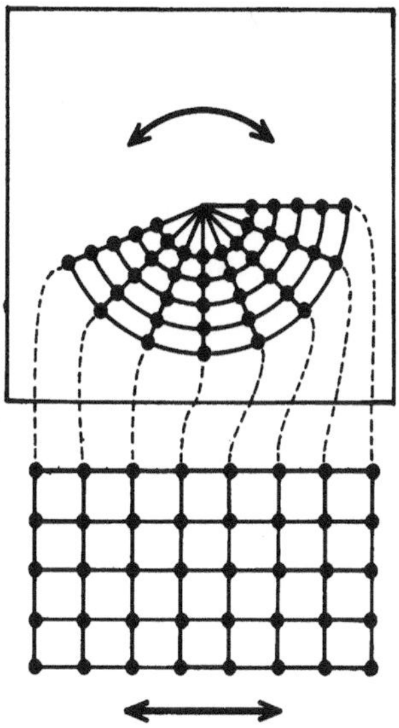

Figure 9.5.4 The radial polar sampling grid mapping rotation into translation; the radial spacings are at the user's disposal.

alignment now outlined, which offers the cross-correlation coefficient between the ACFs or PSs as a function of angle over the full range from 0 to π, with only a single interpolation of the data being necessary.

In this method (Saxton, 1974c), the ACFs or PSs are sampled on a radial polar grid, as shown in Fig. 9.5.4, so that one coordinate of the resulting array measures orientation in the source pictures. Orientation is in effect mapped into a one-dimensional translation. The samples must be evenly spaced in angle, but the radial separation is irrelevant to the alignment, and the radii and number of the sampling arcs may be freely adjusted to suit the particular data as discussed subsequently. It will be noted that the centrosymmetry of ACFs and PSs makes it unnecessary to sample more than one half of the data, so that semicircular rather than circular arcs are used. Also, because the orientational information is duplicated at each radius, only a small number of arcs are usually needed to effect the alignment (typically four) unless the source noise levels are very high, so the interpolation is quite rapid.

Once the interpolated arrays have been obtained, the one-dimensional lateral misalignment between them has to be obtained, this identifying the angle between the source pictures. At one extreme, the data can be summed radially to give a one-dimensional function of angle for each picture which can be cross correlated against each other rapidly; in principle, the orientation can be established by comparing the phases of a single Fourier coefficient of *these* functions (Tanaka, 1972), but in practice the use of such a small amount of information leads to a very unreliable result (P. R. Smith, personal communication). At the other extreme, standard two-dimensional cross correlation can be used, but this is wasteful of computer time as the peak is never displaced radially. What is required is a one-dimensional cross correlation of two-dimensional data with considerable flexibility in the relative weighting of data from different radii—partly because the sample values at higher radii are further apart and so should be weighted at least in proportion to radius, and partly because, although it is the largest radius contributions that offer the finest definition in orientation, the large radius data in ACFs and PSs are commonly small numerically.

The obvious solution is a form of the weighted cross-correlation function (9.4.6); the background level invariance offered by (9.4.4) is not necessary when the source data are ACFs or PSs, as here. We shall require a function of the form

$$\rho_p = \frac{\sum_j \sum_k w_k f_{j-p,k} g_{jk}}{\sqrt{F}\sqrt{G}}, \tag{9.5.1}$$

with

$$F = \sum_j \sum_k w_k f_{j-p,k}^2 \quad \text{and} \quad G = \sum_j \sum_k w_k g_{jk}^2,$$

where f_{jk} and g_{jk} are the interpolated arrays from the ACFs or PSs, the first subscript referring to angle and the second to radius. The weighting function w_k depends on radius only, and while it may be specified at will to suit particular data, may usefully default to the radius itself. Evaluating (9.5.1) involves a one-dimensional cross correlation of f_{jk} and g_{jk} for each value of k, and accumulation of the resulting functions multiplied by the corresponding weight w_k, after which two simple sums are evaluated to give F and G since the former does *not* in fact depend on p, allowing normalization to give ρ_p. Computationally, the process is extremely rapid therefore, even on small computer systems.

An alternative to this method of cross correlation is simply to produce independently normalized one-dimensional cross-correlation functions between corresponding arcs from the two pictures and then take a weighted average of these. This has the advantage of coping automatically with the huge radial dynamic range encountered with some data, particularly PCs, but could become unstable if the arcs extended beyond the regions of significant information in the ACFs or PSs, since the cross correlation of the very small noise values encountered there will contribute to the overall angular correlation function as strongly as any of the lower radius data. Both methods have been used extensively, and while the former is now the normal choice, most of the examples given in this section were calculated with the latter method. There appears to be little difference in practical cases between the two.

It should be noted that the use of the discrete Fourier transform to evaluate the cross-correlation function between corresponding arcs is, for once, exactly the appropriate technique, since it results, in a *circular* correlation, and it is a circular shift of the data for the arc that occurs as a picture rotates. Consequently, the cross-correlation function is as significant at maximum displacement as it is at the center, and there is no difficulty in identifying rotations near $\pi/2$. On the other hand, it is obviously particularly important in this application that the computed ACFs or PSs should be as free as possible from artifacts on the axes, which could bias the orientation determination towards angles of 0 and $\pi/2$. In practice this has proved less troublesome than might have been expected and even without circular masks (see Section 9.3) applied to the original pictures, no difficulty is likely to be experienced.

Fig. 9.5.5 shows the angular correlation function obtained by this third method for two further 64-point square pictures taken from the micrograph of Fig. 9.5.1 (shown above in the figure together with the ACFs and PSs). Curves are shown for both ACF and PS approaches; there is little to choose between them in this particular case. In each case, the contributions from four arcs were averaged, with weighting proportional to radius; the radii were 4, 8, 12 and 16 units; 1.14 sec of computer time was required for the ACF approach, and 0.73 sec for the PS approach.

The reliance on correlation functions confers on the method the high noise tolerance of all correlation methods. Although ACFs are twice as slow to compute as PSs, their use leads to an even higher than normal noise tolerance because the effects on an ACF of source picture noise are largely confined to the neighborhood of the origin, where the noise ACF appears superposed on that of the signal, the fluctuations produced in the rest of the ACF being small (by the result of the preceding section) so that unless sampling arcs with very small radii are used there is likely to be little problem.

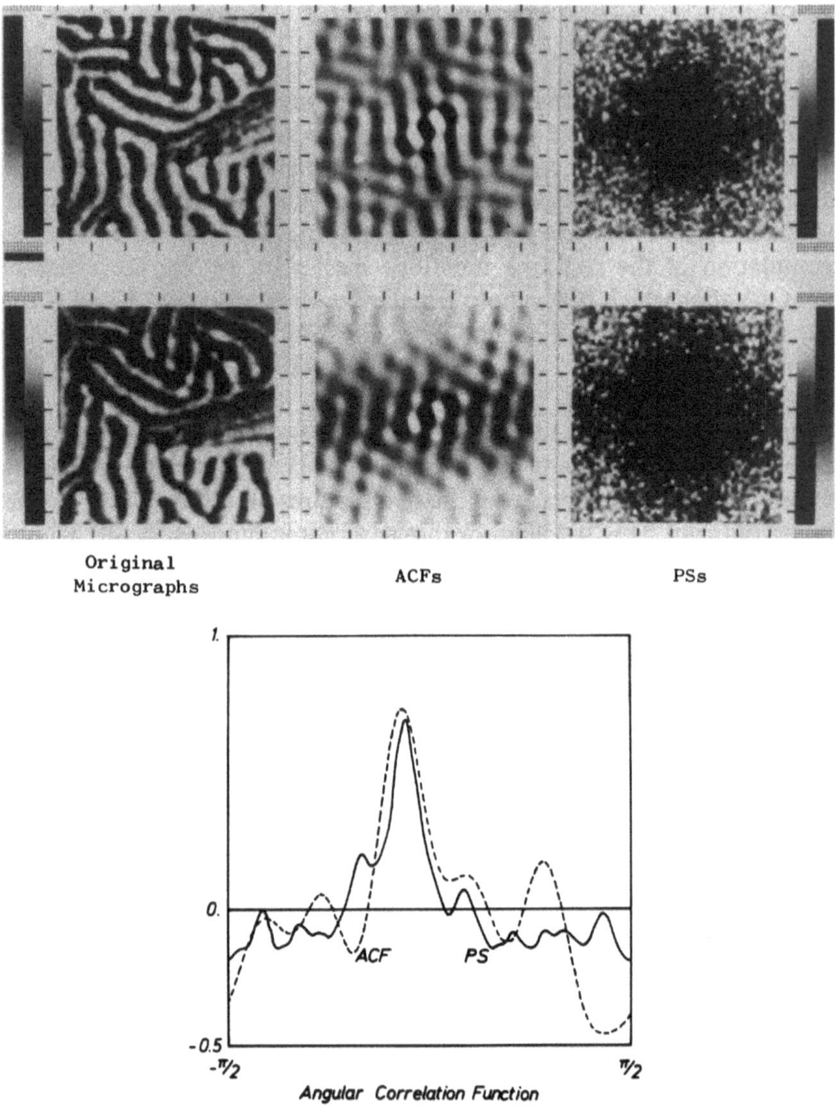

Figure 9.5.5 Orientational alignment with ACFs and PSs (see text for details).

There are other factors affecting the choice between the use of ACFs and PSs, however. The most significant is the level of detail in the original pictures. Whichever approach is chosen, it is the large radius data that provide accurate orientational information, while there are very few sample points available at low radius to define the functional. Consequently, alignment of two pictures with fine detail is best carried out by the PS approach since the PS will extend to a large radius, while the ACF is confined to a few points near the origin. The converse will apply to a picture with predominantly coarse detail. In *principle*, of course, both functionals contain equivalent information, since they are a Fourier transform pair, and indeed the cross-correlation integrals between the ACFs of two pictures should be identical with those between the PSs (Parseval's theorem). However, the sampled nature of the data and the finite range destroy this equivalence for practical purposes.

The dependence of the final angular correlation function of the number of arcs used, the relative weighting of the contributions from the various arcs, and the noise level in the original pictures are illustrated in Fig. 9.5.6, which shows this function calculated in various ways, again from two 64-point square pictures from the micrograph of Fig. 9.5.1, not aligned laterally: (A) uses a single arc with a radius of 12 units; (B) to (D) use 2, 4, and 8 arcs, respectively, with radii evenly spaced between 4 and 16 units, and the arcs being weighted in proportion to their radius; (E) and (F) repeat (B) and (C), but assigning the same weight to each arc; (G) and (H) repeat (C) and (D) but with white noise (uniformly distributed, with a maximum amplitude half that of the micrograph data) superimposed on the pictures. It will be noted that while the peaks are lower in these cases, they are not broader than those in (C) and (D). The accuracy is still as good as the level of detail in the original pictures allows.

Fig. 9.5.7 gives some idea of the variability of the angular correlation function with the level of detail in the original pictures and with the choice of ACF or PS. The pictures here (not aligned laterally) are taken from two optical plates of the radio source 3C465 taken at different wavelengths and scanned on a 25 μm grid. (One plate is reproduced in Fig. 9.5.8.) In Fig. 9.5.7 (A) represents a first trial, using 128-point square pictures taking every data point and using the PS approach—clearly the significant portion of the PS does not extend as far as the arcs used; (B) shows the result of the corresponding ACF version, smooth, but offering poor definition of the peak because the pictures contain predominantly large scale detail. The process was repeated with only 64-point square pictures, taking every fourth data point in each direction; (C) shows the result for the ACF and (D) for the PS.

The dependence of the angular correlation function between two arcs on their radius is shown in Fig. 9.5.9, using the PSs of unmasked pictures, 128-point square, from the micrograph of Fig. 9.5.1, not aligned laterally.

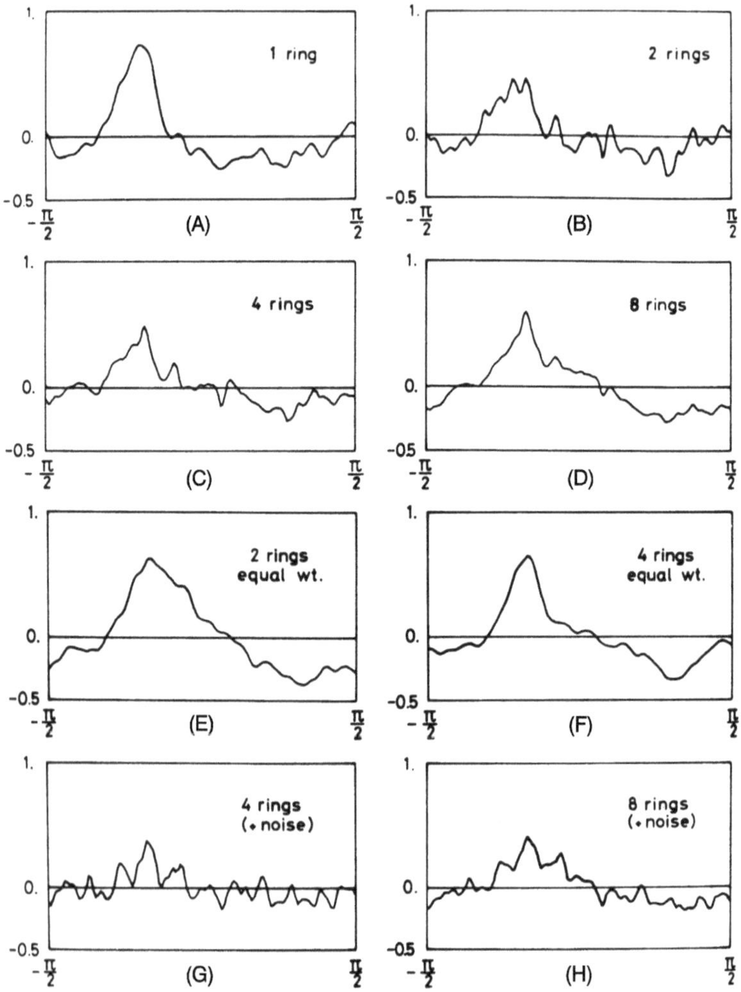

Figure 9.5.6 Eight angular correlation functions for the same data (see text for details). Angles found/rad: (A) -0.614 ± 0.06, (B) -0.568 ± 0.02, (C) -0.572 ± 0.02, (D) -0.568 ± 0.02, (E) -0.539 ± 0.04, (F) -0.566 ± 0.04, (G) -0.553 ± 0.02, (H) -0.558 ± 0.02.

The angular correlation is shown as a function of the two variables angle and radius, as a contour map, the correlation functions at each radius being independently normalized. It is clear that for this set of data at least, there is no merit in using arcs with radii greater than $\frac{3}{4}$ of the maximum possible; indeed, at the maximum radius, the small artificial correlation at 0 and $\pm\pi/2$ is obvious. At the same time, it is also clear that the higher radius arcs offer the best definition of angle.

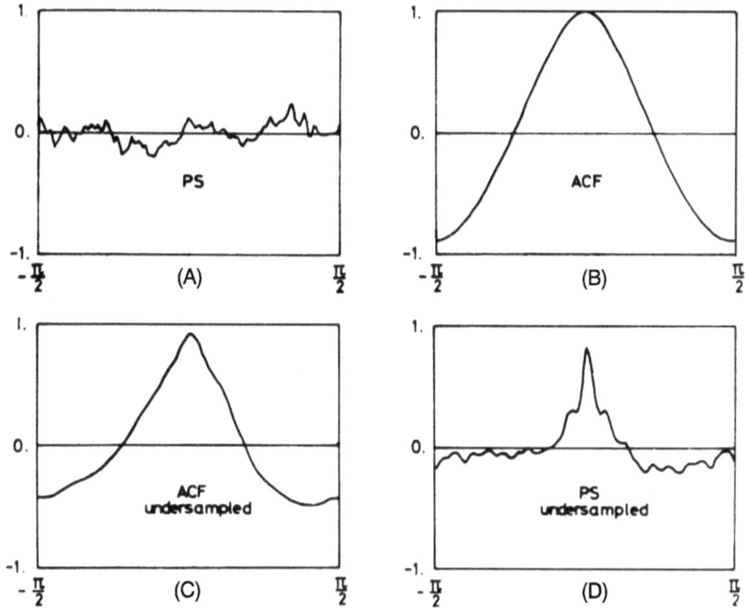

Figure 9.5.7 The dependence of the correlation on the level of detail in the pictures. Angles found/rad: (A) 1.051 ± 0.01, (B) -0.021 ± 0.13, (C) 0.015 ± 0.06, (D) 0.008 ± 0.02.

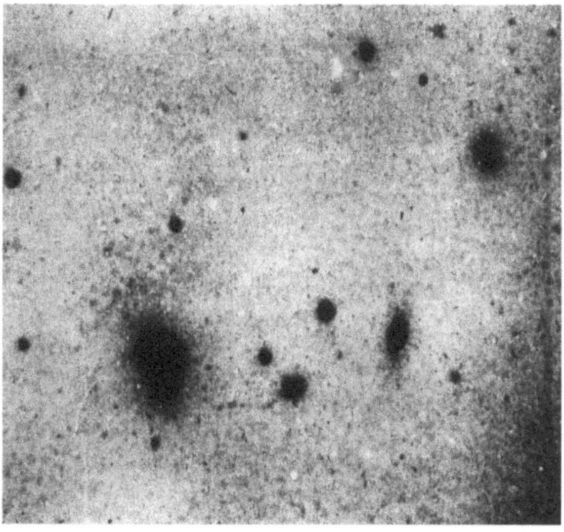

Figure 9.5.8 Optical photograph of the radio source 3C465 (filmwriter reconstruction).

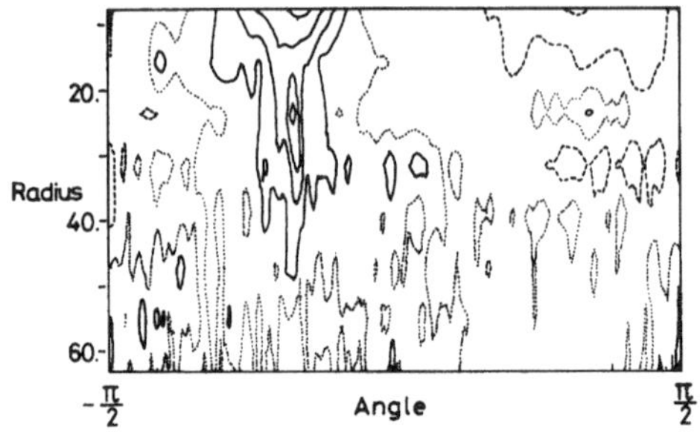

Figure 9.5.9 The radial dependence of angular correlation. Horizontal sections give the angular correlation function between two arcs of given radius. (Positive contours are solid, negative ones broken, and the zero one dotted.)

A final example of angular alignment is given in Fig. 9.5.10 to demonstrate the effectiveness of the alignment even when the original pictures have completely isotropic statistics. The 128-point square pictures to be aligned are portions of a micrograph of an amorphous carbon film misaligned in position and angle. The originals are shown at the top with the ACFs and PSs to either side, and beneath them the angular correlation functions computed using five rings with radii evenly spaced from 0 to $\frac{3}{4}$ of the maximum possible, and weighting proportional to radius.

The number of sampling points used for each arc, incidentally, is twice the linear picture dimension, which means that the points on the largest radius arc ($\frac{3}{4}$ of the maximum) have about the same separation as the data points being interpolated and that the number will usually be a power of 2, convenient for transform and correlation purposes, like the original picture dimensions. Simple bilinear interpolation is normally used (see Section 9.2); this will commonly be fairly accurate for ACFs, but rarely so for PSs. In practice it has not caused any difficulty however, possibly because the distortion will be correlated strongly with the data.

The problem of orientational alignment, then, can be solved reliably by the method just presented. The bulk of the computer time required is devoted to preparing the ACFs or PSs of the original pictures, the interpolation and one-dimensional cross correlation with respect to angle being a very rapid process. Since PSs particularly are readily available at intermediate stages of numerous processing techniques, the cost of establishing or simply verifying orientational alignment is often completely negligible.

The orientational problem having been found so satisfactory a solution, the rest of this section is concerned with an extension of that solution to cope with magnification as well as orientational differences between the

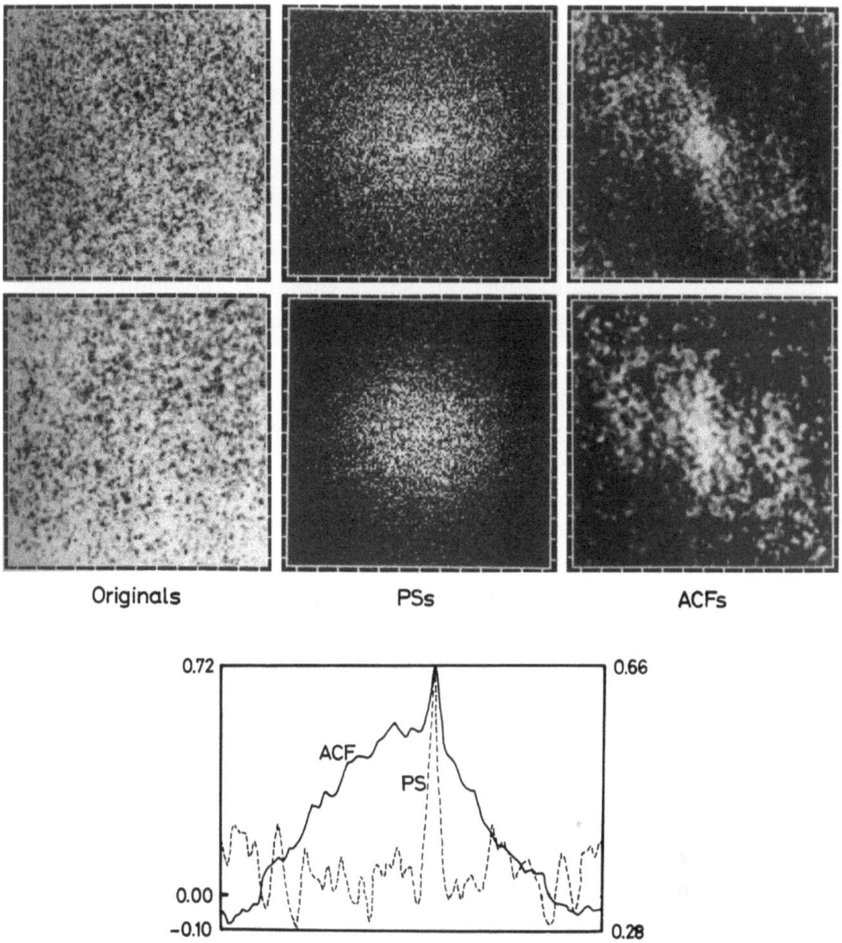

Figure 9.5.10 Angular correlation functions with amorphous subjects. The scale on the left of the graph refers to the ACFs and that on the right to the PSs.

source pictures. Besides reflecting orientation differences, ACFs and PSs both of course reflect magnification differences too, although there is an inversion between them in the sense that the higher magnification picture has the *wider* ACF but the *narrower* PS.

The method is based on a suitable mapping of the ACFs or PSs onto a rectangular grid in such a way that one coordinate measures orientation and the other magnification, after which more or less standard lateral cross correlation can be applied between the interpolated arrays, this time in two dimensions rather than one. The appropriate mapping has recently been pointed out by Hawkes (1976), who extends previous discussion of the use of Mellin transforms for obtaining autocorrelation functions with respect to magnification rather than displacement (Moses & Quesada, 1974) to the case of cross-correlation functions, and shows that the substitution

$$x = \exp(x'), \tag{9.5.2}$$

which interconverts Mellin and Fourier transforms, will convert the magnification cross-correlation function, which cannot be computed rapidly, into a simple displacement cross-correlation function, which *can*. How this is accomplished is easily seen by considering the one-dimensional magnification cross-correlation function

$$c(y) = \int_{-\infty}^{\infty} f(x)g(x\exp(y))\,dx. \tag{9.5.3}$$

The substitution (9.5.2) gives

$$c(y) = \int_{0}^{\infty} f(\exp(x'))g(\exp(x' + y))\exp(x')\,dx'$$

so that if we define new functions

$$\bar{f}(x') = f(\exp(x')),$$

etc., (9.5.3) becomes

$$c(y) = \int_{0}^{\infty} \bar{f}(x')\bar{g}(x' + y)\exp(x')\,dx', \tag{9.5.4}$$

which is a simple (weighted) displacement cross-correlation integral. The weighting factor expresses the fact that for evenly spaced x' values, the spacing of the sampling points in the original functions $f(x)$ and $y(x)$ grows exponentially with x'.

In two dimensions, the appropriate mapping is thus that illustrated in Fig. 9.5.11. The angular separation of sample points is constant as before, but the radial spacing, instead of being free, is now chosen to give a constant ratio between the radii of successive arcs. In this way, a lateral shift of the interpolated arrays in one direction corresponds to rotation of the source pictures, but a lateral shift in the other direction corresponds to a radial scaling of the source pictures by a constant factor, i.e., a change of magnification. If a lateral cross-correlation function is now computed between the two interpolated arrays, one coordinate of the peak will define the relative orientation of the pictures and the other their relative magnification.

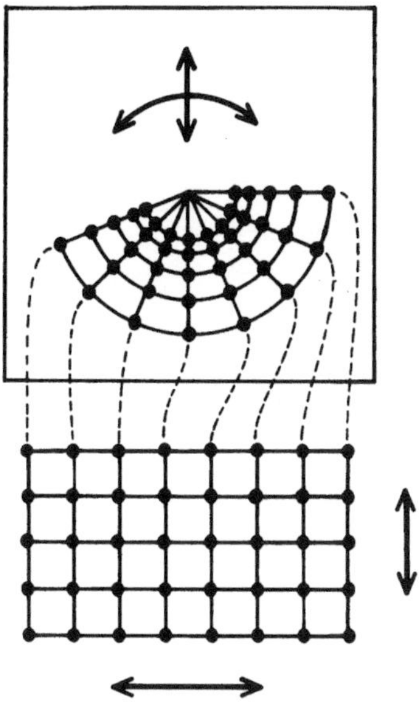

Figure 9.5.11 The radial polar sampling grid mapping rotation and magnification into translation; the ratio of the radii of successive arcs is constant.

Again, the radial decay of ACF or PS values may suggest the use of a stronger weighting still, but a useful default is that appropriate to data without any radial decay, namely weighting in proportion to the square of the radius, one radial factor arising from the uneven spacing of the arcs [cf. (9.5.4)] and the other from the increasing sampling interval around the arcs themselves. We shall require a large number of arcs now instead of a handful. A reasonable choice is half the linear dimension of the picture, which means that the interpolated array requires the same storage as the original ACF or PS if we allow twice as many sampling points around each arc as the linear picture dimension, as previously. If the radii are spread from $\frac{1}{4}$ to $\frac{3}{4}$ of the linear picture dimension, the resulting radial sampling interval ranges from 0.28 to 0.83 of the source sampling interval, so that no information is lost by undersampling.

Again, the cross-correlation function to be computed takes the form of (9.4.6), background level invariance being unnecessary; fortunately the form of the weighting function allows substantial optimization in the calculation of the normalizing function in the denominator. We require

$$\rho_{pq} = \frac{\sum_j \sum_k w_k f_{j-p,k-q} g_{jk}}{\sqrt{F_q}\sqrt{G}}, \tag{9.5.5}$$

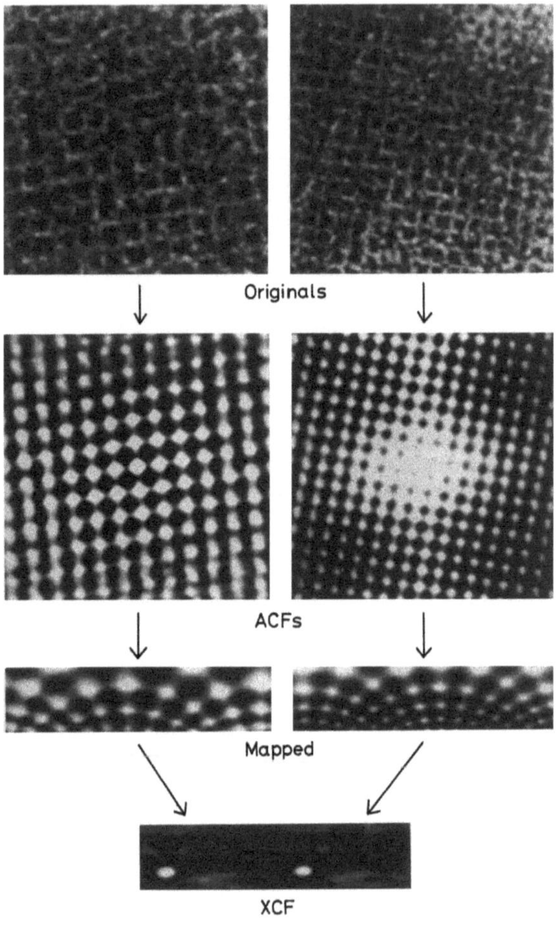

Figure 9.5.12 Alignment of two micrographs in orientation and magnification. Angle found 0.30 ± 0.02 rad; magnification found 1.35 ± 0.03 (micrographs by courtesy of S. M. Salih).

with

$$F_q = \sum_j \sum_k w_k f^2_{j-p,k-q} \quad \text{and} \quad G = \sum_j \sum_k w_k g^2_{jk}.$$

The numerator involves the two-dimensional cross-correlation sum of f_{jk} and $w_k g_{jk}$; F_q, being in fact independent of p, is a one-dimensional cross correlation only of w_k and $\sum_j f^2_{jk}$, while G is a constant.

As a single illustration of the procedure, Fig. 9.5.12 shows the XCF with respect to magnification and orientation calculated in this way for two portions from micrographs of a crystalline protein on a supporting carbon film. (I am indebted to Dr. S. M. Salih for the micrographs, made with specimens supplied by the Basle Biozentrum.) The original misaligned pictures are shown at the top with their ACFs below them, the interpolated arrays extracted from the ACFs being indicated by arrows. The particular choice of data assists in the visualization of the effects of the mapping, and it is not difficult to see that one array is converted into the other by a diagonal displacement, nor indeed that there are *two* horizontally displaced positions for which the arrays match rather than one. The XCF shown below them accordingly exhibits two clearly defined peaks; their separation corresponds to a rotation of $\pi/2$ in one source picture, reflecting the fourfold symmetry of the data, while their positions correctly identify the relative magnification and orientation of the original pictures.

9.6. Averaging repeated structures

If an object structure is repeated in a regular way, either spatially or rotationally, an enormous reduction in the picture noise level is obviously possible by averaging the repeats in large numbers; averaging n repeats will reduce the standard deviation of the noise by a factor of $\sqrt{n}$ while leaving the signal level unchanged, and this improvement is often enough to make intelligible a wealth of detail previously quite inaccessible [e.g. Kuo & Glaeser (1975)].

Spatial periodicity in an image is of course reflected in a localization of transform at the points of a lattice reciprocal to (see Section 2.3) the image plane lattice. As an alternative to direct superposition of unit cell images one upon another, it is therefore possible to pick out the image signal by discarding all nonperiodic contributions to the transform, i.e., all the transform data not in the immediate neighborhood of a reciprocal lattice site. Both approaches have their advantages. Both are capable of optical as well as digital implementation [Markham et al. (1964); De Rosier & Klug (1972); a direct comparison of optical and digital transform plane filtering is given by Aebi et al. (1973)], although it is with the latter that we are (of course) primarily concerned here. The term "structure lattice" will be used to distinguish the lattice of the image periodicity from the "sampling lattice" on which the data are recorded.

In terms of computational speed, the advantage is decidedly with the direct superposition method, which can be very fast even on a small computer system. The small amount of data comprising a single unit cell can be held in core, and a single pass through the picture to be averaged is all that is needed. Each source point is taken in turn and mapped into the unit cell, being accumulated at the nearest sample point (although more elaborate interpolating accumulation schemes can also be employed); a separate count is kept of the number of source points accumulated at each sample point, allowing a final division to produce the averaged unit cell (the count is likely to vary in a random way from point to point, depending on the precise relation of frame positions to lattice vectors). The transform plane method requires at least one Fourier transform, and usually two, besides a pass through the data to perform the masking out of data off the reciprocal lattice sites. The overall speeds of the two approaches are illustrated, for a small computer holding the source data a row at a time on disc, by times of 82 sec for direct averaging of a 256-point square picture and about 20 min for transform plane filtering of the same data.

The next consideration is the effective area over which the source data is averaged. While this is the entire picture frame for the direct superposition method, it is in general substantially less for the transform plane method. The reason for this is the use or discrete data. If the image plane sampling lattice is not an exact submultiple of the structure lattice, then the transform plane structure lattice will not be an exact multiple of the sampling lattice, and the large signal "spots" will fall between, rather than on, the transform plane sampling points. The exact form of the resulting spots is discussed later in this section. For the present, it is enough to note that the spots will be spread over several sampling points so that in filtering out the aperiodic parts of the image it is necessary to pass rather more than one sampling point at each site of the reciprocal structure lattice. Typically, a 3- or 5-point square region might be passed at each site. The filter used to effect the averaging (the function multiplying the transform) is thus an array of such "windows" at each reciprocal structure lattice site, and the averaged image obtained is the convolution of the inverse transform of this with the original data. If the windows were infinitesimally narrow, the inverse transform would simply be an array of delta functions on the image plane structure lattice sites, so that the convolution would produce a superposition of all the unit cells in the picture. The finite width of the windows is described by convolving the "ideal" filter with a single window, and the effect on the inverse transform is of course to multiply it by the inverse transform of a window, thereby attenuating delta functions further from the origin and indeed reversing the sign of some. The image plane convolution thus produces a weighted superposition, in which unit cells further away from a given point contribute progressively less to the average, and may indeed actually detract from it by contributing with reversed contrast.

To be a little more specific, we might suppose that square windows with m points in each direction are passed at each lattice site. For simplicity only we shall take m to be odd. For square pictures with N points in each direction, the inverse transform of such a window is a product of two sums of the form

$$x_k = \sum_{j=-(m-1)/2}^{(m-1)/2} \exp(2\pi ijk/N),$$

one in each direction; the sums are easily evaluated to give

$$x_k = \sin(m\pi k/N)/\sin(\pi k/N). \tag{9.6.1}$$

This has a central lobe extending as far as $|k| = N/m$ so the useful averaging will occur over a region with N/m points in each direction only, rather than over the full picture area, while cells for which $N/m < |k| < 2N/m$ will actually contribute with reversed contrast in the average.

The problem with reducing the window size so as to increase the effectiveness of the averaging is that a small window excludes a significant amount of the signal power at a spot (up to 34%, 27%, 24%, and 23% for 2-, 3-, 4-, and 5-point windows respectively), depending on the relative positions of spot and sampling lattice. The amplitude of such a windowed Fourier component varies across the averaged picture. At the origin, where the best averaged cells are found, the amplitude may be anything from 100–162%, 100–115%, 72–141%, and 94–122% of its true value for 2-, 3-, 4-, and 5-point windows, respectively. The fortuitously narrow range for the 3-point window makes easily the most reliable choice, effecting averaging over the central two-thirds of the picture, with a relative distortion of component amplitudes by no more than ±8%.

There are two ways of curing these problems of the transform plane filtering method completely. One is simply to resample the source data (by the methods of Section 9.2) on a lattice which *is* an exact submultiple of the structure lattice; the signal contributions to the transform are then exactly confined to isolated spots and can be picked out in their entirety; the inverse transform of the window is then unity everywhere, so the averaging extends equally all over the source data; moreover, even the circular nature of the digital convolution, which generally distorts data at the outside of the averaged picture (cf. Fig. 9.6.1), is now ideally suited to the data, since opposite edges of the picture match exactly.

The other method, although requiring more effort to implement initially, is the more flexible in the long run, and also eliminates the need for a second large Fourier transform. The signal spots are "picked out" of the transform one by one and stored in a small array, thus furnishing the (optimally sampled) transform of the unit cell of the image, the unit cell itself being obtainable by a very small (and therefore computationally negligible) Fourier transform.

The picking out of the transform spots involves assessing the transform value at the true spot position from the data at the sample points actually available. This may be done subjectively, on the basis of a small block of printed data around each reciprocal structure lattice site, but an objective approach is not difficult. If the true spot position is (s, t), the corresponding Fourier component of the image structure is

$$x_{km} = c\exp(2\pi i(sk + tm)/N),$$

N being the linear picture dimension and c the required Fourier component amplitude. The transform $\tilde{x}_{jl}$ of this is easily shown to approximate

$$cN^2 \operatorname{sinc}(j - s) \operatorname{sinc}(l - t) \tag{9.6.2}$$

(an exact expression also being easily obtained, but differing little) in the neighborhood of (s, t). Knowledge of (s, t) and the transform value at any nearby sample point is therefore sufficient in principle for the determination of c, but of course in the presence of the high noise levels encountered in the present application, a least-squares fit over a small block of sample points around (s, t) is very much more reliable. Parametrizing the points in such a block by a subscript i, the value of c is obtained as

$$c = \sum_i d_i s_i / \sum_i s_i^2, \tag{9.6.3}$$

in which d_i denotes the value of the transform data, and s_i that of the spot profile [(9.6.2) without the factor c] at the sample points. This expression can be evaluated very rapidly for all the sites of the reciprocal structure lattice even on a small computer.

There are a number of advantages in extracting the signal information from the transform in this way, apart from the elimination of the second large Fourier transform. The reduction in the quantity of data facilitates several other operations, such as transfer function correction (see Section 9.7), the averaging of the data at symmetry related spots, and the incorporation of independent data about the structure, such as the recent use of modulus values taken from electron diffraction patterns to improve the transform values computed from large areas of very low exposure micrographs of radiation-sensitive biological macromolecules (Unwin & Henderson, 1975). These considerations also apply to the unit cell produced by the direct superposition method, of course.

It has been tacitly assumed so far that the lattice vectors for the image structure are known exactly. In practice these may not be easy to determine reliably automatically, although efficient improvement of a first estimate supplied by the user may be possible. The required accuracy grows with the size of the picture to be averaged, obviously, and is the same whether direct superposition or transform filtering is being used, but the latter approach allows substantially easier identification of the structure lattice in the presence of high noise levels than the former, since the signal is concentrated at discrete spots rather than spread out through the picture.

It is a simple matter for a computer user, given a visual display of the transform intensity (on a screen, or as a contour map) to identify the spot positions approximately by eye. The factors that make it less than trivial for the computer to do this for itself are the low signal-to-noise ratio typical at the higher order spot positions (which give the best definition of the lattice vectors), and the strong radial fall-off encountered in many cases, which makes noise values near the origin much larger than signal values further away and necessitates the use of a radial function matched to the specific data if a threshold criterion is to serve for recognizing spots. A promising alternative is to allow the computer to refine the user's estimate of the position of a spot, by allowing s and t in (9.6.2) to vary as well as c in seeking a least squares fit to the data. Better still is to have it vary the lattice vectors so as to maximize the total intensity at the spot sites. The transform values at the estimated sites are obtained by sinc interpolation with a small number of terms, and the problems that might be caused by radial fall-off in the data-noise contributions at low radius points outweighing signal contributions at high radius points can be avoided by treating spots within an annular region only.

As indicated previously, it is rather more difficult to identify the lattice vectors directly from the image. If the noise level is such that a human eye can at least discern the lattice, an estimate can be achieved by measuring a contour map or interactively by having the computer superpose lattice lines on a screen display. A rough estimate can be refined in two ways. First, a relatively small portion from one corner of the picture is cross correlated with similar portions from two other corners (the necessary size being governed by the signal-to-noise ratio of the data as discussed in Section 9.4) to yield a pair of displacement vectors $\mathbf{d}_1$ and $\mathbf{d}_2$ which must lie on the image structure lattice, and can therefore be expressed in terms of its base vectors $\mathbf{x}$ and $\mathbf{y}$ as

$$\mathbf{d}_1 = j_1\mathbf{x} + k_1\mathbf{y}, \qquad \mathbf{d}_2 = j_2\mathbf{x} + k_2\mathbf{y}. \tag{9.6.4}$$

Knowledge of the integers j_1, k_1, j_2, and k_2 would permit solution of (9.6.4) for x and y; the available rough estimates of $\mathbf{x}$ and $\mathbf{y}$ permit estimation of j_1, k_1, j_2, and k_2, however, again by solution of (9.6.4), and the true values may be then found by correction to the nearest integer. Obviously, if the initial uncertainty in the base vectors is such that j_1, k_1, j_2, and k_2 may be more than one from their true values when estimated in this way, it is necessary to build outward from initially much smaller displacement vectors $\mathbf{d}_1$ and $\mathbf{d}_2$.

The second way of refining the estimates of the structure lattice directly from the image data relies on the great speed of direct space averaging, and involves repeated averaging while the lattice base vectors are varied so as to maximize the variance (and therefore signal level) of the resulting unit cell, the point being that incorrect values for the base vectors will cause the signal to disappear on averaging as well as the noise. Again, if the initial uncertainty in the base vectors is fairly large, it will be necessary to perform the maximization on a modest portion of data only and to expand subsequently.

Direct superposition as an averaging technique could be very conveniently carried out on-line with a programmable microdensitometer (see Section 8.2), so it would be very interesting to see whether techniques could be found for recognizing the lattice in these circumstances, perhaps by looking for periodicities in trial single line scans in different directions. This remains conjecture for the present however.

Rotational averaging is far less effective in reducing noise levels than spatial averaging because of the much smaller number of repeats averaged, but is nevertheless useful. Moreover, when the angular period is an even submultiple of 2π, the image signal is real and centrosymmetric, so that its transform shares the property and the transform phases may actually be forced to 0 or π, giving a significant further improvement.

The rotational averaging of an isolated image feature (a disc or helical structure, for example) is easily effected, either directly or by transferring the data to a radial polar grid and filtering the one-dimensional transform with respect to angle, once the center of rotation and the angular period are established. Determining these two are interrelated problems, however. Crowther and Amos (1971) take each possible angular period in turn (the successive submultiples of 2π) and choose for each an origin which minimizes the standard deviation of the phases at the supposedly symmetry-related points of the two-dimensional image transform, after which transfer to a radial polar grid and one-dimensional transformation with respect to angle allows identification of the true angular period through the degree to which the transform intensity is concentrated at its submultiples. It is convenient to work with the image transform rather than the image itself because each origin adjustment then corresponds simply to multiplication by a linear phase factor (2.1.13) and not to a fresh interpolation of the data.

Determination of the center of rotation in this way is also necessary when averaging the symmetry-related points in the unit cells produced from large spatial averages where the cell is known to possess a threefold symmetry axis, or when forcing the transform phases to 0 or π in the case of a two-, four-, or sixfold axis.

A single example of spatial averaging is given in Fig. 9.6.1, which is due to M. Horner and is based on the micrograph used in Fig. 9.5.12. The possibility of extending averaging schemes to aperiodically repeated structures is explored by Frank (1975d) [see also Saxton & Frank (1977)].

9.7. Object reconstruction

With a subject of so broad a scope, it is necessary severely to restrict the material actually covered. Accordingly, only weak objects and the linear image formation theory of Section 1.3 are considered in this section, techniques for the general case having been explored in Chapters 3–7. Section 1.4 has already outlined the principles of the reconstruction. Here the practical details of determining the aberration function γ are considered, and optimal filtering procedures are outlined for obtaining object phase or amplitude components from noisy micrographs.

The power spectrum of the bright-field image of an axially illuminated amorphous weak-phase object, such as a carbon supporting film beneath or near the area of interest, is given by (1.3.24) and shows a series of rings which are the squared product of $\sin\gamma$, the envelopes describing the illumination coherence and the object transform or structure factor. The $\sin^2\gamma$ factor being the most rapidly varying, it can usually be recognized with

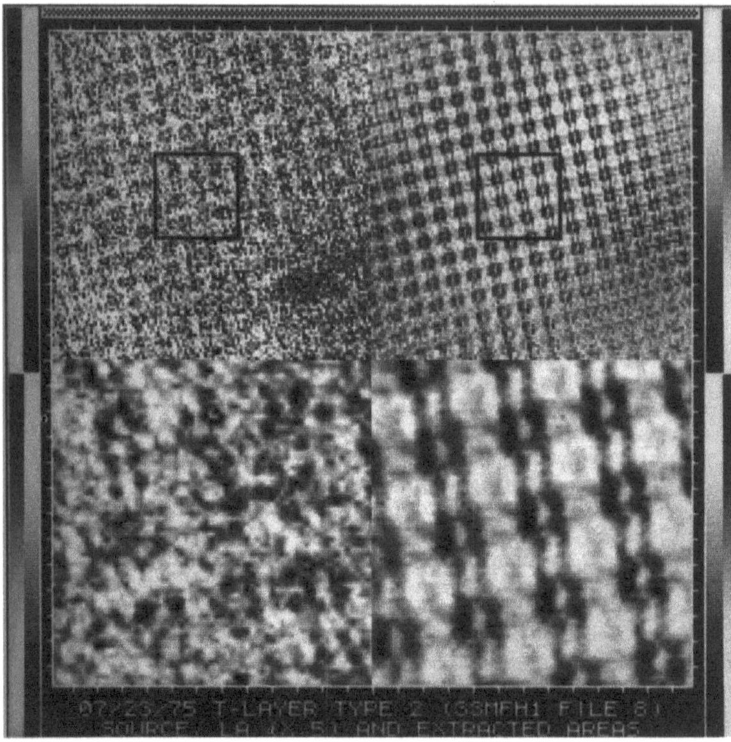

Figure 9.6.1 Spatial averaging over a crystal lattice. Left: unprocessed; right: after averaging by transform plane masking. Enlargements of the marked central regions are shown below each image (by courtesy of M. Horner).

little difficulty in the spectrum, the other factors causing a longer range modulation of it.

The standard optical method of zero fitting for determining defocus and astigmatism involves measuring the positions of minima in a photograph of an optical diffraction pattern of the micrograph and fitting these to a set of standard curves showing the theoretical positions of these as a function of defocus; the positions of maxima can be used as well, though these are less well-defined. The method can also be implemented digitally, the fitting process being performed efficiently by a least-squares program, but the actual measurement of minimum and maximum positions being difficult to automate reliably and probably requiring some manual intervention. If this sort of intervention is reasonably easy, perhaps through the use of a storage tube display, then an effective alternative is for the user to vary the fitting parameters until the computer display of the theoretical diffraction pattern matches that of the actual micrograph; a more rapid compromise is for the computer to superpose on the display of the actual diffraction pattern the predicted minimum/maximum curves.

Accurate transfer function estimation really requires more elaborate procedures, and practical experience so far suggests that the fitting process cannot be made reliable without a degree at least of manual intervention. The basic techniques involved are described by Frank et al. (1970) [see also Frank (1970)], and involve the systematic variation of the parameters to be fitted until the summed squared difference between the predicted and actual diffractograms is minimized. This type of procedure is proof against the problems encountered in simple zero-fitting processes when substantial changes in the transfer function are accompanied by only slight changes in the positions of the zeros (for example, when the generalized defocus is near $1, \sqrt{3}, \sqrt{5} \ldots$), or when only one or two zeros are clearly discernible (which is commonly the case for near optimally focused micrographs at high resolution). There are however two problems associated with it. The first is that predicting the diffractogram involves an assumed object structure factor and assumed illumination coherence conditions. Frank approximates these by a simple $1/|\mathbf{t}|^2$ factor, restricting the sum to an annular region of the spectrum for which the approximation is not intolerably distant; the central region needs excluding in any case because of the contributions there from mechanisms other than phase contrast, and the use of this expression allows reasonable accuracy in the fitted defocus and astigmatism values (about ± 0.05 in the generalized values). The second problem is the one that makes continued manual intervention necessary; the global minimum is usually found only if a fairly close initial estimate of the parameters is made. Fig. 9.7.1 shows the difference function between two spectra of the form $\sin^2 \gamma/|\mathbf{t}|^2$ as a function of the defocus and astigmatism of one spectrum while the parameters of the other remain fixed. It is easy to appreciate the obstacles posed by the high ridges in one direction, though it is possible that an initial exploration of defocus values before minimization is attempted might overcome the problem. A possibility so far unexplored in electron microscopy is the use of the image "power cepstrum," the transform of the logarithm of the spectrum (Cannon, 1976). Since this is essentially dependent on identifying the zero positions it is unlikely to provide a more accurate final value for the defocus, but it might well be helpful in finding a good initial estimate.

The fitting process is greatly assisted if the "speckled" character of a power spectrum calculated from a finite micrograph (see those in Figs. 9.5.5

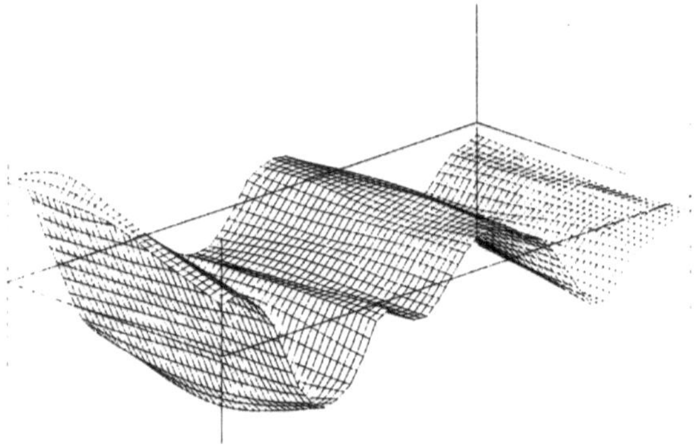

Figure 9.7.1 The difference function between two spectra as a function of the defocus (long axis: range −10 to 730 nm) and astigmatism (short axis: range −35 to 135 nm) of one (by courtesy of M. Horner). The astigmatism direction is correct throughout, so that this represents a relatively easy case with only two variable parameters.

and 9.5.10, for example) is removed. This can be done either by low-pass filtering the spectrum (Frank, 1970) or by rotational averaging yielding smooth radial sections (two perpendicular sections being taken when there is significant astigmatism present), which has the advantage of reducing the fitting process to one-dimensional data, with a corresponding increase in computational speed. Welch (1967) gives a further approach, based on averaging many spectra.

Object restoration by transfer function correction at high resolution is not realistic unless the envelopes describing the illumination coherence are also determined because these affect radically the level of transfer in the higher order rings. Simple zero-fitting procedures are of course of no help at all in this direction since they can determine only the $\sin\gamma$ factor. To include two width parameters (for beam half-angle and energy half-width) in the fitting process generally leads only to instability, unless several members of a focal series are available at once. Frank (1975c) gives a viable if not straightforward optical approach to assessing the width parameters based on the variation with defocus of the highest spatial frequency detectable in the image; the spatial coherence can in fact be very accurately assessed from such a series (Saxton, 1977), but the energy spread determination remains less than wholly satisfactory. Even with modest accuracy in the fitted values for these parameters however, a dramatic improvement is evident in agreement between observed and predicted diffractograms, and a much higher accuracy of defocus determination (to ± 0.02 in the generalized value) becomes immediately possible.

It is important throughout transfer function fitting and the subsequent filtering process to use generalized values for the defocus, astigmatism, and spatial frequency, and to include the power spectrum sampling interval in the fitting process. In this way, there is no possibility of any systematic error being introduced by an incorrect value for the magnification or the spherical aberration. Processing being completed, a scale may be assigned to the restoration by assuming a value either for the magnification or for the spherical aberration and accelerating voltage. The spherical aberration can in fact be deduced from a diffractogram if the magnification is known (e.g., from a calibration micrograph of graphitized carbon fringes) because the ratio of the generalized to actual spatial frequencies t'/t is then determined in the fitting process, and the spherical aberration C_s follows immediately as

$$C_s = \frac{1}{\lambda^3}\left(\frac{t'}{t}\right)^4 \tag{9.7.1}$$

from (1.3.8). Krivanek (1976) gives a similar method, relying solely on zero positions.

Similar methods to the preceding allow estimation of the transfer function in other bright-field imaging modes, such as one-sided imaging (Downing & Siegel, 1973) or tilted beam imaging, but the breakdown of the linear theory prevents any such procedure for dark-field imaging modes, and the processing of such data would need to be based on a bright-field micrograph made immediately after the dark-field one with unchanged instrumental parameters (difficult to achieve in practice).

The remainder of this section presents some of the correction filters and superposition processes that can be applied to images in the linear approximation to obtain object reconstructions. These may be conveniently classified according to whether the object is real or complex, whether it is exactly or overdetermined by the data, and whether or not additional statistical information about the object is assumed. They are presented here in general terms. The images are assumed to be degraded by additive signal-independent noise.

Initially, the case of real objects is treated. The simplest of these is the correction of a single image $i(x)$ of a real object $o(x)$, the system having a point response $p(x)$ and the image noise being $n(x)$. In terms of the transforms of these quantities, the imaging is described by

$$\tilde{i} = \tilde{o}\tilde{p} + \tilde{n}.$$

Dropping tildes for the rest of the section since all quantities will be defined in the Fourier plane, we apply the "inverse" filter $1/p$ to obtain

$$o + n/p.$$

The result is obviously likely to be grossly misleading whenever p is small because of the large values assumed by n/p. In other words there has been an irretrievable loss of information about the object wherever p is small. The least ambitious response to this situation is to multiply i by the filter

$$|p|/p \tag{9.7.2}$$

instead, thereby correcting for the *phase* only of the transfer function; because the factor is unimodular, it cannot cause any increase in the noise level at least. A somewhat better alternative is to apply a filter

$$r = \begin{cases} 1/p & (|p| \geqslant \varepsilon), \\ 0 & \text{(otherwise)} \end{cases} \tag{9.7.3}$$

to i, thereby correcting for the transfer function except where it is *too* small. An empirical choice of ε can be made by inspecting the restored object spectrum, which exhibits high peaks, due to the term n/p, bordering the regions where it vanishes, if ε is too low. A more satisfactory choice might be based on a knowledge of the signal-to-noise ratio, the noise level in the reconstruction being kept lower than the signal level if ε is taken to be the reciprocal of this ratio.

This ratio normally varies with spatial frequency, however, and this leads us to the Wiener filter (Slepian, 1967; Helstrom, 1967; Ansley, 1974) which optimizes the reconstruction on the basis of an assumed knowledge of the signal and noise statistics, as embodied in their (ensemble average) power spectra O and N. Here a reconstructed object transform o' is obtained by applying a reconstructing filter r to the image, r being chosen so that the *ensemble average* of the integrated squared difference between o and o' is minimized. This average may be denoted by

$$\left\langle \int |o - o'|^2 \right\rangle = \left\langle \int |o - r(op + n)|^2 \right\rangle.$$

The minimization condition is simply that the ensemble average integrand be minimal. Differentiating this with respect to r^* gives

$$\langle (o - r(op + n)(o^* p^* + n^*)) \rangle = 0,$$

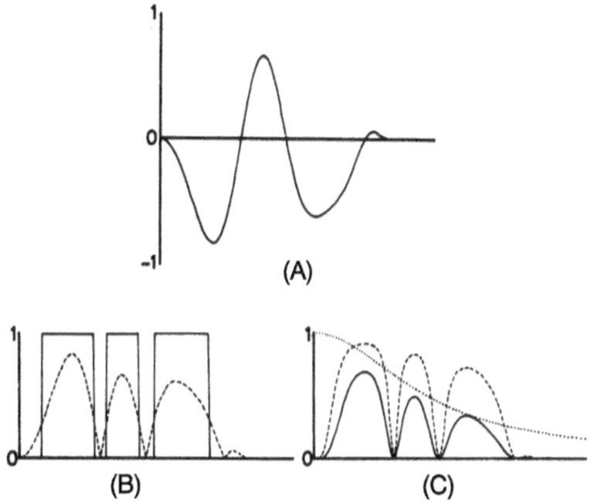

Figure 9.7.2 Three restoration filters compared. Above: the image-forming filter *p*; below left: the filter effective in a reconstruction with the phase filter [(9.7.2); broken] and threshold filter [(9.7.2): solid]; below right: the object spectrum (dotted) and effective filters for Wiener restoration with a white noise spectrum $\frac{1}{4}$ (solid) and $\frac{1}{20}$ (broken) of the object spectrum at zero spatial frequency.

or, recalling that the noise is signal-independent,

$$Op^* - rO|p|^2 - rN = 0,$$

so that the optimal (in this sense) reconstructing filter is

$$r = p^*/(|p|^2 + N/O). \tag{9.7.4}$$

(Differentiating with respect to r gives the complex conjugate condition and the same result for r.) When $N \to 0$, this will be seen to tend to $1/p$, while as $p \to 0$, $r \to 0$ with it, at a rate controlled by the relative signal and noise spectra in the neighborhood.

The three filters presented so far are compared in Fig. 9.7.2 in terms of the product filters pr, being the effective filters converting the actual object into its reconstruction; the nearer these approach unity everywhere the better.

Even with the optimization provided by the Wiener filter, the reconstruction remains unfaithful to the object in the regions where the transfer function is low. The situation can be dramatically improved if more than one image is available, with different transfer functions, so that each image can complement the other in terms of missing object information. In these circumstances, the images are given by

$$i_i = op_i + n_i.$$

If statistical information is not available, we simply choose the object o so as to obtain a least square fit between predicted and observed images (Schiske, 1968), i.e., so as to minimize

$$\sum_i \int |i_i - op_i|^2.$$

Differentiating the integrand with respect to o^* leads at once to

$$o = \sum_i i_i p_i^* \Big/ \Big(\sum_j |p_j|^2\Big), \tag{9.7.5}$$

from which it is clear that no reconstructing filter now has a singularity unless there are places where the transfer function of *every* member of the set vanishes, which is unlikely. If statistical information *is* available (Schiske, 1973), we define a reconstructed object through the superposition filters

$$o' = \sum_i i_i r_i$$

and choose the filters r_i so as to minimize

$$\Big\langle \int |o - o'|^2 \Big\rangle = \Big\langle \int \Big| o - \sum_i r_i(op_i + n_i) \Big|^2 \Big\rangle.$$

Differentiation with respect to r_j^* gives the minimization condition

$$Op_j^* = O\Big(\sum_i r_i p_i\Big) p_j^* + N r_j$$

being a set of linear equations for r_i. The solution can be found conveniently by trying the form $r_i = ap_i^*$ which leads to

$$Op_j^* = Oa\Big(\sum_i |p_i|^2\Big) p_j^* + Nap_j^*,$$

and hence

$$r_i = p_i^* \Big/ \Big(\sum_j |p_j|^2 + N/O\Big), \tag{9.7.6}$$

which may be compared usefully with (9.7.4), the case with a single image, and (9.7.5), the case without the statistical information. In particular, it is worth noting that as the number of images used grows, the term N/O in the denominator of (9.7.6) becomes insignificant, i.e., the statistical information ceases to be valuable.

Now the corresponding filters for a two-component (complex) object are discussed. The two object components o_1 and o_2 are assumed to be transferred into the image by transfer functions p and q, respectively, so that an image transform is described by

$$i = o_1 p + o_2 q + n,$$

and we seek separate restorations of o_1 and o_2. This is patently impossible from a single image if o_1 and o_2 are independent, and two images with different transfer conditions are clearly the minimum number acceptable. Exact solution in the absence of noise would then give

$$\begin{aligned} o_1' &= (q_2 i_1 - q_1 i_2)/(p_1 q_2 - p_2 q_1), \\ o_2' &= (p_1 i_2 - p_2 i_1)/(p_1 q_2 - p_2 q_1), \end{aligned} \qquad (9.7.7)$$

which would be reasonably faithful restorations except near the zeros of the denominator (which occur where the two wave aberration functions γ differ by an integral multiple of 2π, in the electron optical case), where the image noise is severely magnified. Suitable threshold values can be established much as in the real object restoration case (9.7.3).

As in the real object restoration, an improved form of (9.7.7) can be obtained if knowledge of the object and noise spectra O_1, O_2, and N_1 and the correlation spectrum $C = \langle o_1 o_2^* \rangle$ between the two object components are available, while the use of more than two images in the restoration allows the object to be restored even where the denominator of (9.7.7) vanishes. We shall derive the most complicated case only and deduce other useful forms from it. (Note that the statistical information is formulated here in a somewhat different manner from that used by Schiske (1973) and that slightly more information is assumed and exploited.)

We assume, then, a series of images

$$i_i = o_1 p_i + o_2 q_i + n_i$$

available, together with O_1, O_2, C, and N. We define a restored object component o_1' through

$$o_1' = \sum_i r_i i_i,$$

in which r_i is chosen so as to minimize

$$\left\langle \int |o_1 - o_1'|^2 \right\rangle = \left\langle \int \left| o_1 - \sum_i r_i(o_1 p_i + o_2 q_i + n_i) \right|^2 \right\rangle$$

similar expressions applying throughout to the restoration of o_2. Differentiating with respect to r_j^* gives the minimization condition

$$O_1 p_j^* + C q_j^* - \sum_i r_i(O_1 p_i p_j^* + C p_i q_j^* + C^* p_j^* q_i + O_2 q_i q_j^* + N\delta_{ij}) = 0,$$

being a set of linear equations for r_i. The solution is found by trying the form $r_i = a p_i^* + b q_i^*$ which leads to the equation pair

$$\begin{aligned} O_1 &= (O_1 P + C^* X^* + N)a + (O_1 X + C^* Q)b, \\ C &= (CP + O_2 X^*)a + (CX + O_2 Q + N)b, \end{aligned}$$

in which

$$P = \sum_i |p_i|^2 \qquad Q = \sum_i |q_i|^2 \qquad X = \sum_i p_i q_i^*,$$

which may be solved for the coefficients a and b. In terms of the coefficient determinant

$$D = \begin{vmatrix} O_1 P + C^* X^* + N & O_1 X + C^* Q \\ CP + O_2 X^* & CX + O_2 Q + N \end{vmatrix},$$

the restoring filter r_i thus becomes

$$\begin{aligned} r_i = [\{&O_1(CX + O_2 Q + N) - C(O_1 X + C^* Q)\}p_i^* \\ &+ \{C(O_1 P + C^* X^* + N) - O_1(CP + O_2 X^*)\}q_i^*]/D, \end{aligned} \tag{9.7.8}$$

while the corresponding filter s_i restoring o_2 is given by

$$\begin{aligned} s_i = [\{&C^*(CX + O_2 Q + N) - O_2(O_1 X + C^* Q)\}p_i^* \\ &+ \{O_2(O_1 P + C^* X^* + N) - C^*(CP + O_2 X^*)\}q_i^*]/D. \end{aligned}$$

This is the general filter we set out to establish here; various simpler forms are now deducible from it. If the two object components are uncorrelated, so that C vanishes, we have

$$r_i = \frac{O_1(O_2 Q + N)p_i^* - O_1 O_2 X^* q_i^*}{(O_1 P + N)(O_2 Q + N) - O_1 O_2 |X|^2}, \tag{9.7.9}$$

with a similar expression for s_i. If we take N as vanishing, we obtain the filter

$$r_i = (Qp_i^* - X^* q_i^*)/(PQ - |X|^2), \tag{9.7.10}$$

which is the analog of (9.7.5) for the real object case, achieving a least squares fit between observed and predicted images. Restricting the number of images to two does not unfortunately permit much simplification of (9.7.9), although (9.7.10) reduces in this case to (9.7.7). Schiske (1973) points out the convenient forms taken by the filters when the two object component transforms are interrelated by the relations

$$o_1 = o \sin \delta, \qquad o_2 = o \cos \delta,$$

describing an object consisting of a single type of atom for which the scattering factor has a phase δ, and the transfer functions p_i and q_i are the weak-phase and weak amplitude transfer functions $2 \sin \gamma_i$ and $2 \cos \gamma_i$ of Section 1.3: (9.7.8) reduces in this case to

$$r_i = \frac{2 \sin \delta \cos(\delta - \gamma_i)}{4 \sum_j \cos^2(\delta - \gamma_j) + N/O} \tag{9.7.11}$$

and (9.7.10) to the value of this when N vanishes.

The generality of these filters should be emphasized. They should be as useful for making the most of the information in one-sided micrograph pairs/sets or sets made with different illumination tilt angles as they are for exploiting defocus series, provided only that the appropriate functions p_i and q_i can be determined.

Some comment is necessary on the practical determination of the statistical information assumed here. Noise spectra can often be observed directly from images made in the absence of objects; the serious problem is posed by the object spectra. For amorphous objects consisting of a single type of atom, electron scattering factors provide the required information, but more complicated cases will depend on statistical calculation from the assumed object composition and properties, particularly when two components are to be reconstructed. It is sometimes possible to estimate the object spectrum from the image spectrum (Welton & Harris, 1974) if the noise spectrum (known a priori or estimated from regions where the transfer function vanishes) is subtracted from the data, allowance is made for the multiplication of the object spectrum by the squared modulus of the transfer function, and the "speckle" always present in power spectra calculated from finite areas is removed. This is really a reconstruction problem in its own right.

Filters have also been proposed on other criteria than that of minimal ensemble average squared difference. Stockham et al. (1975), considering real objects and transfer functions only, show that the filter

$$1/(|p|^2 + N/O)^{1/2}, \tag{9.7.12}$$

being the geometric mean of the inverse filter $1/p$ and the Wiener filter (9.7.4), forces the power spectra of the restoration to match that of the object (although this must be at the expense of boosting noise where p is small). Hunt (1973) avoids the use of statistical information, showing that the filter maximizing/minimizing the total power in a weighted spectrum of the reconstructed object $|w|^2|o'|^2$, subject to the constraint that the difference between predicted and observed images has a total power matching that of the noise, takes the form

$$p^*/(|p|^2 + \lambda|w|^2). \tag{9.7.13}$$

The weighting function w may be chosen freely to impose some desired property on the restoration—for example, the total power in the restoration is maximized by the choice of unity for w, while choosing w to be the Fourier plane radius minimizes the Laplacean of the object, thereby achieving a *smooth* reconstruction. The parameter λ unfortunately has to be determined iteratively, being varied until the constraint just mentioned is satisfied. Hawkes (1974c) extends Hunt's filter to embrace complex objects. At a rather different level, several authors have considered the minimization of the size of the convolvend corresponding to a restoring filter [see, for example, Frieden (1974)], with the aim of restoring by direct convolution rather than by Fourier plane multiplication. As discussed at the end of Section 9.3, this seems likely to be less attractive in two dimensions than in one, however, except on small input/output bound computer systems.

Examples of practical application of the filters presented in this section may be found, among others, in the papers by Childs and Misell (1973) for the threshold filter (9.7.3) [Stroke et al. (1974) present an optical implementation]; by Frank (1972) for the corresponding complex case (9.7.7) [Burge & Scott (1975) present an optical implementation]; by Hunt (1975) for the Wiener filter (9.7.4) and his own filter (9.7.13); by Stockham et al. (1975) and Cannon (1976) for the filter (9.7.12); restoration of a real object from a defocus series without the use of statistical information, through (9.7.5), is

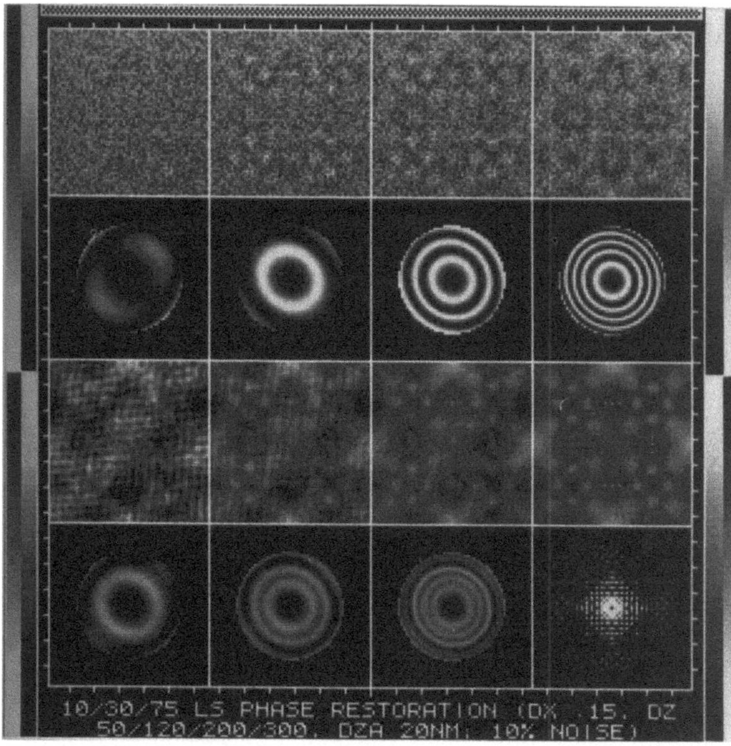

Figure 9.7.3 A series restoration of a weak-phase object. Compare the two right-hand pictures in the third row (see text for details).

illustrated in Fig. 9.7.3, while Frank (1972) uses the corresponding complex object filters (9.7.10). No practical attempts appear to have been published yet using the statistical series restorations, either for real (9.7.6) or complex [(9.7.8) and (9.7.9)] objects, although these are currently being exploited in my own group.

The one illustration given here (Fig. 9.7.3) is a restoration of a weak-phase object from a computed defocus series with a realistic noise level. The test object phase distribution is at the right of the third row, for comparison with the restorations also contained in that row, with the corresponding diffraction pattern beneath it; the maximum phase shift is 0.2 rad, the rms phase shift 0.038 rad, and the field of view 9 nm. Four "micrographs" of this object were computed at defocus values of 50, 120, 200, and 300 nm, respectively, for an instrument operating at 100 kV with 1.8 mm spherical aberration coefficient, 20 nm residual axial astigmatism, and an object aperture cut-off angle corresponding to 0.45 nm at the object; the contrast in these had a maximum of about 12% and an rms value of about 3% (rather less for the first micrograph). Noise was added simulating the recording of the micrographs at an object exposure of about 200 electrons $Å^{-2}$, which gives an rms noise value twice that of the signal. The resulting simulated micrographs are shown in the top row of the figure, with the square of the corresponding transfer function p_i beneath each. Three restorations are shown in the third row, made by applying the superposition filters (9.7.5), respectively to the first two, the first three, and to all four micrographs; beneath each is shown the summed squared transfer functions of the contributing micrographs (the denominator of (9.7.5)), dark bands revealing regions where *all* the contributing transfer functions are low, and the reconstruction may remain unreliable. The dramatic improvement in interpretability of the restorations over the original micrographs is evident.

One may ask whether this kind of restoration scheme constitutes *resolution* enhancement. The answer must depend on what one means by resolution. If one means the smallest object period passed by the objective aperture, the answer is clearly "no"; if one means the radius of the point spread function, as determined by the instrumental optics, it can be "yes." By removing contrast attenuation and reversal effects, and by separating different object components, the limit is lowered on the size of detail that can be interpreted visually in the image, and this can be said to constitute an improvement in the *useful* resolution. Where however the limiting factor is the critical dose for the destruction of the specimen, rather than the electron optical properties (cf. Section 1.6), a posteriori transfer function correction is of no value at all (though this is not necessarily true of its a priori correction in the microscope), and efforts must be directed at reducing the image noise level in the first instance by averaging the images of many objects (cf. Section 9.6) if this is possible. Genuine resolution enhancement, in the sense of making visible detail too small to be passed by the objective aperture must depend on additional a priori information about the object, such as the crystallographic constraint that the object consists of an array of point atoms, or that it is positive, or that it is of finite extent. These have been studied by many authors (e.g., Harris, 1964; Barnes, 1966; Biraud, 1969; Frieden, 1972; Ables, 1973; Hoenders & Ferwerda, 1973; Gerchberg, 1974) but the high noise levels typical of electron microscopy at high resolution have so far prevented their being exploited usefully in this context.

CHAPTER TEN

Improc[☆,☆☆]

W. Owen Saxton
Murray Edwards College, University of Cambridge, Cambridge, United Kingdom
e-mail address: wos1@cam.ac.uk

Contents

10.1. The objectives and environment

"Improc" is an unpretentious acronym subsequently applied to an image processing language designed to provide a basic set of facilities for digital manipulation of micrographs in an easily used form, without at the same time sacrificing the flexibility essential to the more sophisticated user. It is currently in regular use in the Electron Microscopy group of the Cavendish Laboratory, Cambridge, as a research tool. This chapter describes its structure and the technique used to implement it in the off-line multiuser environment of a central University computer (Saxton, 1974b).

It is arguable that it is too early to define the operations that should be provided within such a language; it is certainly clear from the foregoing chapters that many processing techniques are still in their infancy. On the other hand, it is possible to list a number of operations that have emerged as necessary preliminaries to more advanced techniques, and these are sufficient to form the basis of a language for manipulating pictures. The list

☆ Reprinted from Advances in Electronics and Electron Physics, Supplement 10, 1978.
☆☆ I am grateful to Dr. J. Frank and M. Horner for help and advice throughout the development of the Improc System.

ISSN 1076-5670
https://doi.org/10.1016/bs.aiep.2020.04.010

includes microdensitometer input, filmwriter (and other forms of) output, the scaling, translation, rotation and alignment of pictures, the suppression of background level variations, transformation, and the calculation of correlation functions. It is this list which is represented in the language as presented in this chapter (in fact a "frozen" form now dating from 1974). Subsequent developments are mentioned in Section 10.6.

The Improc system runs on an IBM 370/165 (operating under OS with MVT) on-line access being effectively limited to the generation of jobs for subsequent off-line processing. Input is by means of a rotating drum microdensitometer (a Joyce–Loebl/Tech-Ops "Scandig") and output to a rotating drum filmwriter (an Optronics "Photowrite"), both instruments being interfaced with magnetic tape decks independent of the computer. Up to 400 K 8-bit bytes are available to an individual user, for times of the order of minutes per week (central processing time). Immediate execution of submitted jobs is possible only for small jobs, and most image processing jobs are deferred to overnight turnround. Permanent disc file space is severely limited.

A number of processing routines had been developed before any attempt was made to bind them into a cohesive whole. These all stored pictures as doubly-subscripted arrays in core rather than on backing store, the limitation in picture sizes (to about 256 elements square) being traded for convenience and speed. The picture values were held as 32-bit floating point numbers (the mantissa being 24 bits long excluding the sign bit): 32-bit integers offered no advantage in speed or storage space, and the next lower unit that can be easily manipulated, a 16-bit integer, presented too many problems in precision and dynamic range for general use. The routine used for Fourier transformation (the Fortran subroutine FOUR2 from the IBM Contributed Program Library) required picture dimensions to be powers of 2 (cf. Section 9.3). The interpolation routines effecting rotation assumed the sampling lattice to be square. The coordinate system used originated off the top left of the picture, the first subscript increasing from one across the picture, and the second down it. All these features became the inheritance of the Improc system when it developed. Other features of the routines that are less obvious to the Improc user, but are important nevertheless, include the sharing of purely local working space (in a single labeled COMMON block), the calculation of transforms, correlation function, etc. in situ rather than in independent storage, the internal initialization/closing of any input/output units used, the use of multiple routine entry points to provide variants on a single facility, multiple return addresses to assist error trapping, the provision of default values allowing the optional omission of arguments from the routine calling sequence, and the avoidance of double subscripting within routines in the interests of efficiency. Nearly all the routines were written in Fortran (exploiting the IBM subset of the various common extensions), but tape transfers are handled by assembler code subroutines allowing greater efficiency and flexibility than the Fortran language permits, and some other functions (mainly data reformatting) are also coded at assembler level.

The Improc system developed as an attempt to unify these routines into a convenient whole, with an easily grasped central language replacing the complicated subroutine argument lists and data storage allocations. However, the requirement of flexibility remained the most important and was the strongest influence on the method of implementation selected. The systems developed elsewhere have necessarily restricted the facilities at the immediate disposal of the user to a sharply defined set for which routines were already available (cf. Section 9.1), a substantial amount of effort being entailed in even the most trivial extensions. Here it was felt vital that the more sophisticated user should be able to extend the language much more readily; this is reflected in the inserted Fortran facility, and the MACRO statement described in the following sections. It was also hoped that the same command language would be used interactively on a dedicated PDP 8/E computer addressing itself to the same image processing problems with very different hardware resources; this is reflected in the conversational style of the language and the lack of forward references, though in the event, the PDP 8/E system developed independently with a command language not very dissimilar from, but not identical to, Improc (Horner, 1976).

Within such a language system, the distinction disappears between arrays representing optical density (electron intensity) and those representing, for example, wave phase, or difference images; for Improc purposes, all such arrays are referred to as "pictures."

10.2. The implementation technique

Improc is a superset of Fortran, and while it would overstate the case to say that it offers the microscopist the advantages over Fortran that Fortran offers over an assembler code, the advantages are of the same character. In the first instance, an Improc program is translated into Fortran, and the Fortran is then compiled, loaded together with precompiled library routines, and executed in the usual way. The fact that Fortran is an intermediate step in the compilation means that given a basic knowledge of how Improc organizes the data the user can easily insert Fortran into an Improc program to perform operations not provided by Improc statements. Statements that merely duplicate Fortran statements have been deliberately omitted from Improc in the belief that the possibility of intermixing Improc and Fortran is a strength, rather than a weakness, of the system. By storing the library routines without their external references satisfied, considerable economy in disc space is achieved; the system operates with only 15 tracks (about 200,000 8-bit bytes) of permanent disc storage. Core space is also used economically since only those routines actually needed by a given job are loaded.

Specifically, the translation from Improc into Fortran is performed by macrogeneration. Contexts known as "macros" are recognized in the Improc and systematically replaced by a "replacement text," while the rest of the Improc source is passed unaltered. The macrogeneration was programmed in ML/I, a language developed by Brown (1967), which permits quite sophisticated context replacement to be programmed entirely in a high-level language. In a given job, a set of macro definitions which comprise the Improc language are passed through an ML/I processor, and followed by the user's Improc source program in which the macros defined are recognized and replaced, the output from the processor being a Fortran main program with the necessary storage allocations, type declarations, input/output, and subroutine calls to perform the operations requested by the Improc source. The combination of the ML/1 processor with the Improc macro definitions is subsequently referred to as the "Improc processor."

Development of a language system implemented in this way is particularly easy. The Improc processor may be used with as little as a single macro defined, the user writing his program almost entirely in Fortran. As new macros are added one by one, the source programs contain progressively fewer Fortran and more Improc statements until the range of the latter is such that they constitute a new language in their own right.

It is instructive to consider the Improc statement SET (see Appendix A10.1 and the specimen programs of Section 10.4) as an example of the flexibility of the system; by taking advantage of the Fortran compiler for syntactical analysis, it offers facilities for point by point arithmetical operations involving arbitrary numbers of pictures (scaling, subtraction, squaring, logarithm calculation, etc.) which none of the alternative approaches of Section 9.1 can match.

A certain time overhead is of course attached to the translation of Improc into Fortran—typically 1.5 sec at present, which is not serious. This figure could be halved if the version of ML/I available could be stored with the Improc macros already defined; currently these must be redefined with each job. Processing time in ML/I rises linearly with the number of macros defined, so that it is useful to organize Improc in such a way that similar operations are grouped together as a single macro. It is probable that the time overhead is in any case outweighed by the dramatic reduction in user errors due to the simple syntax of the Improc statements.

Before the language structure is discussed in detail in the next section, a brief description is given of some concepts used in ML/I and some facilities it offers, which are relevant to Improc for the benefit of any reader not already familiar with the language.

In ML/I, text is processed in "atoms" consisting either of alphanumeric words (demarcated by spaces) or single nonalphanumeric characters. Any one of these, or a juxtaposition of several, may serve as a "delimiter," an entity recognized in its own right. A "macro" consists of a sequence of delimiters, the first being the "macro name" and the character strings (if any) between the delimiters being the "arguments" of the macro. The process of defining a macro involves specifying a structure of delimiters and replacement text to be output instead of an "instance" of the macro in the text being processed. The delimiter structure is very flexible, its definition including specification of which delimiters may follow a given delimiter, so that in one instance of the macro, some delimiters might appear in a different order from that in which they occurred in a second instance, or indeed might not appear at all. The replacement text can be varied widely according to the actual delimiters and arguments present in the particular instance of the macro being processed, the other macro definitions in force at the time of replacement and the value of a number of internal numerical variables which may be set, manipulated, and tested at any time.

10.3. The language structure

The structure of Improc is obviously derivative from both Fortran and ML/I. The Fortran convention that statements begin in the seventh character position, optionally preceded by a numerical label and terminated by a "new line," applies; Fortran type conventions apply, except that pictures are always REAL; program flow is controlled by Fortran statements, though the DO statement has been somewhat extended, and Fortran syntax is used in arithmetical and logical expressions. Indeed a program consisting entirely of standard Fortran is legal, and would be copied without alteration by the Improc processor.

The form of the peculiarly Improc statements reflects the fact that they are macros; they begin with a mnemonic describing the operation to be performed (the macro name), followed by operands of the user's choice (the arguments) separated by keywords that are part of the statement syntax (the delimiters). The statement

READIN BFP FROM FILE 3 SAMPLING 2,2

for example, is an instance of the macro READIN, with delimiters FROM, FILE, SAMPLING, and "new line," with the arguments BFP, 3, and 2,2 intervening. (It specifies that a picture BFP is to be read into core from the third file of a microdensitometer tape, reading every other point in both directions.) The distinction between arguments and delimiters in fact is not always obvious or relevant to the user.

Most statements have several possible forms with different delimiters, useful defaults being provided wherever it is possible to omit a specification. Other forms of the preceding statement, for example, identify different source tapes or specify an offset in the position within the file from which the picture is to be read, while the SAMPLING specification may be omitted if all points are to be read. Variables may always replace literal numbers in Improc statements, provided they are of the appropriate (Fortran) type.

The principal operands are pictures, coded as digitized values on a sampling lattice that is usually square (though some of the facilities in the language are applicable to general lattices), the number of points in each direction being the same and currently restricted to be a power of two for Fourier transformation purposes but otherwise of the user's choice; they are identified, like all other Improc operands, by an alphabetic (*not* alphanumeric) name of up to five letters (in fact, to one less than the number recognized by the Fortran compiler used in the subsequent step). Picture origins are at the top left rather than the center—again for Fourier transformation purposes (see Section 2.3). One-dimensional arrays are also supported, partly to allow the use of *sets* of parameters, but also to permit one-dimensional modeling of two-dimensional procedures with very large numbers of points. Many Improc statements accordingly accept one-dimensional arrays as alternative arguments to pictures (e.g., those requesting display, shifting, Fourier transformation, and cross correlation). Simple variables may be used as in Fortran, and may be given as arguments to Improc statements.

Most Improc statements naturally refer to operations on pictures and arrays: a small number, however, are provided to assist with other aspects of programming, such as simple variable input/output, the trapping of error returns from processing routines, program looping, and the monitoring of processor time. These are not essential features of an image processing language in any sense but are natural and useful extensions of the Improc processor.

The recursive nature of text evaluation by ML/I makes it possible for the user to define at the beginning of his program new Improc statements of his own, effective for the duration of his job, thereby avoiding the repetition of frequently occurring sets of statements. The Improc statement MACRO allows this definition to be effected rather more easily than it could be directly in ML/I. An indefinite number of arguments are permitted, delimited by commas, and replacement text may be given either in Improc or Fortran. While this facility is obviously analogous to Fortran subroutines in a sense, the analogy is not exact, since the replacement text of such a MACRO statement will be inserted complete in the output Fortran on each occasion when it is called, whereas the code for a subroutine appears once only. Certain restrictions follow from this: for example, lines may not be labeled within the replacement text for a MACRO statement.

A further consequence of the use of a macro-generation process is that such words as Improc statement names cannot be used freely except in the appropriate context because they might initiate unintentional macrogeneration, resulting in error diagnostics (about missing delimiters) and incorrect output Fortran. Such words are called "reserved names." The full list includes all statement names and delimiters, together with a small number of macro names internal to the Improc processor or to ML/I itself. They can be used nevertheless if they are enclosed in "literal brackets," pairs of angle brackets, which denote material to be copied *without* evaluation by the Improc processor. It is good practice, though not essential, to enclose all inserted Fortran in literal brackets because this reduces the amount of text the Improc processor has to scan in detail.

A special restriction applies to DO loops in Improc; if the labeled line on which a loop terminates generates more than one line of Fortran, the label will refer to the first of these rather than the last; consequently, it is usually necessary to close such loops on CONTINUE statements. Similar considerations apply to Fortran logical IF statements which should not normally involve an Improc statement. Labels used should be in the range 0–9999 as the Improc processor generates labels of its own above 9999.

Appendix A10.1 gives a description of the Improc statements defined in the ML/I program listed in Appendix A10.2. Limitations of space have imposed a somewhat unsatisfactory format, and this should be regarded more as an indication than as a definitive statement of the syntax. It will however serve to explain the examples given in the following section of complete specimen programs.

10.4. Specimen programs

The scope of a system such as the Improc implementation being described obviously depends on the range of Fortran subroutines at its disposal, though certain operations are performed at the level of the main program. The four programs described in this section reflect the range of operations available, applied to various situations. Reference to Appendix A10.1 will supplement the comments below on the functions of each. The Fortran generated by the examples is listed in Appendix A10.3 and gives an idea of the distribution of processing tasks between the main program and the precompiled subroutines.

Table 10.4.1 shows a program which examines the spatial frequency dependence of the structural damage in a specimen exposed to the electron beam of the microscope. Two micrographs, one made with a minimum exposure of a fresh specimen and the other with a similar exposure but after the action of the electron beam has caused a significant structural deterioration, have been digitized in the same orientation and approximately aligned laterally. The SETUP statement (line 3) defines two pictures, MINEX and DAM, each 128 points square, to be used in the program. These are read from the microdensitometer tape and copied to temporary disc storage (lines 4–6). The XCORR statement (line 7) computes in DAM the cross-correlation function between the two micrographs, the position of the peak (M,N) defining the relative displacement of the two micrographs. Following a shift of origin from the top left to the center (line 8), the cross-correlation function is displayed on the line printer with the title CCF (line 9). The correlation process having overwritten the pictures, they are retrieved from disc (line 10) and the second is shifted to match the first (line 11). The HISTOGRAM statement (line 12) sets, among other variables, MU to the average of the picture MINEX; this value being noted, a further HISTOGRAM statement finds the average of the picture DAM. The difference between the pictures is now computed, scaling DAM to account for a possible difference in exposure between the plates (line 15). In line 16, the difference picture, stored in MINEX, is masked with a circle (see Section 9.3); its power spectrum is calculated in DAM (line 17) and displayed logarithmically in a contour map (line 18); finally, MINEX is replaced by its autocorrelation function, which is also displayed (lines 19–20).

Table 10.4.2 shows a program for analyzing a pair of micrographs made with complementary halves of the diffraction plane obscured (see Section 3.5) in an aberration-free, weak-scattering situation, on the assumption that they are already aligned. The pictures are only 64 points square here, in default. The scanning grid having been chosen rather smaller than was necessary, only alternate points are read from the microdensitometer data. The alignment data read by the INPUT statement (line 5) specify the offset within the second file of the portion that corresponds to the first file (line 6). Superposition of the two micrographs yields the real part of the full scattered wave (lines 7–8); the picture HPA is restored (line 9), and both

Table 10.4.1 Specimen Improc program.

```
 1  #  COMPARISON OF FRESH WITH DAMAGED SPECIMEN
 2  #
 3     SETUP MINEX/128, DAM/128
 4     READIN MINEX FROM FILE 1
 5     READIN DAM FROM FILE 2
 6     DISC SAVE MINEX, DAM
 7     XCORR MINEX, DAM GIVING RHO,M,N
 8     SHIFT DAM
 9     SKETCH DAM/'CCF'
10     DISC FETCH MINEX, DAM
11     SHIFT DAM BY M,N
12     HISTOGRAM MINEX
13     RM = MU
14     HISTOGRAM DAM
15     SET MINEX = MINEX – DAM * RM/MU
16     MASK MINEX BEYOND 43.
17     PWRSPEC MINEX TO DAM
18     PLOT DAM LOG/'DIFFERENCE PS'
19     ACF MINEX
20     PLOT MINEX/'DIFFERENCE ACF'
21     END
```

Table 10.4.2 Specimen Improc program.

```
1  #  PHASE FROM ALIGNED HALF-PLANE PICTURES
2  #
3     SETUP HPA, HPB
4     READIN HPA FROM FILE 1 SAMPLING 2,2
5     INPUT IOFF,JOFF
6     READIN HPB FROM FILE 2 AT IOFF,JOFF SAMPLING 2,2
7     SET HPA = HPA + HPB
8     SKETCH HPA/'RECONSTRUCTED AMPLITUDE'
9     SET HPA = HPA − HPB
10    HILBERT HPA
11    HILBERT HPB
12    SET HPA = HPA − HPB
13    SKETCH HPA/'RECONSTRUCTED PHASE'
14    END
```

Table 10.4.3 Specimen Improc program.

```
1  #  ABERRATION-FREE PROBE FOR GAUSSIAN SOURCE
2  #
3     SETUP A/128,B/128,$C/65;B = C
4     INPUT BFPR,SWID
5     SET A = 1.
6     MASK A BEYOND BFPR TO 0.
7     PWRSPEC A
8     SHIFT A
9     <W = 2.*SWID*SWID>
10    SET B = EXP(−((I−65)**2 + (J−65)**2)/W)
11    XCORR A,B GIVING RHO,M,N
12    PLOT C/'RADIAL SECTION THROUGH PROBE'
13    END
```

pictures are Hilbert transformed to yield the corresponding imaginary parts of the scattered wave (lines 10–11), the sign change necessary in one case being postponed until the two components are added (line 12).

Table 10.4.3 shows the language in a modeling rather than processing application, investigating the effect of a finite source size on the shape of a scanning microscope probe formed by a diffraction limited lens with a given aperture size. The aperture radius and the half-width of the Gaussian source (referred to the plane of the probe) are taken as data (line 4); the probe for a point source is calculated as the squared modulus of the Fourier

Table 10.4.4 Specimen Improc program.

```
 1 #        ALIGNMENT AND STRUCTURE DETECTION
 2 #
 3 MACRO    READFLAT
 4          READIN %A 1. FROM TAPE %A2. FILE %A3.
 5          REMOVE RAMP,DUST FROM %A1.
 6          HISTOGRAM SKETCH %A1.
 7 END
 8          SETUP STAIN/128,UNSTN/128
 9          READFLAT STAIN,2,1
10          WRITEOUT STAIN
11          SKETCH STAIN /'STAINED SPECIMEN'
12          MASK STAIN BEYOND 43.
13          DISC SAVE STAIN
14          CPUTIME
15          DO 1 WHILE TIME.LT.15.
16          READFLAT UNSTN,1,COUNT
17          MASK UNSTN BEYOND 43.
18          <IF (COUNT.EQ.l) GOTO 2>
19          DISC FETCH STAIN
20 2        ORIENT (ACF) STAIN,UNSTN GIVING THETA,M,N
21          SET STAIN = UNSTN
22          EXTRACT UNSTN FROM STAIN ROTATED BY THETA
23          MASK UNSTN BEYOND 43.
24          WRITEOUT UNSTN
25          DISC FETCH STAIN
26          XCORR STAIN,UNSTN GIVING RHO,M,N /PRINT,PLOT
27          CPUTIME
28 1        CONTINUE
29          END
```

transform (power spectrum) of a circle of radius BFPR (lines 5-8). Following a line of inserted Fortran (line 9), a Gaussian of the appropriate width is generated (line 10) and convolved with the point source probe ((line 11) for centro-symmetric data, convolution and correlation are equivalent). The resultant probe shape is left in picture B, with its origin top left; the first few points of the array therefore constitute a radial section through the probe, and so a one-dimensional array C comprising these locations (line 3) is plotted as the final output.

Table 10.4.4 shows a rather longer program which attempts to detect periodic structure in a series of micrographs of an unstained, preparation of a specimen by cross correlating them with one of a stained preparation available on a separate microdensitometer tape. It will obviously be impossible in this case to align the micrographs by eye, either in angle or position, so the orientation of the micrographs is regarded as unknown initially. A new Improc statement READFLAT is defined for the job (lines 3–7) embracing reading a microdensitometer file, removing background variation and extreme density values, and a histogram display of the corrected data. The program begins by reading the stained micrograph into core, writing a corrected version to a further tape for subsequent filmwriter display (line 10), displaying and masking this, and copying it to temporary disc storage since it is overwritten in each cycle of the subsequent program loop. This loop now commences (line 15), with as many pictures being processed as 15 sec of processor time permits. TIME is set by the CPUTIME statements. In each cycle, an unstained micrograph is read, the stained one being retrieved from disc as necessary; the two are aligned in orientation THETA, the positional information (M,N) being ignored (line 20); the unstained micrograph is rotated appropriately, the stained one being used as working space (lines 21–22). The corners of the rotated picture will of course be meaningless, but these are immediately masked off (line 23). The aligned unstained micrograph is output for filmwriter display (with maximum contrast) and cross correlated with the stained one (retrieved from disc once more), the correlation function being plotted in contour representation (line 26) and revealing by an array of small peaks any periodic structure in the micrograph of the unstained preparation that correlates with that of the stained preparation.

10.5. The Improc macro definitions

This section discusses the techniques adopted in translating Improc into Fortran and the methods of data management used; a rather more detailed knowledge of ML/I is assumed here than has been necessary to appreciate the broader principles discussed previously.

The entire set of macro definitions that comprise Improc will be found listed in Appendix A10.2, and line numbers given with the following comments refer to this listing. The version of ML/I available used "new line" as the final delimiter of most operation macros, NL representing new line in delimiter structure representations.

The initial environment is set up in lines 1–11; line 9 defines literal brackets, and line 10 an end-of-line skip which also serves as the comment facility in Improc, while line 11 makes it possible for text strings enclosed in single quote marks to use reserved names freely.

As mentioned previously, pictures in Improc are held in doubly subscripted Fortran REAL arrays. These have the same name as the corresponding picture. The restriction to alphabetic names of up to five characters for Improc operands allows the Improc processor to define additional unique variables by attaching digits to picture names; thus an integer variable PIC1 is created to hold the linear dimension of a picture PIC, while a two-element integer array PIC2 holds both dimensions (the elements in fact being equal but the array being required by the Fourier transform routine FOUR2); and an integer PIC3 identifies the picture numerically for the temporary disc storage macro DISC described below. A one-dimensional array PIC0 is also created, slightly larger than PIC itself, to allow the extra space necessary for Fourier transformation, and is equivalenced to PIC. (The conjugate symmetry of the transforms means that only the columns numbered 0 to PIC1/2 need be stored, so that the total storage required exceeds that of the original array by only two columns.) Similar variables are created for one-dimensional arrays. The use of one-dimensional equivalenced arrays simplifies the calculation of the necessary offsets for EQUIVALENCE statements generated by Improc directives to allocate one series of pictures to store shared with another. These various declarations are made by the SETUP macro (lines 14–50) which also defines global macros of the form $RPIC, set to the rank of each array (lines 24 and 45) for the use of the macros evaluating subsequent Improc statements; the macro $DPIC of line 29, storing the array size for equivalencing purposes, lapses at the conclusion of the SETUP macro.

The macro which follows, ACF (lines 51–53), is one of the simplest, and it is examined in detail to illustrate the principles of the translation process. Line 51 specifies the permissible sequence of delimiters, and hence the statement syntax, the two permitted forms being

ACF PICA

ACF PICA TO PICB.

The statement is replaced by a simple call of the routine ACF (the name being enclosed in literal brackets to prevent its being recognized as a further instance of the macro which would cause a program loop, the word CALL being output indefinitely). The arguments of this call are the source array name (%A1.), its linear dimension (%A1.1, the variable being assigned by the SETUP macro), and a destination array name (%AT1., for the *last* argument present, since T1 is initialized to the number of arguments present in the particular instance of a macro being evaluated) which will be the same as the source array for the first form of the statement.

The SKETCH macro (lines 54–68) illustrates a reference to the SR macros created by SETUP, in order to distinguish pictures from one-dimensional arrays (line 55). Because the SR is at a different textual level from the array name %A1., the macro must be evaluated twice, and the macro MCW (defined in lines 12 and 13) serves this purpose. According to the result of the test, and the delimiters present, the macro may call the routines SKETCH, FLSKET, or GRAPH, with appropriate arguments in each case. Lines 66 and 67 show the generation of a unique label for a FORMAT statement providing a title.

Lines 77–91 define the SET macro; a one- or two-dimensional loop is generated on the subscripts I and J as necessary; the delimiters of the macro are the various possible arithmetical operators, and the expression to be calculated is evaluated argument by argument, using the ML/I variable T1 as a pointer, subscripts being added to any which prove to be arrays of the same rank as the destination array.

The DISC macro (lines 93–103) allows efficient use of temporary disc storage. READ or WRITE statements are generated in the main program involving the array name only, which is much more efficient than using statements involving implied or actual DO loops, which would be necessary if a subroutine was used for this purpose. A subroutine DISCIO is used to maintain records of the contents of the disc unit, the pictures being identified by the numerical values assigned by the SETUP macro (PIC3 etc.), and rewind the units as necessary.

The macro PWRSPEC (lines 139–144) illustrates recursive use of the macros; if the destination picture differs from the source picture, a SET macro is used to generate a loop transferring the picture across.

Lines 204–206 of the macro XCORR illustrate the output of a FORMAT statement that does not change with the instance of the macro, an ML/I variable P21 is used to note the fact of its being output and is tested subsequently to avoid duplication.

The INPUT macro (line 217) has an alternative name, INPUTDATA —in fact an earlier name being phased out.

The PRINTOUT macro (lines 225–245) adopts the rather tedious expedient of scanning its arguments twice; by testing whether they are of class I (alphanumeric), it assigns them to the list of variables or to the FORMAT statement. The nonstandard Fortran G format is used for output, providing simultaneously for INTEGER or REAL data of any magnitude.

The ORIENT macro (lines 262–283) is probably the most sophisticated, calling the routines MSECPS or MSECAC to align pictures with unknown orientations. As used here, the routine overwrites its data, so these are first saved by a call of the DISC macro; on return, the PS or ACF of one of the aligned pictures is displayed while it is available, and then the data are restored from disc to allow the routine MSECFC to pursue the lateral alignment with the orientation now established.

The implementation of the MACRO statement is remarkably simple (lines 297–299). All that is necessary is to specify a delimiter structure (an indefinite sequence of commas, and a final "new line") and to copy the replacement text given. The use of %WA2. rather than %A2. defers the evaluation of the latter until the newly created macro is being called.

The routines which have error returns use the ML/I variable P2 to identify the error return label (e.g., line 303); consequently, the only function of the TRAP macro (line 317) is to reset this variable. It is necessary to complete the output line with a CONTINUE statement since the succeeding line would otherwise be doubly indented. (Line 2 initializes P2 to 10000 which is included in the replacement text of the END macro (lines 69–76).)

Lines 330–334 contain a short heading output with all programs; the user's program is concatenated with the macro definitions at this point.

A major weakness of the present structure is the poor diagnostics available; a missing (or mistyped) delimiter can cause the Improc processor to search for it as far as the end of the program. Diagnostics when they do appear are extremely verbose, the ML/I processor giving a full traceback of the macros being evaluated at the time of the error. It is possible, with the MCNOTE operation macro, for the Improc processor to generate its own diagnostics; these are also, however, followed by a full traceback, and the extra processor time involved in systematic exploitation of the facility has not been considered worthwhile so far.

It should incidentally be made clear that the lack of adequate diagnostics refers to the Improc processor only and not to the run-time code. The library of Fortran subroutines is provided with a reasonably good set of diagnostics covering meaningless data or arguments and input/output tape error conditions and takes sensible corrective action where possible.

10.6. Recent extensions

Improc has developed continuously if slowly since the preceding sections were first assembled, and a number of extensions have been made to both its facilities and its structure.

Provision for convenient access to COMPLEX pictures (Fourier transforms) has been made through a suitably dimensioned COMPLEX array equivalenced to each picture. The confusion occasioned by the fact that origins for Fourier transform purposes were top left has been removed; REAL picture origins are now always central, and transform origins are half-way down the first column (only the right-hand half-plane data being stored), the Improc processor generating appropriate shifting routine calls before and after transformation, and those Fortran routines involving transforms having been adjusted internally. Many statements (e.g., SET, DISC, MASK, ACF, PWRSPEC, and XCORR) accept COMPLEX pictures as alternatives to REAL operands, the word "COMPLEX" being added at the end of the Improc statement in such cases.

An effective compromise has been made in the direction of picture storage on disc without a full disc-based set of routines being provided. The DISC statement maintains disc-held pictures in the form of relatively short records (4K bytes); routines called by new forms of the SET statement read these a record at a time and, without therefore requiring significant additional storage space, allow a core-held picture to be replaced by the sum, difference, product, or quotient of itself and a disc-held picture. Even this relatively limited use of disc storage makes possible the implementation with full-size pictures of all the filters described in Section 9.7. It seems unlikely that provision will ever be made for *complete* flexibility between core and disc storage, but several further operations on disc-held pictures, such as rotation and cross correlation, will certainly be added in due course.

New statements provide for the application of weak phase/amplitude contrast transfer functions, including the effect of finite source sizes and energy spreads (see Section 1.3); for the simulation of shot noise limited versions of pictures; for the rotational averaging (within an arbitrary sector) and radial weighting (with an arbitrary weighting function) of pictures and transforms; for the conversion of pictures to the format used by the PDP 8/E system, which involves a change of floating point representation by a suitable assembler code routine; for efficient use of magnetic tape for long term picture storage; for interpolation on nonsquare sampling grids; and for lattice averaging [cf. Section 9.6]. The restriction to powers of 2 for picture sizes has been lifted through the optional use of an alternative Fourier transform routine.

The type (REAL or INTEGER) of numbers or variables used as arguments to statements now matters only if the variable is to be reset by the statement. In place of the argument list ..3,X,Y,J.. the Improc processor, in order to force the first two arguments to be REAL and the second two to be INTEGER, would now generate ..3 + 0.,X + 0.,IFIX(Y + 0.),IFIX(J + 0.)... An optional run-time "trace" facility is now available which causes the Improc source statements to be printed as they are executed; this is effected by the use of the "start line" character as a macro, taking the rest of the line as its argument, which generates WRITE and FORMAT statements to print its argument and then evaluates the argument as usual. Titling facilities for line printer and plotter displays have been improved through the use of the PRINTOUT macro to process the relevant section of the statements, thus allowing the value of variables to be included in titles. A titling facility has also been provided for filmwriter output (WRITEOUT), with the same flexibility, but much less easily. An assembler code routine modifies the Fortran system input/output package so as to divert the title from an output unit to a core buffer, and suitable output blocks are generated from this by reference to a table of 6 by 5 matrix graphics for the standard 64 character ASCII set. Since very little processing is now done with pictures smaller than 128 elements square, this figure is now the default picture size in place of 64.

In the hope of reducing dramatically the time involved in macrogeneration, a version of the Improc processor was prepared that exploited the "warning" facility of ML/I. In warning mode, macros are only recognized following special warning characters, and so although these have to be included in the source introducing each statement, there are no longer any reserved names or literal brackets in the language. The reduction in processing time proved disappointing however, being by no more than 20%.

The standard operating system of the IBM 370/165 does not allow the generation of system control statements at run time, and this was previously the source of some inconvenience as the user was obliged to supply full specifications for all input/output units required by a job, which were by no means brief. The use of the Improc processor to generate a fresh job, complete with the necessary control statements, for resubmission to the computer was considered, although it was not a particularly attractive solution because of the turnround problems in a multiaccess environment. In the event, this problem was removed by the provision of a much more powerful on-line control language for the computer (also, incidentally, implemented largely by macrogeneration) which has made it possible to generate the input/output specifications from simple parameters included in the job submission command.

Since no special hardware provisions are made for the Improc system, it is relatively easy to transfer the language from one machine to another. Certain adjustments would inevitably be necessary in the Fortran routines to conform with local software, but no such changes would be needed in the Improc macroprocessor. All available ML/I processors have themselves been macrogenerated from a single master source version and are therefore syntactically identical. For the same reason, it is a comparatively simple matter to provide an ML/I processor for a machine which does not already possess one, any one working implementation being capable of generating suitable code for any other machine (Brown, 1969). The only essential requirements of a suitable machine are that it have a Fortran system and that it be large.

In conclusion, Improc has proved itself a reliable and useful tool for digital image processing in a flexible frame of reference. Saxton (1974c), Frank et al. (1974), Frank (1975a), Frank and Al-Ali (1975), and Al-Ali (1976a) all report work carried out with its aid, and it is also used outside the field of electron microscopy. The hard work of developing new standard procedures is still carried out in Fortran, but an invaluable interface is available between the proved routines and the user. Additional statements are very easily added to the language as new routines become available, so there is no obvious limit to its potential scope. The techniques of implementation described here may also find application in developing specialized languages in other fields.

Appendix A10

A10.1 Specification of Improc statements

The specification below, in view of the limited space available, relies to a large extent on the self-explanatory character of the statements, making no pretense to give precise definitions of syntax or function. It is included to give an impression of the scope of the system in 1974 and to provide a framework for the discussion in Section 10.3–10.5.

The statement names are underlined; alternative items appear one beneath the other; items not necessarily required appear in braces; items optionally repeatable, usually with commas intervening, are enclosed in square brackets. The names PICA and PICB denote any picture names; ARRA and ARRB denote any one-dimensional array names; isolated letters denote simple variables or literal numbers; 13 denotes any Fortran label; and an ellipsis marks a break necessitated here by the page width, and not in fact part of the statement syntax.

1. $$\underline{\text{SETUP}}\left[\begin{matrix}\text{PICA}\\ \text{\$ARRA}\end{matrix}\;\{/\text{M}\}\right]\left\{\left[;\begin{matrix}\text{PICA}\\ \text{ARRA}\end{matrix}=\left[\begin{matrix}\text{PICB}\\ \text{ARRB}\end{matrix}\right]\right]\right\}$$

   Initiates program, declaring pictures and ID arrays (the latter being identified by \$). M sets the dimension. Equivalencing is specified in the latter part of the statement, groups of pictures being allocated sequential store shared with a given picture.

2. <u>END</u>

   Concludes program; includes a label 10000.

3. $$\underline{\text{READIN}}\ \text{PICA FROM}\left\{\text{TAPE}\begin{matrix}1\\ 2\end{matrix}\right\}\text{FILE N}\ \{\text{AT I,J}\}\{\text{SAMPLING...}$$
   $$\text{...K, L}\}$$

   Reads pictures from microdensitometer tapes.

4. $$\underline{\text{WRITEOUT}}\ \text{PICA}\ \{\text{SIZE N}\}\left\{\begin{matrix}\text{TRUNCATING F}\\ \text{DIRECTLY}\end{matrix}\right\}$$

   Generates a filmwriter output file, N points square, scaling so as to truncate a fraction F of the histogram, unless DIRECTLY is specified.

5. $$\underline{\text{DISC}}\ \begin{matrix}\text{SAVE}\\ \text{FETCH}\\ \text{DELETE}\end{matrix}\left[\begin{matrix}\text{PICA}\\ \text{ARRA}\end{matrix}\right]$$

   Manages temporary disc storage: FETCH retrieves the most recently SAVEd version; DELETE releases the disc unit (rarely necessary).

6. $$\begin{matrix}\underline{\text{SKETCH}}\ \text{PICA}\\ \underline{\text{PLOT}}\ \ \text{ARRA}\end{matrix}\ \{\text{LOG}\}\{\text{FULL}\}\{/\text{'TEXT'}\}$$

   Generates line printer or plotter displays with optional title. LOG and FULL refer to pictures only; the latter forces multipage displays.

7. <u>INPUT</u>[V]

   Simple variable input: any type, free format.,. Also reads one-dimensional arrays.

8. PRINTOUT[V]

Simple variable output: any type. The list may also contain text strings in quotes, or '/' to force a line throw.

9. SHIFT $\begin{array}{c}\text{PICA}\\ \text{ARRA}\end{array}$ {BY I{,J}}

Circular shift by quoted vector: omission of BY implies scrambling (see Section 9.3).

10. EXTRACT PICA FROM PICB $\left\{\begin{array}{c}\text{AT I,J}\\ \text{\{ROTATED BY T\} \{ABOUT I,J\}}\end{array}\right\}$

Extraction of portions with given position and orientation: AT refers to top left of portion, ABOUT to center.

11. SUPERPOSE ± PICA ON PICB ABOUT I,J

The added portion may be negated.

12. HISTOGRAM {SKETCH} $\begin{array}{c}\text{PICA}\\ \text{ARRA}\end{array}$ {RANGE B TO T}{/DUST}

The variables TOP, BOTTM, MU, and SIGMA are set, but the actual display is optional, and the range may be forced. DUST suppresses extreme values.

13. MASK PICA {BEYOND R} {TO C}

Circular masking (to picture average, in default).

14. REMOVE [{RAMP}{, }{DUST}] FROM PICA

RAMP removes linear background variation (see Section 9.2) and DUST suppresses extreme values as in the HISTOGRAM statement.

15. SET $\begin{array}{c}\text{PICA}\\ \text{ARRA}\end{array}$ = Expression.

Point by point arithmetic: any Fortran arithmetic expression is valid, optionally involving picture or array names.

16. FFT ± $\begin{array}{c}\text{PICA}\\ \text{ARRA}\end{array}$.

Discrete Fourier transformation in direction given by sign. One-half only of Fourier plane is stored, with the origin at the (top) left.

17. PWRSPEC PICA {TO PICB}.

Power spectrum: overwrites unless TO is present.

18. ACF PICA {TO PICB}.

Autocorrelation function: overwrites unless TO is present.

19. HILBERT PICA {TO PICB}.

Hilbert transformation: overwrites unless TO is present.

20. XCORR $\begin{matrix}\text{PICA}\\ \text{ARRA}\end{matrix}, \begin{matrix}\text{PICA}\\ \text{ARRB}\end{matrix}$ GIVING R,I{J}{/PRINT{,PLOT}}.

PICA (ARRA) returns the cross-correlation function, R the peak height, and I, J (I) the displacement vector. PRINT and PLOT cause displays of the correlation function.

21. ORIENT $\begin{bmatrix}\text{PS}\\ \text{ACF}\end{bmatrix}$ PICA, PICB GIVING T,I,J ...

... {USING N RINGS{,ARRA {,ARRB}}}.

T returns the angle between PICA and PICB, and I,J the center of rotation in PICB. ARRA and ARRB optionally specify radii and weights for the rings (see Section 9.5).

22. TRAP 13.

Diverts to quoted label subsequent error returns which otherwise jump to END.

23. DO 13 $\begin{matrix}\text{WHILE}\\ \text{UNTIL}\end{matrix}$ Condition.

The condition is a Fortran logical value, possibly involving the loop counter, COUNT. A standard Fortran DO statement is legal too.

24. CPUTIME

Prints and sets TIME to processor time elapsed since first use of the statement.

25. #

Introduces comment lines that are ignored.

26. MACRO Name.

Replacement text.

END

Defines new statements: any alphanumeric name; replacement text in Improc or Fortran, with arguments written %Al., %A2., etc.

A10.2 The Improc macro definitions

```
MCSET P1=1
MCSET P2=10000
MCSET P20=0
MCSET P21=0
```

```
MCSET P22=0
MCSET P23=0
MCSET P24=0
MCINS % .
MCSKIP MT, <>
MCSKIP # NL
MCSKIP DT, ' '
MCDEF MCW / SSAS<MCDEF <TEMP> AS %A1.
TEMP>
MCDEF SETUP N1 OPT $ N3 OR N3 / N2 OR N2 , N1
OR N4 ; = N4 OR + N4 OR NL ALL
SSAS<MCSET T1=0
MCGO L10
%L1.    %L10.MCSET T1=%T1+1.
MCGO L8 IF %WDT1.=$
MCSET T2=64
MCGO L2 UNLESS %WDT1.=/
MCSET T2=% WAT1+1.
%L2.MCSET T3=%T2*T2+2*T2.
MCDEFG $ WITH R%WAT1. AS 2
REAL %WAT1.(%T2.,%T2.),%WAT1.0(%T3.)
    EQUIVALENCE (%WAT1.(1,1),%WAT1.0(1))
    INTEGER %WAT1.1/%T2./,%WAT1.2(2)/2*%T2./,%WAT1.3/%P1./
%L4.MCSET P1=%P1+1.
MCDEF $ WITH D%WAT1. AS %T3.
MCGO L3 UNLESS %WDT1.=/
MCSET T1=%T1+1.
%L3.MCGO L1 IF %WDT1.=,
%L9.MCGO L0 UNLESS %WDT1.=;
MCSET T1=%T1+1.
MCSET T2=1
MCSET T3=%T1.
%L7.MCSET T1=%T1+1.
    EQUIVALENCE (%AT1.0(1),%AT3.0(%T2.))
MCSET T2=%T2+MCW $D%AT1./.
MCGO L7 IF %WDT1.=+
MCGO L9
%L8.MCSET T1=%T1+1.
MCSET T2=%WAT1+1.
MCSET T3=%T2+2.
MCDEFG $ WITH R%WAT1. AS 1
DIMENSION %WAT1.(%T2.),%WAT1.0(%T3.)
    EQUIVALENCE (%WAT1.(1), %AT1.0(1))
    INTEGER %WAT1.1/%WAT1+1./,%WAT1.3/%P1./
MCGO L4
>
MCDEF ACF OPT TO N1 OR N1 NL ALL
```

```
SSAS<CALL <ACF>(%A1.,%A1.1,%A1.1,%AT1.)
>
MCDEF SKETCH OPT LOG N1 OR N1 FULL N2 OR N2 / N3 OR N3 NL ALL
SSAS<MCGO L2 IF MCW $R%A1./=1
CALL MCGO L3 IF %WD1.=FULL
MCGO L5 IF %T1.=1
MCGO L3 IF %WD2.=FULL
%L5.<SKETCH>MCGO L4
%L3.<FLSKET>%L4.(%A1., %A1.1,%A1.1#
MCGO L1 UNLESS %WD1.=LOG
,6,LOG#
MCGO L1
%L2.CALL GRAPH(%A1.,%A1.1,1%L1.)
MCGO L0 UNLESS %WDT1-1.=/
    WRITE (6,%T2+10000.)
%T2 +10000. FORMAT (1H0,%AT1.)
>
MCDEF END WITH NL SSAS<CONTINUE
10000 MCGO L1 IF %P23.=0
CALL UWEOF
    %L1.WRITE (6,10005)
10005 FORMAT (/' IMPROC VERSION 8.4.74')
    STOP
    <END>
>
MCDEF 4 VARS SET = N1 OPT + N1 OR - N1 OR * N1 OR / N1 OR , N1
OR ( N1 OR ) N1 OR NL ALL
SSAS<MCSET T3=%T2+10000.
MCSET T4 = MCW $R%A1./
MCDEF $ WITH S AS(I#
MCGO L4 IF %T4.=1
,J%L4.)
MCGO L3 IF %T4.=1
<DO> %T3. J=1,%A1.1
    %L3.<DO> %T3. I = 1, %A1.1
%T3. %A1.$S=MCSET T1=1
$L1.MCSET T1=%T1+1.
%AT1.MCGO L2 UNLESS MCW $R%AT1./=%T4.
$S%L2.%WDT1.MCGO L1 UNLESS %WDT1.=<
>
>
MCDEF DISC OPT SAVE OR FETCH OR DELETE ALL OPT , N1 OR N1 NL ALL
SSAS<<CALL DISCIO(&>%P2.,NUNIT1,%A2.3,MCGO L1 UNLESS %WD1.=SAVE
<1)
    WRITE>%L2.< (NUNIT1) >%A2.MCGO L4 UNLESS %WD2.=,
,%A3.%L4.
MCGO L0
```

```
%L1.MCGO L3 IF %WD1.=DELETE
<2)
   READ>MCGO L2
%L3.3)
>
MCDEF PLOT OPT LOG N1 OR N1 / N2 OR N2 NL ALL
SSAS<MCGO L2 IF MCW $R%A1./=1
CALL CNTPLT(%A1.,%A1.1,%A1.1#
MCGO L1 UNLESS %WD1.=LOG
,10,8,0,LOG#
MCGO L1
%L2.CALL FIXPLT(%A1.,%A1.1,1%L1.)
MCGO L0 UNLESS %WDT1-1.=/
   WRITE (8, %T2+10000.)
%T2+10000. FORMAT (1H ,%AT1.)
>
MCDEF SUPERPOSE OPT + OR - ALL ON OPT ABOUT N1 OR N1 NL ALL
SSAS<CALL PATCH(%A3.,%A3.1,%A3.1,%A2.,%A2.1,%A2.1,%WD1.1.#
MCGO L1 UNLESS %WD3.=ABOUT
,%A4.%L1.)
>
MCDEF REMOVE N1 OPT , N1 OR FROM ALL NL
SSAS<MCSET T2=1
%L1.MCGO L2 IF %WAT2.=RAMP
CALL STAT(%AT1.,%AT1.1,%AT1.1,MU,SIGMA,BOTTM, TOP)
   CALL DUST(%AT1.,%AT1.1,%AT1.1)
MCGO L3
%L2.CALL TREND(%AT1.,%AT1.1,%AT1.1)
%L3.MCSET T2=%T2+1.
MCGO L0 IF %T2.=%T1.
   MCGO L1
>
MCDEF SHIFT OPT BY N1 OR N1 NL ALL
SSAS<<CALL SHIFT>MCGO L1 IF MCW $R%A1./=1
2(%A1.,%A1.1#
MCGO L2
%L1.1(%A1.%L2.,%A1.1#
MCGO L3 UNLESS %WD1.=BY
,%A2.%L3.)
>
MCDEF PWRSPEC OPT TO N1 OR N1 NL ALL
SSAS<MCGO L1 UNLESS %WD1.=TO
SET %A2. = %A1.
   %L1.CALL FOUR2(%AT1.,%AT1.2,2,-1,0)
   CALL CMPLT(%AT1.,%A1.1,%A1.1,%AT1.)
>
MCDEF MASK OPT BEYOND N1 OR N1 TO N2 OR N2 NL ALL
```

```
SSAS<CALL CRCMSK(&%P2.,%A1.,%A1.1#
MCGO L1 UNLESS %WD1.=<
>
)
MCGO L0
%L1.MCGO L2 IF %WD1.=TO
,%A2.MCGO L3 IF %WD2.=TO
)
MCGO L0
%L2.,0.%L3.,%AT1.)
>
MCDEF EXTRACT FROM OPT AT N1 OR ROTATED WI THS BY N2 OR N2 ABOUT
N1 OR N1 NL ALL
SSAS<CALL MCGO L1 IF %WD2.=ROTATED BY
EXTRCT(%A2.,%A2.1,%A2.1,%A1.,%A1.1,%A1.1#
MCGO L2 IF %WD2.=<
>
,%A3.MCGO L2 IF %WD2.=AT
,1%L2.,&%P2.)
MCGO L0
%L1.ROTSQU(%A2.,%A2.1,%A2.1,%A1.,%A1.1,%A1.1,
   *%A3.MCGO L2 IF %WD3.=<
>
,%A1.1,%A1.1,%A4.,&%P2.)
>
MCDEF READIN FROM OPT TAPE N1 OR N1 FILE ALL OPT AT , N2 OR N2
SAMPLING , N3 OR N3 NL ALL
SSAS<CALL SCRD#
MCSET T1=3
MCGO L1 IF %WD2.=FILE
MCSET T1=4
%A3.MCGO L2
%L1.1%L2.(&%P2.,%A1.,%A1.1,%A1.1,%AT1.#
MCGO L3 UNLESS %WDT1.=<
>
)
MCGO L0
%L3.MCGO L4 IF %WDT1.=SAMPLING
MCSET T1=%T1+2.
,%AT1-1.,%AT1.#
MCGO L5 UNLESS %WDT1.=<
>
)
MCGO L0
%L4.,1%L5.,%AT1+1.,%AT1+2.)
>
MCDEF XCORR , GIVING OPT / N1 OR N1 NL ALL
```

```
SSAS<CALL XCR#
MCGO L2 IF MCW $R%A1./=1
2S(%A1.,%A1.1,MCGO L3
%L2.1S(%A1.,%L3.%A1.1,%A1.,%A1.)
    CALL XCR#
MCGO L4 IF MCW $R%A1./=1
2C(%A2.,%A1.1,MCGO L5
%L4.1C(%A2.,%L5.%A1.1,%A2.,%A1.,
    *%A3.MCGO L1 UNLESS %WD3.=/
,%WA4.%L1.)
    WRITE (6,10002) %A3.
MCGO L0 IF %P21.=1
10002 FORMAT (/' CORRELATION PEAK',F7.3,'  AT',2I5)
MCSET P21=1
>
MCDEF DO OPT WHILE N1 OR UNTIL N1 OR N1 NL NO ALL
AS<<DO> %A1. MCGO L0 IF %WD1.=<
>
MCDEFG %A1. WITHS CONTINUE AS <%WD0.
>%T2 +10000.< CONTINUE>
L%T2.=1,100000
    COUNT=L%T2.
    IF (MCGO L1 IF %WD1.=UNTIL
.NOT.%L1.(%A2.)) GOTO %T2+10000.>
MCDEF OPT INPUTDATA OR INPUT ALL NL
SSAS<<READ (5,*)  >%A1.<
    WRITE (6,10006)
    WRITE (6,*) >%A1.
MCGO L0 IF %P24.= 1
<10006 FORMAT (/' DATA:')>
MCSET P24=1
>
MCDEF 4 VARS PRINTOUT N1 OPT , N1'OR NL ALL
SSAS<<WRITE> (6,%T2+10000.) MCSET T1=0
MCSET T3=0
MCSET T4=1
%L4.MCSET T1=%T1+1.
MCGO LT4 IF %AT1.BC I
MCGO LT4+1
%L1.MCGO L8 IF %T3.=0
,%L8.%AT1.MCSET T3=1
%L2.MCGO LT4+2 IF %WDT1.=<
>
MCGO L4
%L3.
%T2+10000. FORMAT ('0'MCSET T1=0
MCSET T4=5
```

```
MCGO L4
%L5.,G12.5#
MCGO L2
%L6.,%AT1.MCGO L2
%L7.)
>
MCDEF HISTOGRAM OPT <SKETCH> N1 OR N1 RANGE TO N3 OR N3 / N4 OR
N4 NL ALL
SSAS<CALL MCGO L1 IF %WD1.=<SKETCH>
STAT(%A1.,%A1.1,MCGO L6 IF MCW $R%A1./=1
%A1.%L6.1,MU,SIGMA,BOTTM,TOP)
MCSET T2=1
MCGO L2
%L1.HISTG(%A2.,%A2.1,MCGO L4 IF MCW $R%A2./=1
%A2.%L4.1,MU,SIGMA#
MCSET T2=2
MCGO L3 UNLESS %WD2.=RANGE
,%A3.,%A4.%L3.)
%L2.MCGO L0 UNLESS %AT1. = DUST
   CALL DUST(%AT2.,%AT2.1,MCGO L5 IF MCW $R%AT2./=1
%AT2.%L5.1)
>
MCDEF ORIENT WITHS () , GIVING , OPT USING OPT RINGS OR RING ALL
N1 OR N1 NL ALL
SSAS<DISC SAVE %A2.,%A3.
MCDEF MCSEC AS %A2.,%A3.,%A2.1,%A2.,%A3,.%A4.,%A5.
   CALL MSEC#
MCSUB(%WA1.,1,2)(MCSEC#
MCGO L1 IF %WD5.=<
>
,%A6.%A7.%L1.)
MCGO L2 IF %WA1.=PS
   SKETCH %A2.
MCGO L3
%L2.        SKETCH %A2. LOG
%L3.        WRITE (6,10003)
       DISC FETCH %A2.,%A3.
       CALL MSECFC(MCSEC)
       WRITE (6,10004) %A5.
MCGO L0 IF %P22.=1
10003 FORMAT (/' ACF/PS OF FIRST MICROGRAPH')
10004 FORMAT (/' CENTRE OF ROTATION',I5,',',I4)
MCSET P22=1
>
MCDEF CPUTIME NL SSAS<MCGO L1 IF %P20.=1
<CALL SCLOCK
   TIME=0.
```

```
10001 FORMAT (/' CPU TIME',F6.2,' SECONDS')
>MCSET P20=1
MCGO L2
%L1.<CALL RCLOCK(ITIME1)
   TIME=FLOAT(ITIME1)/100.
>%L2.<   WRITE (6,10001) TIME>
>
MCDEF HILBERT OPT TO N1 OR N1 NL ALL
SSAS<<CALL HILBRT (>%A1.,%A1.1,%A1.1,%AT1.)
>
MCDEF MACRO NL END WITH NL
SSAS<MCDEFG %WA1. N1 OPT # N1 OR NL ALL AS%WA2.
>
MCDEF WRITEOUT OPT SIZE N1 OR N1 DIRECTLY N2 OR N2 TRUNCATING
N3 OR N3 NL ALL
SSAS<MCSET P23=1
CALL PWRITE(&%P2.,%A1.,%A1.1,%A1.1#
MCGO L2 UNLESS %WD1.=<
>
%L1.)
MCGO L0
%L2.MCGO L3 IF %WD1.=SIZE
,256,256%L4.MCGO L5 UNLESS %WDT1-1.=DIRECTLY
,-1.)
MCGO L0
%L5.MCGO L1 UNLESS %WDT1-1.=TRUNCATING
,%AT1.)
MCGO L0
%L3.,%A2.,%A2.MCGO L4
>
MCDEF TRAP NL AS<MCSET P2=%A1.
CONTINUE
>
MCDEF FFT OPT + OR - ALL NL
SSAS<CALL FOUR2(%A2.,MCGO L1 IF MCW $R%A2./=1
%A2.2,2#
MCGO L2
%L1.%A2.1,1%L2.,%D1.1#
MCGO L3 IF %D1.=+
,0)
MCGO L0
%L3.,-1)
>
<C   IMPROC VERSION 8.4.74
C
   DATA PI/3.1415927/
   INTEGER COUNT
```

```
334     REAL MU>
```

A10.3 Fortran generated from Improc programs (Section 10.4)

From Table 10.4.1

```
C       IMPROC VERSION 8.4.74
C
        DATA PI/3.1415927/
        INTEGER COUNT
        REAL MU
        REAL MINEX(128, 128),MINEX0(16640)
        EQUIVALENCE (MINEX(1.1),MINEX0(1))
        INTEGER MINEX1/128/,MINEX2(2)/2*128/,MINEX3/1/
        REAL DAM(128,128),DAM0(16640)
        EQUIVALENCE (DAM(1,1),DAM0(1))
        INTEGER DAM 1/128/,DAM2(2)/2*128/,DAM3/2/
        CALL SCRD1(&10000,MINEX,MINEX1,MINEX1,1)
        CALL SCRD1(&10000,DAM,DAM1,DAM1.2)
        CALL DISCIO(&10000,NUNIT1,MINEX3,1)
        WRITE (NUNIT1) MINEX,DAM
        CALL XCR2S(MINEX,MINEX1,MINEX1,MINEX,MINEX)
        CALL XCR2C(DAM,MINEX1,MINEX1,DAM,MINEX,
        *RHO,M,N)
        WRITE (6,10002) RHO,M,N
10002   FORMAT (/' CORRELATION PEAK',F7.3,'   AT',2I5)
        CALL SHIFT2(DAM,DAM1,DAM1)
        CALL SKETCH(DAM,DAM1,DAM1)
        WRITE (6,10107)
10107   FORMAT (1H0,'CCF')
        CALL DISCIO(&10000,NUNIT1,MINEX3,2)
        READ (NUNIT1) MINEX,DAM
        CALL SHIFT2(DAM,DAM1,DAM1,M,N)
        CALL STAT(MINEX,MINEX1,MINEX1,MU,SIGMA,BOTTM,TOP)
        RM=MU
        CALL STAT(DAM,DAM1,DAM1,MU,SIGMA,BOTTM,TOP)
        DO 10153 J=1,MINEX1
        DO 10153 I=1,MINEX1
10153   MINEX(I,J)=MINEX(I,J)-DAM(I,J)*RM/MU
        CALL CRCMSK(&10000,MINEX,MINEX1,43.)
        DO 10200 J=1,DAM1
        DO 10200 I=1,DAM1
10200   DAM(I,J)=MINEX(I,J)
        CALL FOUR2(DAM,DAM2,2,-1,0)
        CALL CMPLT(DAM,MINEX1,MINEX1,DAM)
        CALL CNTPLT(DAM,DAM1,DAM1,10,8,0,LOG)
        WRITE (8,10220)
10220   FORMAT (1H , 'DIFFERENCE PS')
```

```
      CALL ACF(MINEX,MINEX1,MINEX1,MINEX)
      CALL CNTPLT(MINEX,MINEX1,MINEX1)
      WRITE (8,10230)
10230 FORMAT (1H , 'DIFFERENCE ACF')
      CONTINUE
10000 WRITE (6,10005)
10005 FORMAT (/' IMPROC VERSION 8.4.74')
      STOP
      END
```

From Table 10.4.2

```
C     IMPROC VERSION 8.4.74
C
      DATA PI/3.1415927/
      INTEGER COUNT
      REAL MU
      REAL HPA(64,64),HPA0(4224)
      EQUIVALENCE (HPA(1,1),HPA0(1))
      INTEGER HPA1/64/,HPA2(2)/2*64/,HPA3/1/
      REAL HPB(64,64),HPB0(4224)
      EQUIVALENCE (HPB(1,1),HPB0(1))
      INTEGER HPB1/64/,HPB2(2)/2*64/,HPB3/2/
      CALL SCRD1(&10000,HPA,HPA1,HPA1,1,1,1,2,2)
      READ (5,*) IOFF,JOFF
      WRITE (6,10006)
      WRITE (6,*) IOFF,JOFF
10006 FORMAT (/' DATA:')
      CALL SCRD1(&10000,HPB,HPB1,HPB1,2,IOFF,JOFF,2,2)
      DO 10080 J=1,HPA1
      DO 10080 I=1,HPA1
10080 HPA(I,J)=HPA(I,J)+HPB(I,J)
      CALL SKETCH(HPA,HPA1,HPA1)
      WRITE (6,10108)
10108 FORMAT (1H0,'RECONSTRUCTED AMPLITUDE')
      DO 10120 J=1,HPA1
      DO 10120 I=1,HPA1
10120 HPA(I,J)=HPA(I,J)-HPB(I,J)
      CALL HILBRT (HPA,HPA1,HPA1,HPA)
      CALL HILBRT (HPB,HPB1,HPB1,HPB)
      DO 10150 J=1,HPA1
      DO 10150 I=1,HPA1
10150 HPA(I,J)=HPA(I,J)-HPB(I,J)
      CALL SKETCH(HPA,HPA1,HPA1)
      WRITE (6,10178)
10178 FORMAT (1H0,'RECONSTRUCTED PHASE')
      CONTINUE
10000 WRITE (6,10005)
```

```
10005   FORMAT (/' IMPROC VERSION 8.4.74')
        STOP
        END
```

From Table 10.4.3

```
C       IMPROC VERSION 8.4.74
C
        DATA PI/3.1415927/
        INTEGER COUNT
        REAL MU
        REAL A(128,128),A0(16640)
        EQUIVALENCE (A(1,1),A0(1))
        INTEGER A1/128/,A2(2)/2*128/,A3/1/
        REAL B(128,128),B0(16640)
        EQUIVALENCE (B(1,1),B0(1))
        INTEGER B1/128/,B2(2)/2*128/,B3/2/
        DIMENSION C(65),C0(67)
        EQUIVALENCE (C(1),C0(1))
        INTEGER C1/65/,C3/3/
        EQUIVALENCE (C0(1),B0(1))
        READ (5,*) BFPR,SWID
        WRITE (6,10006)
        WRITE (6,*) BFPR,SWID
10006   FORMAT (/' DATA:')
        DO 10096 J=1,A1
        DO 10096 I=1,A1
10096   A(I,J)=1.
        CALL CRCMSK(&10000,A,A1,BFPR,0.)
        CALL FOUR2(A,A2,2,-1,0)
        CALL CMPLT(A,A1,A1,A)
        CALL SHIFT2(A,A1,A1)
        W=2.*SWID*SWID
        DO 10120 J=1,B1
        DO 10120 I=1,B1
10120   B(I,J)=EXP(-((I-65)**2+(J-65)**2)/W)
        CALL XCR2S(A,A1,A1,A,A)
        CALL XCR2C(B,A1,A1,B,A,
        *RHO,M,N)
        WRITE (6,10002) RHO,M,N
10002   FORMAT (/' CORRELATION PEAK',F7.3,'   AT',215)
        CALL FIXPLT(C,C1,1)
        WRITE (8,10256)
10256   FORMAT (1H ,'RADIAL SECTION THROUGH PROBE')
        CONTINUE
10000   WRITE (6,10005)
10005   FORMAT (/' IMPROC VERSION 8.4.74')
```

```
      STOP
      END
```

From Table 10.4.4

```
C       IMPROC VERSION 8.4.74
C
        DATA PI/3.1415927/
        INTEGER COUNT
        REAL MU
        REAL STAIN(128,128),STAINO(16640)
        EQUIVALENCE (STAIN(1,1),STAINO(1))
        INTEGER STAIN1/128/,STAIN2(2)/2*128/,STAIN3/1/
        REAL UNSTN(128,128),UNSTNO(16640)
        EQUIVALENCE (UNSTN(1,1),UNSTNO(1))
        INTEGER UNSTN1/128/,UNSTN2(2)/2*128/,UNSTN3/2/
        CALL SCRD2(&10000,STAIN,STAIN1,STAIN1,1)
        CALL TREND(STAIN,STAIN1,STAIN1)
        CALL STAT(STAIN,STAIN1,STAIN1,MU,SIGMA,BOTTM,TOP)
        CALL DUST(STAIN,STAIN1,STAIN1)
        CALL HISTG(STAIN,STAIN1,STAIN1,MU,SIGMA)
        CALL PWRITE(&10000,STAIN,STAIN1,STAIN1)
        CALL SKETCH(STAIN,STAIN1,STAIN1)
        WRITE (6,10098)
10098   FORMAT (1H0,'STAINED SPECIMEN')
        CALL CRCMSK(&10000,STAIN,STAIN1,43.)
        CALL DISCIO(&10000,NUNIT1,STAIN3,1)
        WRITE (NUNIT1) STAIN
        CALL SCLOCK
        TIME=0.
10001   FORMAT (/' CPU TIME ',F6.2,' SECONDS')
        WRITE (6,10001) TIME
        DO 1 L123=1,100000
        COUNT=L123
        IF (.NOT.(TIME.LT.15.)) GOTO 10123
        CALL SCRD1(&10000,UNSTN,UNSTN1,UNSTN1,COUNT)
        CALL TREND(UNSTN,UNSTN1,UNSTN1)
        CALL STAT(UNSTN,UNSTN1,UNSTN1,MU,SIGMA,BOTTM,TOP)
        CALL DUST(UNSTN,UNSTN1,UNSTN1)
        CALL HISTG(UNSTN,UNSTN1,UNSTN1,MU,SIGMA)
        CALL CRCMSK(&10000,UNSTN,UNSTN1,43.)
        IF (COUNT.EQ.1) GOTO 2
        CALL DISCIO(&10000,NUNIT1,STAIN3,2)
        READ (NUNIT1) STAIN
        CALL DISCIO(&10000,NUNIT1,STAIN3,1)
        WRITE (NUNIT1) STAIN,UNSTN
        CALL MSECAC(STAIN,UNSTN,STAIN1,STAIN,UNSTN,THETA,M,N)
        CALL SKETCH(STAIN,STAIN1,STAIN1)
```

```
       WRITE (6,10003)
       CALL DISCIO(&10000,NUNIT1,STAIN3,2)
       READ (NUNIT1) STAIN,UNSTN
       CALL MSECFC(STAIN,UNSTN,STAIN1,STAIN,UNSTN,THETA, M,N)
       WRITE (6,10004) M,N
10003  FORMAT (/' ACF/PS OF FIRST MICROGRAPH')
10004  FORMAT (/' CENTRE OF ROTATION',I5,',',I4)
       DO 10192 J=1,STAIN1
       DO 10192 I=1,STAIN1
10192  STAIN(I,J)=UNSTN(I,J)
       CALL ROTSQU(STAIN,STAIN1,STAIN1,UNSTN,UNSTN1,UNSTN1,
       *THETA,&10000)
       CALL CRCMSK(&10000,UNSTN,UNSTN1,43.)
       CALL PWRITE(&10000,UNSTN,UNSTN1,UNSTN1)
       CALL DISCIO(&10000,NUNIT1,STAIN3,2)
       READ (NUNIT1) STAIN
       CALL XCR2S(STAIN,STAIN1,STAIN1,STAIN,STAIN)
       CALL XCR2C(UNSTN,STAIN1,STAIN1,UNSTN,STAIN,
       *RHO,M,N,PRINT,PLOT)
       WRITE (6,10002) RHO,M,N
10002  FORMAT (/' CORRELATION PEAK',F7.3,'   AT',2I5)
       CALL RCLOCK(ITIME1)
       TIME=FLOAT(ITIME1)/100.
       WRITE (6,10001) TIME
1      CONTINUE
10123  CONTINUE
       CONTINUE
10000  CALL UWEOF
       WRITE (6,10005)
10005  FORMAT (/' IMPROC VERSION 8.4.74')
       STOP
       End
```

References

Ables, J. (1973). *Astron. Astrophys. (Suppl.), 15*, 383.

Ade, G. (1975). *Optik, 42*, 199.

Aebi, U., Smith, P. R., Dubochet, J., Henry, C., & Kellenberger, E. (1973). *J. Supramol. Struct., 1*, 498.

Al-Ali, L. S. (1976a). In J. A. Venables (Ed.), *Developments in electron microscopy and analysis* (p. 225). London and New York: Academic Press.

Al-Ali, L. S. (1976b). *Optical and computer analysis of electron micrographs* (Thesis). Univ. of Cambridge.

Allpress, J. G., & Sanders, J. V. (1973). *J. Appl. Cryst., 6*, 165.

Ansley, D. A. (1974). *Optik, 40*, 400.

Arndt, U. W., Crowther, R. A., & Mallett, J. F. W. (1968). *J. Sci. Instrum. (Ser. 2), 1*, 510.

Arsenault, H., & Lowenthal, S. (1974). *C. R. Acad. Sci. Paris B, 269*, 518.

Barnes, C. W. (1966). *J. Opt. Soc. Am., 56*, 575.

Bates, R. H. T. (1969). *Monthly Not. Roy. Astr. Soc., 142*, 413.

Baumeister, W., & Hahn, M. (1974). *J. Microsc., 101*, 111.

Billingsley, F. C. (1971). *Adv. Opt. Electron Microsc., 4*, 127.

Biraud, Y. (1969). *Astron. Astrophys., 1*, 124.

Blachman, N. M. (1974). *Proc. IEEE, 62*, 346.

Boas, R. P. (1954). *Entire functions*. New York: Academic Press.

Brown, P. J. (1967). *Comm. ACM, 10*, 618.

Brown, P. J. (1969). *Comp. J., 12*, 327.

Budinger, T. F., & Gullberg, G. T. (1974). *IEEE Trans., NS-21*, 2.

Burge, R. E. (1973). *J. Microsc., 98*, 251.

Burge, R. E. (1976). In M. A. Hayat (Ed.), *Principles and techniques of electron microscopy: biological applications, vol. 6* (p. 85). Princeton, New Jersey: Van Nostrand-Reinhold.

Burge, R. E., & Scott, R. F. (1975). *Optik, 43*, 301.

Burge, R. E., Fiddy, M. A., Greenaway, A. H., & Ross, G. (1974). *J. Phys. D*, 7, L65.

Burge, R. E., Dainty, J. C., & Thom, J. (1976). In J. A. Venables (Ed.), *Developments in electron microscopy and analysis* (p. 221). London and New York: Academic Press.

Burge, R. E., Fiddy, M. A., Greenaway, A. H., & Ross, G. (1976). *Proc. R. Soc. London, 350*, 191.

Cannon, M. (1976). *IEEE Trans., ASSP-24*, 58.

Chapman, J. N. (1975). *Phil. Mag., 32*, 527, 541.

Chien, R. T., & Snyder, W. E. (1975). *IEEE Trans., CAS-22*, 541.

Childs, P. A., & Misell, D. L. (1973). *J. Phys. D, 6*, 1653.

Connor, R. C., Brainard, R. C., & Limb, J. O. (1972). *Proc. IEEE, 60*, 779.

Cooley, J. W., & Tukey, J. W. (1965). *Math. Comp., 19*, 297.

Cowley, J. M. (1953). *Proc. Phys. Soc., 66B*, 1096.

Cowley, J. M. (1975). *J. Phys. D, 8*, L77.

Cowley, J. M., & Iijima, S. (1972). *Z. Naturforsch., 27a*, 445.

Crowther, R. A., & Amos, L. A. (1971). *J. Mol. Biol., 60*, 123.

Crowther, R. A., De Rosier, D. J., & Klug, A. (1970). *Proc. R. Soc. London A, 317*, 319.

Dallas, W. J. (1975). *Optik, 44*, 45.

De Rosier, D. J., & Klug, A. (1968). *Nature (London), 217*, 130.
De Rosier, D. J., & Klug, A. (1972). *J. Mol. Biol., 65*, 469.
De Rosier, D. J., & Moore, P. B. (1970). *J. Mol. Biol., 52*, 355.
De Santis, P., & Gori, F. (1975). *Opt. Acta, 22*, 691.
Dekkers, N. H., & De Lang, H. (1974). *Optik, 41*, 452.
Doetsch, G. (1950). *Handbuch der Laplace Transformationen*. Basel: Birkhäuser.
Downing, K. H., & Siegel, B. M. (1973). *Optik, 38*, 21.
Downing, K. H., & Siegel, B. M. (1975). *Optik, 42*, 155.
Drenth, A. J. J., Huiser, A. M. J., & Ferwerda, H. A. (1975). *Opt. Acta, 22*, 615.
Efron, E. (1968). *Photogramm. Eng., 34*, 1058.
Engel, A. (1974). *Optik, 41*, 117.
Engel, A., Wiggins, J. W., & Woodruff, D. C. (1974). *J. Appl. Phys., 45*, 2739.
Erickson, H. P. (1971). *Adv. Opt. Electron Microsc., 5*, 163.
Erickson, H. P., & Klug, A. (1970). *Ber. Buns. Phys. Chem., 74*, 1129.
Farnell, G. C., & Flint, R. B. (1975). In M. A. Hayat (Ed.), *Principles and techniques of electron microscopy: biological applications, vol. 5*. Princeton, New Jersey: Van Nostrand-Reinhold.
Fowle, E. N. (1964). *IEEE Trans., IT-10*, 61.
Frank, J. (1969). *Optik, 30*, 171.
Frank, J. (1970). *Untersuchungen von Elektronenmikroskopischen Aufnahmen hoher Auflösung mit Bilddifferenz- und Rekonstruktionsverfahren* (Thesis). München: Tech. Univ.
Frank, J. (1972). *Biophys. J., 12*, 484.
Frank, J. (1973a). In J. K. Koehler (Ed.), *Advanced techniques in biological electron microscopy* (p. 215). Berlin and New York: Springer.
Frank, J. (1973b). *Optik, 38*, 519.
Frank, J. (1973c). In P. W. Hawkes (Ed.), *Image processing and computer aided design in electron optics* (p. 196). London and New York: Academic Press.
Frank, J. (1974). *Optik, 38*, 582.
Frank, J. (1975a). *J. Phys. E, 8*, 582.
Frank, J. (1975b). *Optik, 43*, 25.
Frank, J. (1975c). *Optik, 44*, 379.
Frank, J. (1975d). *Ultramicroscopy, 1*, 131.
Frank, J., & Al-Ali, L. S. (1975). *Nature (London), 256*, 376.
Frank, J., Buszler, P. H., Langer, R., & Hoppe, W. (1970). *Ber. Buns. Phys. Chem., 74*, 1105.
Frank, J., Salih, S. M., & Cosslett, V. E. (1974). In *Proc. int. congr. electron microsc., 8th., vol. 2* (p. 678).
Frieden, B. R. (1972). *J. Opt. Soc. Am., 62*, 511.
Frieden, B. R. (1974). *J. Opt. Soc. Am., 64*, 682.
Gabor, D. (1948). *Nature (London), 161*, 777.
Gerchberg, R. W. (1972). *Nature (London), 240*, 404.
Gerchberg, R. W. (1974). *Opt. Acta, 21*, 709.
Gerchberg, R. W., & Saxton, W. O. (1971). *Optik, 34*, 275.
Gerchberg, R. W., & Saxton, W. O. (1972). *Optik, 35*, 237.
Gerchberg, R. W., & Saxton, W. O. (1973a). In P. W. Hawkes (Ed.), *Image processing and computer aided design in electron optics* (p. 66). London and New York: Academic Press.
Gerchberg, R. W., & Saxton, W. O. (1973b). *J. Phys. D, 6*, L31.
Glaeser, R. W. (1975). In B. M. Siegel, & D. R. Beaman (Eds.), *Physical aspects of electron microscopy and microbeam analysis* (p. 205). New York: Wiley.
Glaser, W. (1952). *Grundlagen der Elektronenoptik*. Vienna: Springer.

Glaser, W. (1956). In S. Flügge (Ed.), *Handbuch der Physik, vol. 33* (p. 123). Berlin and New York: Springer.

Goldberger, M. L., Lewis, H. W., & Watson, K. M. (1963). *Phys. Rev.*, *132*, 2764.

Gonsalves, R. A. (1976). *J. Opt. Soc. Am.*, *66*, 961.

Gordon, R., & Herman, G. T. (1974). *Int. Rev. Cytol.*, *38*, 111.

Gordon, R., Bender, R., & Herman, G. T. (1970). *J. Theor. Biol.*, *29*, 471.

Grünbaum, A. (1974). *Z. Wahrschein. Theor.*, *29*, 345.

Hanszen, K.-J. (1969). *Z. Naturforsch.*, *24a*, 1849.

Hanszen, K.-J. (1971). *Adv. Opt. Electron Microsc.*, *4*, 1.

Hanszen, K.-J., & Ade, G. (1975). *Optik*, *42*, 1.

Hanszen, K.-J., & Morgenstern, B. (1965). *Z. Angew. Phys.*, *19*, 215.

Hanszen, K.-J., & Trepte, L. (1971a). *Optik*, *32*, 519.

Hanszen, K.-J., & Trepte, L. (1971b). *Optik*, *33*, 182.

Harris, J. L. (1964). *J. Opt. Soc. Am.*, *54*, 931.

Hawkes, P. W. (1973a). In P. W. Hawkes (Ed.), *Image processing and computer aided design in electron optics* (p. 129). London and New York: Academic Press.

Hawkes, P. W. (1973b). In P. W. Hawkes (Ed.), *Image processing and computer aided design in electron optics* (p. 2). London and New York: Academic Press.

Hawkes, P. W. (1974a). *Optik*, *40*, 539.

Hawkes, P. W. (Ed.). (1974b). *Catalogue of computer programs in electron optics*. Cambridge, England: Cavendish Laboratory.

Hawkes, P. W. (1974c). *Optik*, *41*, 64.

Hawkes, P. W. (1976). *Optik*, *45*, 427.

Hegerl, R., & Hoppe, W. (1972). In *Proc. eur. reg. conf. electron. microsc., 5th* (p. 628).

Helstrom, C. W. (1967). *J. Opt. Soc. Am.*, *57*, 297.

Herrmann, K.-H., Krahl, D., & Rindfleisch, V. (1972). *Siemens Forsch. Entwickl. Ber.*, *1*, 167.

Herrmann, K.-H., Krahl, D., Rust, H.-P., & Ulrichs, O. (1976). *Optik*, *44*, 393.

Hibi, T., & Takahashi, S. (1969). *Z. Angew. Phys.*, *27*, 132.

Hirsch, P. B., Howie, A., Nicholson, R. B., Pashley, D. W., & Whelan, M. J. (1965). *Electron microscopy of thin crystals*. London: Butterworths.

Hoenders, B. J., & Ferwerda, H. A. (1973). In P. W. Hawkes (Ed.), *Image processing and computer aided design in electron optics* (p. 220). London and New York: Academic Press.

Hofstetter, E. M. (1964). *IEEE Trans.*, *IT-10*, 119.

Högbom, J. A. (1974). *Astron. Astrophys. (Suppl.)*, *15*, 417.

Hoppe, W. (1969). *Acta Crystallogr. A*, *25*, 495, 508.

Hoppe, W. (1970). *Acta Crystallogr. A*, *26*, 414.

Hoppe, W. (1971a). *Z. Naturforsch.*, *26a*, 1155.

Hoppe, W. (1971b). *Phil. Trans. Roy. Soc.*, *261*, 71.

Hoppe, W. (1972). *Z. Naturforsch.*, *27a*, 919.

Hoppe, W. (1974). *Naturwiss.*, *61*, 534.

Hoppe, W., & Strube, G. (1969). *Acta Crystallogr. A*, *25*, 502.

Hoppe, W., Langer, R., & Thon, F. (1970). *Optik*, *30*, 538.

Hoppe, W., Kostler, D., Typke, D., & Hunsmann, N. (1975). *Optik*, *42*, 43.

Horner, M. (1976). In J. A. Venables (Ed.), *Developments in electron microscopy and analysis* (p. 209). London and New York: Academic Press.

Huiser, A. M. J., & Ferwerda, H. A. (1976). *Opt. Acta*, *23*, 445.

Huiser, A. M. J., Drenth, A. J. J., & Ferwerda, H. A. (1976). *Optik*, *45*, 303.

Hunsmann, N., Buszler, P., & Hoppe, W. (1972). In *Proc. eur. reg. conf. electron. microsc., 5th* (p. 654).

Hunt, B. R. (1973). *IEEE Trans.*, *C-22*, 805.

Hunt, B. R. (1975). *Proc. IEEE*, *63*, 693.

Iijima, S. (1971). *J. Appl. Phys.*, *42*, 5891.

Iijima, S. (1973). *Acta Crystallogr. A*, *29*, 18.

Imura, T., Saka, H., Todokoro, H., & Ashikawa, M. (1971). *J. Phys. Soc. Jpn.*, *31*, 1849.

Jackson, P. D. (1961). In G. R. Screaton (Ed.), *Dispersion relations*. Edinburgh: Oliver and Boyd.

Khalfin, L. A. (1958). *Sov. Phys.-JETP*, *6*, 1053.

Krakow, W. (1976). *Ultramicroscopy*, *1*, 203.

Krakow, W., & Siegel, B. M. (1975). *Optik*, *42*, 245.

Krakow, W., Ast, D. G., Goldfarb, W., & Siegel, B. M. (1976). *Phil. Mag.*, *33*, 985.

Krakow, W., Welles, K. B., & Siegel, B. M. (1976). *J. Phys. D*, *9*, 175.

Krivanek, O. (1976). *Optik*, *45*, 97.

Kuo, I. A. M., & Glaeser, R. M. (1975). *Ultramicroscopy*, *1*, 53.

Langer, R., Frank, J., Feltynowski, A., & Hoppe, W. (1970). *Ber. Buns. Phys. Chem.*, *74*, 1120.

Lannes, A. (1976a). *Contributions aux Formalismes de Diffusion et de Transfert en Microscopie Electronique par Transmission Applications* (Thesis). Toulouse: Paul Sabatier Univ.

Lannes, A. (1976b). *J. Phys. D*, *9*, 199.

Lenz, F. (1954). *Z. Naturforsch.*, *9a*, 185.

Lenz, F. (1971). In U. Valdre (Ed.), *Electron microscopy in material science* (p. 540). London and New York: Academic Press.

Lohmann, A. W. (1974). *Optik*, *41*, 1.

Lynch, D. F., Moodie, A. F., & O'Keefe, M. A. (1975). *Acta Crystallogr. A*, *31*, 300.

Markham, R., Hitchburn, J. H., Hills, G. J., & Frey, S. (1964). *Virology*, *22*, 342.

McFarlane, S. C., & Cochran, W. (1975). *J. Phys. C*, *8*, 1311.

Misell, D. L. (1973a). *J. Phys. D*, *6*, L6.

Misell, D. L. (1973b). *J. Phys. D*, *6*, 2200, 2217.

Misell, D. L. (1973c). *Adv. Electron. Electron Phys.*, *32*, 63.

Misell, D. L., & Burge, R. E. (1973). In P. W. Hawkes (Ed.), *Image processing and computer aided design in electron optics* (p. 168). London and New York: Academic Press.

Misell, D. L., & Burge, R. E. (1975). *J. Microsc.*, *103*, 195.

Misell, D. L., & Greenaway, A. H. (1974). *J. Phys. D*, 7, 832. 1660.

Misell, D. L., Burge, R. E., & Greenaway, A. H. (1974). *J. Phys. D*, 7, L27.

Möllenstedt, G., & Düker, H. (1956). *Z. Phys.*, *145*, 377.

Moses, H. E., & Quesada, A. F. (1974). *J. Math. Phys.*, *15*, 748.

Mulvey, T. (1973). *J. Microsc.*, *98*, 232.

Munch, J. (1974). *Optik*, *43*, 79.

Muskhelishvili, N. I. (1953). *Singular integral equations*. Gröningen: Noordhoff.

Nagata, F., Matsuda, T., & Komoda, T. (1975). *Jpn. J. Appl. Phys.*, *14*, 1815.

Nathan, R. (1971). *Adv. Opt. Electron Microsc.*, *4*, 85.

Nevanlinna, R. (1953). *Eindeutige Analytische Funcktionen*. Berlin and New York: Springer.

Nevanlinna, R. (1969). *Introduction to complex analysis*. Reading Massachusetts: Addison-Wesley.

Paley, R. E. A. C., & Wiener, N. (1934). *Fourier transforms in the complex domain. Am. math. soc. coll. pub., vol. XIX*.

Papoulis, A. (1962). *The Fourier integral and its applications*. New York: McGraw-Hill.
Peřina, J. (1972). *Coherence of light*. Princeton, New Jersey: Van Nostrand-Reinhold.
Rabiner, L. R., & Rader, C. M. (Eds.). (1972). *Digital signal processing*. New York: IEEE Press.
Ramachandran, G. N., & Lakshminarayanan, A. V. (1971). *Proc. Nat. Acad. Sci. U.S.*, *68*, 2236.
Reichenbach, H. (1944). *Philosophic foundations of quantum mechanics*. London and New York: Cambridge Univ. Press.
Reimer, L., & Gilde, H. (1973). In P. W. Hawkes (Ed.), *Image processing and computer aided design in electron optics* (p. 138). London and New York: Academic Press.
Riddle, H. N., & Siegel, B. M. (1971). In *Ann. meet. electron microsc. soc. am., 29th* (p. 226).
Roman, P., & Marathay, A. S. (1963). *Nuovo. Cimento*, *30*, 1452.
Roos, B. W. (1969). *Analytic functions and distributions in physics and engineering*. New York: Wiley.
Rose, A. (1948). *Adv. Electron.*, *1*, 131.
Rose, H. (1975). *Optik*, *42*, 217.
Saxton, W. O. (1974a). *J. Phys. D*, *7*, L63.
Saxton, W. O. (1974b). *Comp. Graph. Im. Proc.*, *3*, 266.
Saxton, W. O. (1974c). In *Proc. int. congr. electron microsc., 8th., vol. 1* (p. 314).
Saxton, W. O. (1975). *Computer techniques for image processing in electron microscopy* (Thesis). England: Univ. of Cambridge.
Saxton, W. O. (1976). In J. A. Venables (Ed.), *Developments in electron microscopy and analysis* (p. 191). London and New York: Academic Press.
Saxton, W. O. (1977). *Optik*, *49*, 51.
Saxton, W. O., & Frank, J. (1977). *Ultramicroscopy*, *2*, 219.
Schiske, P. (1968). In *Proc. eur. reg. conf. electron. microsc., 4th, vol. 1* (p. 145).
Schiske, P. (1973). In P. W. Hawkes (Ed.), *Image processing and computer aided design in electron optics* (p. 82). London and New York: Academic Press.
Schiske, P. (1974). *Optik*, *40*, 261.
Schiske, P. (1975). *J. Phys. D*, *8*, 1372.
Siegel, B. M., & Hertel, R. J. (1968). In *Proc. eur. reg. conf. electron. microsc., 4th, vol. 1* (p. 219).
Slepian, D. (1967). *J. Opt. Soc. Am.*, *57*, 918.
Stenn, K., & Bahr, G. F. (1970). *J. Ultrastruct. Res.*, *31*, 526.
Stockham, T. G., Cannon, T. M., & Ingebretson, R. B. (1975). *Proc. IEEE*, *63*, 678.
Stroke, G. W. (1972). *IEEE Spectrum*, *9*, 24.
Stroke, G. W., Halioua, M., Thon, F., & Willasch, D. (1974). *Optik*, *41*, 319.
Tanaka, K. (1972). *Pattern Recog.*, *4*, 251.
Thomson, M. G. R. (1973). *Optik*, *39*, 15.
Titchmarsh, E. C. (1937). *Introduction to the theory of Fourier integrals*. London and New York: Oxford Univ. Press.
Titchmarsh, E. C. (1939). *The theory of functions*. London and New York: Oxford Univ. Press.
Toll, J. S. (1956). *Phys. Rev.*, *104*, 1760.
Toms, N., Chapman, J. N., & Ferrier, R. P. (1972). In *Proc. eur. reg. conf. electron. microsc., 5th* (p. 422).
Unwin, P. N. T. (1972). *Proc. R. Soc. London A*, *329*, 327.
Unwin, P. N. T. (1974). *Z. Naturforsch.*, *29a*, 158.
Unwin, P. N. T., & Henderson, R. (1975). *J. Mol. Biol.*, *94*, 425.

Vainshtein, B. K. (1971). *Sov. Phys. Crystallogr.*, *15*, 781.

Valentine, R. C. (1966). *Adv. Opt. Electron Microsc.*, *1*, 180.

Van der Lugt, A. (1974). *Proc. IEEE*, *62*, 1300.

Van Toorn, P., & Ferwerda, H. A. (1976). *Opt. Acta*, *23*, 457, 469.

Voelcker, H. B. (1966). *Proc. IEEE*, *54*, 340.

Wahl, H. (1974). *Optik*, *39*, 585.

Walther, A. (1963). *Opt. Acta*, *10*, 41.

Welch, P. D. (1967). *IEEE Trans.*, *AU-15*, 70.

Welton, T. A., & Harris, W. W. (1974). In *Proc. int. cong. electron. microsc., 8th, vol. 1* (p. 318).

Whittaker, E. T., & Watson, G. N. (1950). *A course of modern analysis*. London and New York: Cambridge Univ. Press.

Willasch, D. (1975). *Optik*, *44*, 17.

Williams, R. C., & Fisher, H. W. (1970). *J. Mol. Biol.*, *52*, 121.

Wolf, E. (1962). *Proc. Phys. Soc.*, *80*, 1269.

Zeitler, E. (1968). *Adv. Electron. Electron Phys.*, *25*, 277.

Zeitler, E., & Thomson, M. G. R. (1970). *Optik*, *31*, 258, 359.

Index

R

S

T

U

V

W

FSC
www.fsc.org
MIX
Papier | Fördert gute Waldnutzung
FSC® C083411

Kolibri 360 GmbH
Ferdinand-Jühlke-Straße 7
99095 Erfurt, Deutschland
produktsicherheit@kolibri360.de